THE SCOTTISH SHALE OIL INDUSTRY
&
MINERAL RAILWAY LINES

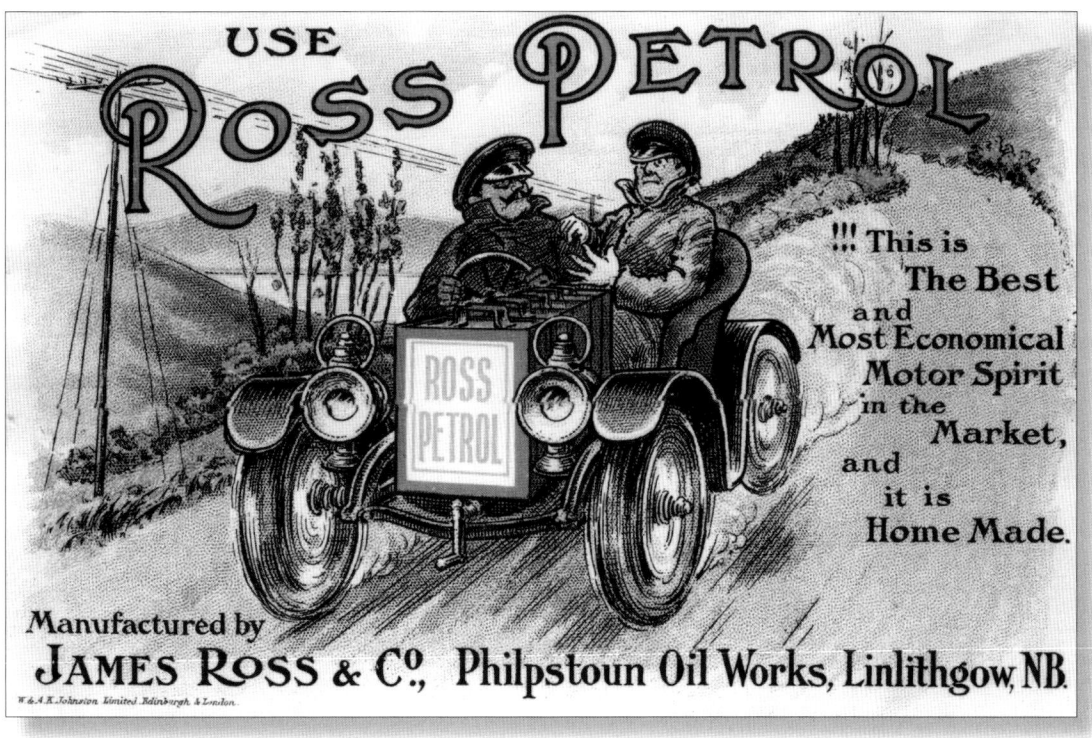

An early advert for James Ross & Son's new motor spirit, produced at Philpstoun Oil Works.

Pumpherston oil works and refinery, showing the benches of pot coking stills. On the skyline, two shale bings at other refineries can just be seen behind the chimneys. The bing on the left is most likely that at Broxburn Stewartfield and the one in the right distant the bing at Winchburgh. On the right hand side are the wooden 'Bryson' cooling towers associated with the ammonia compression refrigeration process of paraffin wax production, peculiar to Pumpherston. At lower right is an ex-NBR Class 'A' 0-6-2 tank engine (L&NER Class N15) of Bathgate shed, busy shunting in the sidings with a range of wagons in the foreground. The oil tank wagons belong to Pumpherston Oil Company. Identifiable private owner wagons include Ebbw Vale, Crump (Forest of Dean), Butterley (Derbyshire), Monckton (Royston near Barnsley), William Harrison Ltd (Cannock), Florence Coal & Iron (Stoke on Trent), Haydock (St Helens) and Berry Hill Collieries (Stoke on Trent).

© Harry Knox and Lightmoor Press 2013
Designed by Nigel Nicholson
British Library Cataloguing-in-Publication Data. A catalogue record for this book is available from the British Library
ISBN 9781 899889 73 0
All rights reserved. No part of this publication may be reproduced, stored in a retrieval system or transmitted in any form or by any means, electronic, mechanical, photocopying, recording or otherwise, without the written permission of the publisher

LIGHTMOOR PRESS

Unit 144B, Lydney Trading Estate, Harbour Road,
Lydney, Gloucestershire GL15 5EJ

www.lightmoor.co.uk

Lightmoor Press is an imprint of
Black Dwarf Lightmoor Publications Ltd.
Printed by Berforts Information Press, Eynsham, Oxford.

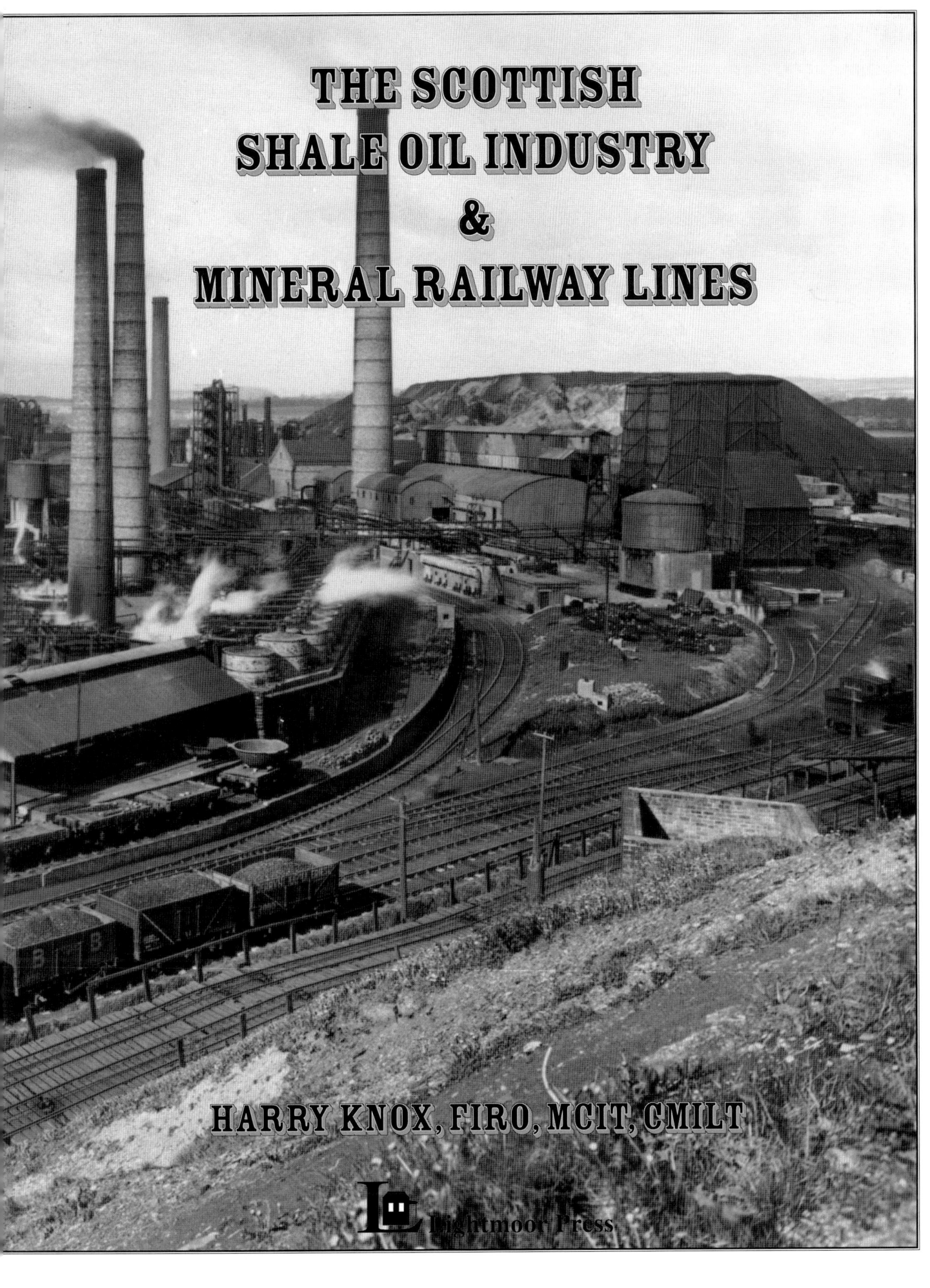

THE SCOTTISH SHALE OIL INDUSTRY & MINERAL RAILWAY LINES

HARRY KNOX, FIRO, MCIT, CMILT

Lightmoor Press

CONTENTS

INTRODUCTION	7
PART ONE THE SCOTTISH SHALE OIL INDUSTRY	
1 THE GEOGRAPHY	13
2 THE GEOLOGY	15

    Geological Time Periods ♦ Oil Shale Described
    Geological Position of the Oil Shale Measures
    Chemical Composition of Oil Shale

3 THE SHALE FIELDS IN DETAIL	19

    Southern Area: Cobbinshaw to Uphall
    Northern Area: Uphall to the River Forth
    Other Scottish Shale Workings ♦ The Occurrence of Natural Oil and Gas

4 THE HISTORY OF THE INDUSTRY	33

    The Early Years ♦ The Growth Years
    The Years of Decline ♦ Renaissance (but not Shale)

5 THE MINING OF OIL SHALE	45

    Mining Techniques ♦ Longwall Working
    Stoop and Room Working
    Opencast Working ♦ Mining Equipment

6 MINING: SAFETY, HEALTH AND ACCIDENTS	61

    Safety ♦ Health ♦ Accidents ♦ Summary

7 THE PITS AND MINES	69
8 EXTRACTION AND REFINING OF SHALE OIL	75

    Crushing the Shale ♦ The Shale Retort ♦ Retorting the Shale
    Naphtha and Sulphate of Ammonia Separation
    Refining the Crude Oil: First Distillation
    Refining the Crude Oil: Second Distillation
    Crude Solid Paraffin ♦ The Refined Products
    Production of Sulphuric Acid

9 THE MAJOR SHALE OIL COMPANIES AND THEIR WORKS	95

    E.W. Binney & Co., Bathgate ♦ Broxburn Oil Company
    The Glasgow Oil Company (Broxburn) Ltd ♦ The Uphall Mineral Oil Company
    Young's Paraffin, Light & Mineral Oil Company ♦ Linlithgow Oil Company
    J. Ross & Sons (Philpstoun Oil Company)
    Hermand Oil Company ♦ West Calder Oil Company
    Mid Calder Oil Company ♦ Bathgate Oil Company
    West Lothian Oil Company ♦ Tarbrax Oil Company Ltd
    Oakbank Oil Company ♦ Dalmeny Oil Company
    Pumpherston Oil Company ♦ Clippens Oil Company
    Holmes Oil Company ♦ Burntisland Oil Company
    Other Oil Companies in the Lothians
    The Final Years ♦ The Legacy of the Shale Bings

10 THE SOCIAL ISSUES	117

    Pay and Conditions ♦ Housing
    Health ♦ Education ♦ Leisure

PART TWO   THE MINERAL RAILWAYS OF THE SCOTTISH OIL INDUSTRY

11 An Introduction to the Railways — 141

12 Non-standard Gauge Railways — 143
Underground Haulageways   ◆   Surface Haulageways

13 The Winchburgh Tramway — 149

14 The Caledonian Railway — 155
Tarbrax Junction (Cobbinshaw) to Tarbrax Oil Works
Benhar Junction to Benhar Oil Works and Polkemmet Colliery
Muldron Junction to Muldron Pit   ◆   Levenseat Junction to Handaxwood Oil Works
The Caledonian West Calder Loop   ◆   Addiewell Oil Works
Mid Calder Junction to Oakbank Oil Works   ◆   Camps Junction to Raw Quarry and Camps Goods

15 The North British Railway — 171
Linlithgow to Bridgend Oil Works (Linlithgow Oil Company)   ◆   Philpstoun Oil Works
Bathgate to Addiewell Goods and Addiewell Oil Works
Bathgate (Polkemmet Junction) to the Torbane Pits and Armadale
Fauldhouse Crofthead to Levenseat and Handaxwood Oil Works
Bathgate (Starlaw) to Cousland (Seafield) Oil Works   ◆   Bathgate to Deans and Boghall Mines
Bathgate to Deans Oil Works   ◆   Uphall Junction
Uphall Junction to Hopetoun Oil Works
Uphall (Drumshoreland) to Holygate Goods, Broxburn and Albyn Oil Works
Uphall (Drumshoreland) to Roman Camp Oil Works
Uphall Junction to East Calder Goods and Camps Raw Quarry
Uphall Junction to Pumpherston No. 4 Mine   ◆   Uphall Junction to Pumpherston Oil Works
Uphall Junction to Pumpherston No. 5 Mine
Uphall Junction to Holmes Oil Works   ◆   Newliston Junction to Newliston No. 29 Pit
Broxburn Junction to Broxburn Albyn Oil Works
Broxburn West Junction to Niddry Castle Oil Works
Queensferry Junction to South Queensferry and Port Edgar
Queensferry Junction to Ingliston Pits
Ingliston Pits to Almondhill Paraffin Works, Dalmeny Oil Works, South Queensferry and Port Edgar
Edinburgh to Loanhead and Straiton Oil Works (Pentland)
Kinghorn to Binnend Oil Works and Mines (Burntisland Oil Company)

16 Internal Standard Gauge Mineral Lines — 199
Addiewell Oil Works to Burngrange, Baads Mine and Fraser Pit
Addiewell Oil Works to the Gavieside and Polbeth Pits
Philpstoun Oil Works to the Philpstoun and Ochiltree Mines
Hopetoun Oil Works to Various Shale Mines
Albyn Oil Works to Various Shale Mines
Oakbank Oil Works to Mid Calder Mines

17 Oakbank Overhead Rope System — 201

APPENDICES

A Locomotives — 205
Allocation of Steam Locomotives
Allocation of Diesel Mechanical Locomotives   ◆   Allocation of Electric Locomotives

B Known Oil Companies and Works Post-1850 — 223

C Glossary of Mining Terms — 225

Bibliography — 227

Index — 229

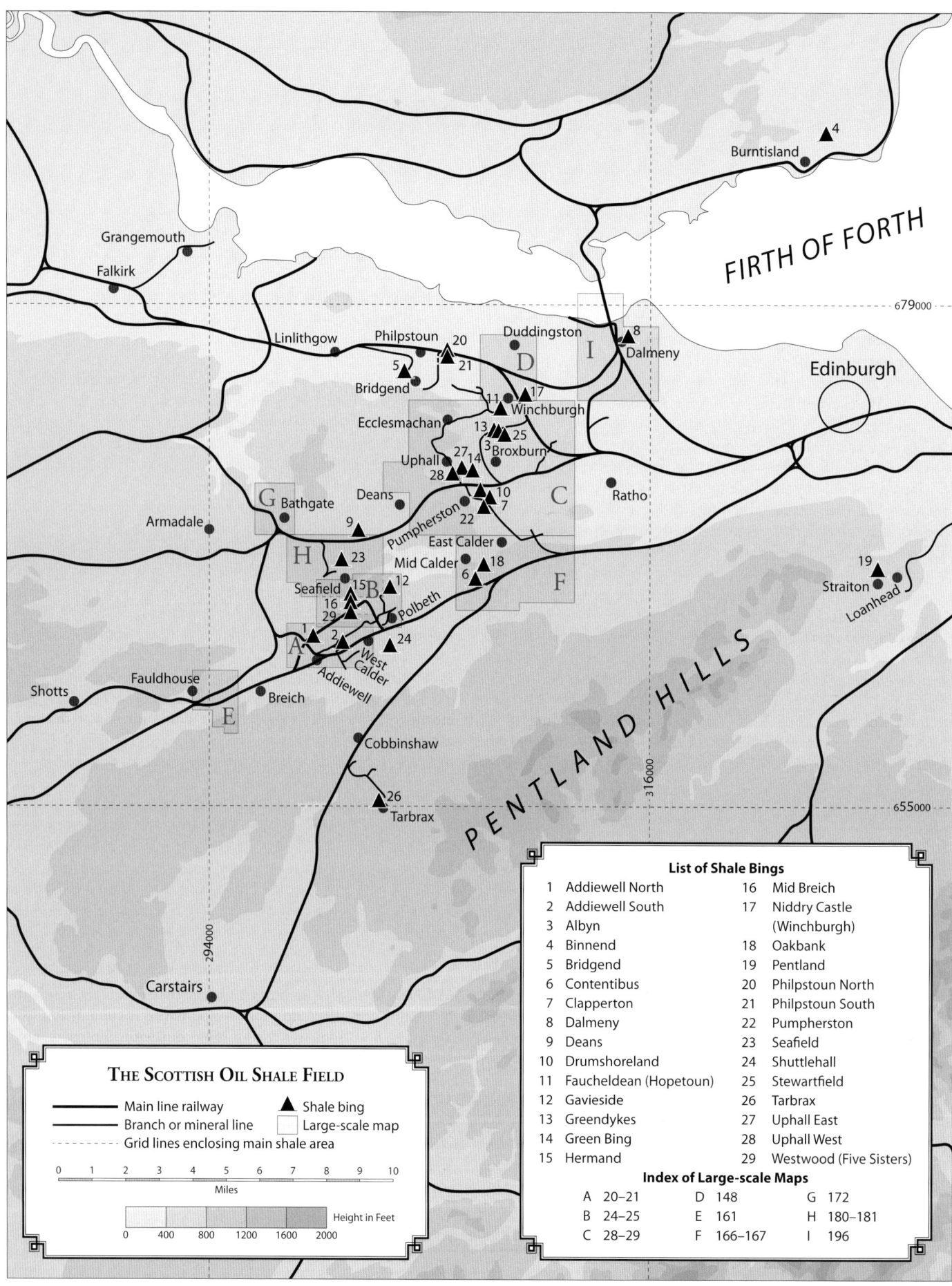

INTRODUCTION

This book has been a long-term ambition. I was born in West Calder at the beginning of the Second World War. Growing up in the village, one was very much exposed to the shale oil industry, the industry around which village life revolved. As a young boy, whilst my father was not employed in the industry, many of my friends had fathers who were miners and we inevitably had neighbours who were miners. In the family home in Hartwood Road it was possible of an evening to hear the shots going off under the house as the mining of the raw material continued apace, night and day. The shale oil industry completely dominated life and was not to be ignored.

This was the area with its dominating industry into which I was born in 1940. As time went by and I grew older, I came to be fascinated not only by the shale industry as a whole but also by the train workings which served the pits and the oil works, as well as the works complexes themselves. The industry was, without doubt, highly significant to the UK economy. What is also without doubt is the fact that whilst working practices were governed by such things as Mining Regulations and other safety standards of the day, there was sometimes a fairly cavalier attitude to strict compliance. In the chapters which follow I comment upon the fact that, prior to the Second World War, the shale miners were required to purchase their own explosives from the local Co-operative ironmongery department! The upshot was that it was not unknown for miners to hold explosives at home, and not necessarily in small quantities.

One of my early recollections was the evening of Friday 10th January 1947, around 8.30pm. The street, normally quiet at that time of the evening, was busy, with an extraordinary number of people out and about. The following morning, we woke to the news of the terrible underground disaster which had occurred, just close by, at Burngrange Nos 1 & 2 pit. I was taken to the pithead later that Saturday morning and my memories, even now as clear as day, are of the many people, especially women, standing silently around in small groups, the snaking coils of fire hoses running to the shafts and the caged canaries waiting their turn to go underground with the Mines Rescue Service. It was a sad and tragic time in the village communities of West Calder and Seafield, the latter being where many of the men came from, villages where everyone knew everyone else and where everyone was touched in one way or another. Fifteen men perished in the accident, the worst single incident ever recorded in the history of the industry.

When I was thirteen, I was taken down the same Burngrange pit by my second cousin, one David Brown, who was oversman there and who had been awarded the Edward Medal (later to be exchanged for the George Cross) for bravery during the pit disaster. I had the opportunity to visit a working face and had the methods of shale extraction explained.

Later, when in my fifth year at West Calder High School and undertaking a higher geography project covering the geology of the shale oil industry, through the good offices of Mr John MacArthur, the Mines Manager, two colleagues and I were afforded the opportunity to again go underground. Initially it was agreed that we should go down Hermand No. 4 mine on a designated Saturday morning but, sadly, owing to a roof fall there which had killed one miner earlier in the week, we were diverted to Burngrange pit where we were given the grand underground tour of the pit right down to a working face and there experienced shots being prepared and then fired.

Miners placed much store on education, most hoping that sons would not follow them underground. In this respect we were extremely fortunate, since West Calder High School, one of the four senior secondary schools in the County of Midlothian at that time, and my *Alma Mater*, was a seat of learning *par excellence*. We also enjoyed an excellent Carnegie Trust library in West Calder, where all children received much encouragement to read books and learn, from the incumbent librarian, Miss Blackwood, or, as we referred to her (but never in her hearing), 'wee wheest'! Should anyone have been imprudent enough to make a noise whilst in these hallowed premises, a loud '*wheest*' would emanate from the front desk. Similar quality learning and educational establishments existed at Bathgate, Linlithgow and Broxburn.

Whilst the coming of the shale oil industry was to change the way of life in West and Mid Lothian, and expand or create new communities with a burgeoning population, it was also to change the physical face of both counties significantly. The mining and retorting of raw shale to extract the crude oil meant that there was a large quantity of spent (burnt) shale to dispose of and this waste in fact equated in volume to the shale mined, that is, every cubic yard of shale mined meant a cubic yard of spent shale to be disposed of; this was done by the expedient of just creating a waste tip, or, as it is called in Scotland, a 'bing', adjacent to the retorts. Much good farming land was quickly swallowed up under the ever-growing bings and at Addiewell at least one farm, complete with steadings, named 'Clash me Doon' was soon to be swallowed up by the immense Addiewell (South) bing. These unsightly red bings littered the West Lothian and Midlothian landscapes and were, and many still are, visible for miles.

When the early bings were formed, the retorting process was producing somewhat improperly burnt shale, with the result that the residue was black and prone to spontaneous combustion (such as happened at Addiewell North bing) but with improvements in retort technology, so all constituent properties were then removed leaving a red, inert, slightly alkaline blaes.

Following the extraction of raw shale, so the problem of surface subsidence arose as abandoned underground workings slowly began to fall in on themselves. This subsidence could be, and often was, gradual and insidious, occurring over a period of time, but on occasions it could be surprisingly quick.

Around 1947, new playing fields were created for my school, as well as new pre-fabricated annexes to provide additional classroom accommodation. The annexes were constructed so that the effects of subsidence could be addressed, but the playing fields, laid out

A panoramic view of Winchburgh village and Niddry Castle Oil Works taken from Hopetoun Oil Works bing (Faucheldean). Niddrie Castle Oil Works lies to the upper right whilst the Hopetoun Rows in the village can clearly be seen from left to centre rear. The Union Canal bisects the picture.

on ground lying above the workings and within clear sight of Burngrange pit, became almost unusable in just a short space of time because of subsidence and the creation of large, deep pools of surface water which gathered and could not drain away. Truly a case of short-sighted planning!

In 1924 the Government of the day introduced a Tax Preference which gave Scottish hydrocarbon oil products an advantage over imported fuels. This preference was guaranteed for ten years in 1934 but, only four years later, in 1938, was further extended and the rate increased. In 1962, the Government announced the ending of the tax preference and Westwood, the last remaining oil works, closed in 1962. The Scottish shale oil industry was dead.

Not a lot remains in 2012 of the industry which once was all-important to the area. Strangely, whilst it was always thought that the gigantic red bings would serve for posterity as a physical reminder, they themselves are now disappearing due to (commendable) landscaping projects and the recovery of the spent shale blaes for infill in major motorway and other civil engineering works. Addiewell (South) bing was probably one of the largest of the many bings in the area, but this too has all but disappeared. Indeed,

recently, after a visit to West Calder, I returned home using the old and very familiar road via Tennant's March and Addiewell village to Bathgate, which ran round behind what had been Addiewell bing and … I actually missed the bing! West Lothian District Council are to be commended for having had the Westwood bings, now known as the Five Sisters, preserved with three others as national historic monuments to an industry which was cutting edge in its day and which affected so closely the lives of many thousands over so many years.

I suppose it was inevitable – since both my grandfathers were mining engineers – that I inherited an interest from them in what was the primary industry around which my younger days revolved. Whilst my working career took me to the railway industry, my first and life-long love, nevertheless I continued to take an interest in the local industry and was, with many others, very sad when the shale oil industry, and a way of life, was swept away in such a very short time.

After all these years, with spare time now a reality, I have set about trying to describe, in layman's terms, just how the oil was won, where the many pits and oil works, now long disappeared, actually

INTRODUCTION

stood and where the myriad internal mineral railway lines servicing this industry ran. Happily, many signs are still there to be found, and it is hoped that this book might encourage others to go out and explore, to find the traces, the industrial archaeology, of what was the making of our home county.

This book, I must stress, deals with the shale oil industry in and around West and Mid Lothian and purposely ignores other oil refining activities which were common-place throughout Scotland. It is written in several parts, tracing the geology of shale, the extent of the individual shale fields, the mining of shale, the refining of shale oil and just how the raw materials were transported from source to oil works and the final products from oil works to consumer. It also looks at the social aspects of shale life. However, I must also highlight the fact that, whilst the book traces in some depth the history and other aspects of this industry, it is written more as an overview and it is not, and never was intended to be, a detailed technical treatise. The technical details of all the procedures described herein can be readily found elsewhere by the inquiring reader. Neither can it be considered a definitive work since much has been lost over the intervening years, but some sterling work is now being undertaken by the Scottish Shale Museum (see below) in putting together a complete and detailed history of the industry as far as is now possible. Being a historical work of the life and times of a now largely forgotten industry, it is accepted that the modern-day reader can only interpret the contents within the context of their own circumstances. Nevertheless, I think it behoves any historian, amateur or not, to reflect relativism in a way which could not be expected of contemporary participants, and this I have tried to do in this work.

This book was also written against the background that, within the UK, oil from shale was not confined to the Lothians but around the 1890s was also being obtained by the Kimmeridge Oil & Carbon Company, when the exposed seams of Kimmeridge Jurassic Shales (Blackstone) at Clavell's Head in Kimmeridge Bay, Dorset, were being worked. The company reported that there were 5,000 feet of underground workings *in situ* at that time. The Blackstone was a very rich mineral shale consisting of some 70 per cent organic material (kerogen) and producing, under pyrolysis, some 120 gallons of oil per ton, an immense quantity in anyone's book. The company did not survive for long thereafter, although the Kimmeridge shale is now, once more, being closely studied. Nevertheless, Scotland, and West Lothian, was to be the world leader in shale oil production right up until the 1930s.

However, one must take pride from the fact that it was the Scottish shale oil industry, certainly the largest in the world up to 1921, which was at the forefront of pushing the development of the technology for this purpose and which then pursued the extraction of oil by this process for just over 100 years.

The industry is also recalled in a fine display of artefacts contained at the Scottish Shale Museum, part of the Almond Valley Heritage Centre which incorporates the old Livingston Mill Farm (complete with mill); located centrally within Livingston, West Lothian, it is well worth a visit. I am indebted to Dr Robin Chesters and Karen Bell of that establishment, and also to John Holt who oversaw the Shale Villages Project 2010, for all help and encouragement freely given, and for allowing me to access the most valuable and comprehensive archives held there, a collection now quite correctly considered as being of national interest to Scotland. The unique Almond Valley Heritage Centre photographic collection surely records all the facets of the Scottish shale industry for posterity, and all photographs included in this book are, unless otherwise acknowledged, reproduced by the kind permission of Dr Chesters on behalf of AVHC. It is here at Livingston that the ongoing work mentioned above is currently being undertaken. Thanks also go to the BP Archives at Warwick University, Coventry.

I must specially mention the late William (Bill) Maxwell of Polbeth, an employee of Scottish Oils and a fund of information, who both welcomed and fostered my early interest in the industry, and Lynda Maxwell for help in dispelling my doubts when confronted by the sheer enormity of what I had proposed to take on, and for ongoing support. To Hamish Stevenson, my sincere thanks, again for assistance given and photographs kindly loaned, and to HMRS for permission to use images of shale activities and oil tank wagons. The National Archives of Scotland (and, from 1st April 2011, The National Records of Scotland) in Charlotte Square, Edinburgh was, as always, a valuable reference source and I am indebted to the duty archivists there for their unfailing help and support; also to Andrea Massey at the Map Room of the NLS in Edinburgh, my grateful thanks for maps supplied, and to the Trustees of the National Library of Scotland for their kind permission to make use these maps. Thanks to Douglas Yuill, Ed McKenna, Charles Davidson, Douglas Blades and Graham Todd for all the railway information freely given. To Jim Wilson, my thanks for all his assistance given in locating various pits/mines on the OS maps. The West Lothian Local History Library in Blackburn (now relocated to Linlithgow in 2011) under Head Librarian Sybil Cavanagh, with her almost encyclopaedic knowledge, proved to be the source of much information regarding shale and the industry. Without all named, this book would not have been possible. I am also grateful for the assistance provided by the Grangemouth Heritage Trust in sourcing some additional, but most valuable photographs. Finally, I am indebted to Neil Parkhouse and Nigel Nicholson of Black Dwarf Lightmoor Publishing for making this book a reality. Whilst every effort has been made to ensure that the contents are as accurate a record as possible, it may be that some discerning readers may spot inaccuracies. If this is the case then the responsibility for any such omission or error is mine, and mine alone.

The book itself is, quite simply, dedicated to both the industry which dominated my early life and to the many men (and women) who gave a lifetime (and indeed, in some instances, life itself) in its service. It is my sincere hope that it will evoke memories of the days, now long gone, amongst readers who, like myself, fondly remember those times and also perhaps stimulate an interest amongst those who know nothing of the Scottish shale oil industry.

Harry Knox, FIRO MCIT CMILT
Linlithgow, 2012

A view of the longwall face in the Fraser pit showing hutches being loaded by hand. Note the wooden sleeper 'pack' filled with waste material supporting the roof to the left of the picture. The inclination of the shale seam being worked is very obvious, a feature which made the roof very difficult to shore up securely. A short-lived coal cutting machine lies ready to make a start undercutting the face.

PART ONE
THE SCOTTISH SHALE OIL INDUSTRY

The Oakbank Oil Company aerial ropeway passing over Glebe Farm to the south-west of Mid Calder village. Note the safety screening to protect against spillage.

1

THE GEOGRAPHY

In geographical terms, the whole of the central belt of modern Scotland is a clearly delineated rift valley, lying between the Highland Fault Line (Northern Boundary running between Helensburgh and Montrose) and the Southern Uplands Fault Line (Southern Boundary stretching from Stranraer to Cockburnspath). In this central area lay the major mineral deposits – coal, oil shale, sandstone, limestone, fireclay and iron stone – formed over many millions of years; they were here discovered, and worked, to serve the great industrial revolution occurring in the eighteenth, nineteenth and early twentieth centuries. There is little doubt, then, as to why the greater part of the total Scottish population can be found in this relatively small area of Scotland.

The area containing the oil-bearing shales worked by the oil extraction industry in the period from circa 1850 until 1963 is, to all intents and purposes, contained within one single sheet of the OS Landranger 1:50,000 series of maps, sheet 65; the precise area bounded to the north by grid line 679000, to the east by grid line 316000, to the south by grid line 655000 and to the west by grid line 294000. Some mining of shale and crude oil production took place just beyond these boundaries, in Fife at Burntisland, outside Edinburgh to the south at Loanhead, and in the west in Ayrshire and Renfrewshire, but these activities in overall terms were relatively insignificant and short-lived; the main and most significant concentration was in the area described.

Oil-bearing shale was to be the making of the then county of West Lothian (Linlithgowshire) and to a lesser extent, Midlothian (Edinburghshire). Before the advent of the shale industry, working of the coal seams which were a continuation of the Lanarkshire coalfields had occurred in certain localities, particularly on the exposed Monklands plateau and in and around Armadale and Bathgate where coal, canneloid coal, ironstone and fireclay had been found in some considerable quantity. Indeed, in even earlier days, the mining of silver had taken place in the Bathgate Hills.

Below the rich soil of the Almond Valley lay, largely undiscovered until the mid-nineteenth century, various rich seams of oil-bearing shale. This shale field, finally surveyed in 1857, was found to measure some 75 square miles in total. Whilst mainly lying within the West Lothian county boundaries, it also encroached into Midlothian on the southern and western borders lying along Breich Water and the River Almond, these two geographical features marking the old county marches before the regional boundary changes under the Local Government (Scotland) Act, 1973 which reformed local authorities in Scotland. The new West Lothian District created by these boundary changes in 1975 now more or less covers the complete area being described.

The main villages (some mere hamlets) and towns to be changed almost beyond recognition by this developing industry were, from west to east, Bathgate, West Calder, Mid Calder, Pumpherston, Uphall, Broxburn, Winchburgh and to a lesser extent, Dalmeny. Some almost entirely new communities were also created within this area. The West Lothian county town of Linlithgow lay just outside the western edge of the shale fields, although a company called the Linlithgow Oil Company was to have a short existence, working and retorting shale no more than one mile east of the town at Bridgend.

From the early 1960s, the development of Livingston New Town commenced within the area around the Almond Valley, north-east of Bellsquarry and north-west of Mid Calder, and the town, now much expanded, sits almost completely in the centre of the shale fields and is bounded by Bathgate and West Calder in the west, Uphall and Broxburn to the north-east, and East Calder to the south.

It is the shale oil activities carried out within this area upon which this book will now primarily focus.

Almondell viaduct carrying the NBR branch from Uphall Junction to Camps across the River Almond. The river formed the former county march between West Lothian (left bank) and Midlothian (right bank).

A view of Broxburn Oil Works looking west. The Union Canal runs from lower right to left.

2

THE GEOLOGY

Geological Time Periods

The creation of the land mass which was to become Scotland, whilst extremely interesting, is largely beyond the scope of this book. Suffice to simply say that the splitting of ancient super-continents such as Laurusia (Euramerica) or the Old Red Sandstone Continent, the movement (drift) of the resultant land masses occurring over many millions of years, and the collisions between these, were to form the Scotland we now know. This is a country which was more closely associated with the North American land mass (Laurentia) – until a collision with another drifting land mass (Avalonia) upon which lay what was to be England – caused the formation of the land mass now known as Britain. Over these eons of time, the drift of the Laurentian land mass northwards, from the South Pole up through the equatorial regions to the present position, gave rise over the long process to Scotland's unrivalled geo-diversity.

The creation of these mineral deposits can be traced back over some five geological time periods of the Paleozoic Era as described below.

Old Red Sandstone Period (400 to 360 million years ago)

This was the period most significant in Scotland's rock formation and included a suite of sandstone rocks, the red colouring coming from the presence of iron oxide. The rock formations also included conglomerates, limestone, mudstones and siltstones. The central belt was one of the four major basinal areas where this rock formation was to be found.

During this period, the rift valley began to be filled with detritus deposited from the newly developing Caledonian Mountains. There was also, within this period, some extensive volcanic activity which, amongst other features, resulted in the formation of the Pentland Hills on the south-eastern edge of the valley south of Edinburgh.

Carboniferous Period (360 to 285 million years ago)

This period followed the Old Red Sandstone Period. During it the land mass which was to become Scotland lay on the equator, close to the southern margins of the ancient super-continent, and experienced a tropical climate. This was a period of some glaciations, low sea levels and the formation of mountains.

Further volcanic activities occurred towards the end of this era, resulting in the rock formations and the creation of significant volcanic plugs such as Arthur's Seat and the Salisbury Crags in Edinburgh, the Bass Rock, Taprain Law and North Berwick Law in East Lothian, Binnie Craig, Dechmont Law, Cockleroy and other volcanic plugs and sills in West Lothian.

Between these periods of volcanic activity, this rift valley filled with warm water in one vast tropical lagoon and limestones developed from the coral reefs. This lagoon was retrospectively to be known as Lake Cadell, so named after the famous Scottish geologist H.M. Cadell who did so much to record the geology of the central belt and the shale fields. There are some differing opinions offered as to the actual water content of this body of water and whether it was salt (sea) water or fresh water. The latter is the generally accepted view, although it is highly likely that the sea did encroach into the lake on regular occasions.

During the Lower Carboniferous Dinatian and Early Namurian periods (Heerlin classification), the many amphibians of the period lived and died in this lake, and were to be preserved as fossils in the layers of the fine silt and plant debris deposited by the tides. The organic (vegetable) content of these layers (kerogen) gradually decomposed and accumulated to eventually form an oil-bearing shale. Around the edge of this lagoon lay a swamp of primitive plants which, over time, also formed great coal seams, lying under a great thickness of sandstone; these were to become the rich and productive coal fields of Ayrshire, Lanarkshire, Midlothian, Fife, Stirling and Dunbartonshire. Thus, the Lothian oil shales were laid down in water ('lagoonal' or 'esturine' shale) whilst the coal reserves were laid down on the surrounding drier ground, but both as a consequence of the layering of decaying vegetation.

Towards the end of the Carboniferous Period, the rock strata, including the shale seams of the Central Belt valley, became severely folded with a series of folds running on a roughly north to south axis. Owing to the pressures exerted by the newly developing mountains of the north, these folds were then the subject of considerable and complex east to west faulting and produced very complicated outcropping patterns which were later to be identified and recorded by Cadell. This complexity led to the rather random and piecemeal extraction of the shale during the life of the industry.

Permian, Triassic and Tertiary Periods (245 to 5 million years ago)

This was the period when the great land masses began to split up and which saw the evolution of diverse life forms. The mass of land (Laurentia) from which Scotland was formed now began the slow drift northwards, and new sandstone, developing from the dry sand dune formations created during the Tertiary Period, became deposited in the folds of the central belt valley. Thereafter, three major glacial advances occurred within the 50 million to 10,000 years period; the detritus resulting from glacial tills and scouring by glaciers, plus a tilting to the east of the whole land mass, left a diverse mix of rocks and minerals across the central belt. The areas which became West and Mid Lothian thus inherited, from west to east, considerable bands of coal, fireclay, cement limestone, sandstone, limestone and oil shale, all covered by a rich glacial till, and was to be blessed with fertile soil in addition to rich mineral wealth.

Oil Shale Described

The term 'shale' comes for the Old English word *scealu*, meaning shell. Oil shale is, as we have seen, a sedimentary rock of great antiquity (350 million years) occurring in smaller bands within the thicker seams of shale, itself also sedimentary and based on clay. The oil shale is, therefore, in simple terms, the compressed and hardened mud or silts of the geological periods described, containing large quantities of organic and vegetable matter (sapropel fuels) rich in fatty or waxy substances. There is no firm geological definition for oil-bearing shale, neither does it have a specific chemical formula; instead, oil shale is a general term adopted to describe this fine-grained sedimentary rock which yields significant quantities of oil upon decomposition by heating to high temperatures (pyrolysis).

The palaeontological evidence (fossils) discovered in the shale measures, and the distribution thereof, were to prove of immense value to the geologists in assisting with the dating process. Numerous examples of marine fossils were identified in both Raeburn and Pumpherston shale beds. As a further point of interest, the oldest known fossilised lizard in the world (reptiliomorph), from the Lower Carboniferous Period, was discovered in East Kirkton quarry near Bathgate in 1989. This fossil was found in finely grained limestone and the blue-grey (non oil-bearing) shale measures (Little Cliffe shale) existing there, and was named Westlothiana Lizziae. Lizzie, as it is more affectionately known, can now be seen in the Royal Scottish Museum in Edinburgh.

Geological Position of the Oil Shale Measures

The noted Scottish geologist Sir Archibald Geikie first mapped the portions of West Lothian and Midlothian on behalf of the Geological Survey in 1857, detailing various exposures of what he was to describe as 'bituminous shales'. He also referred to other significant minerals of the same group such as the two-foot coal, Houston marls and Binny sandstone, all of which were of great stratigraphical significance in assisting with the subsequent identification of the geological structure of the region. From Geikie's maps it is readily apparent that he fully recognised the scale and extent of the folding which had occurred in the strata of West Lothian.

The subsequent and rapid development of the oil shale fields required a more detailed survey of the area and shale fields, and this was undertaken between 1884 and 1887 by H.M. Caddell, who was attached to the staff of the Geological Survey. Through his work it was confirmed that the oil shale measures upon which the oil industry was later to depend formed part of the calciferous sandstone series of West and Mid Lothian and the southern coastline of Fife between Burntisland and Inverkeithing. The shale occurred in a mixture of mudstones, sandstones, marls, marine limestone, freshwater limestones and thin coal seams. The Carboniferous system in Scotland was thereafter to be based on the most valuable works of Cadell and another geologist, J.S. Grant Wilson, and arranged in four main divisions and in a descending order:

- *Coal Measures.* These consisting of red sandstone, shales and marls, overlaying white and grey sandstone; containing numerous valuable coal seams and ironstones.
- *Millstone Grit.* Thick, coarse sandstones, fireclay, thin coals, ironstones and thin limestones.
- *Carboniferous Limestone Series.* Arranged in three sub-divisions: the highest containing limestones; then, in the centre, thick beds of sandstone and some valuable seams of coal and ironstone; finally the lowest, beds of marine limestone, sandstone, shales, some coals and ironstones.
- *Calciferous Sandstone Series.* Consisting two sub-divisions. The upper, the Oil Shale Group, was over 3,000 feet in thickness in the Lothians; in its upper part it contained beds of coal of mainly inferior quality, and below that a series of oil shale seams, with about seven significant seams intermixed with beds of sandstone, shale, fireclay, marl and estuarine limestones. In the lower, the Cementation Group, no oil shales were found, but this group consisted of white sandstones, grey, green and red shales, clays, marls and other sandstones.

As already stated, the West and Mid Lothian shale fields covered approximately 75 square miles, being located mainly in a belt of territory about 3 miles broad, stretching over a distance of some 25 miles. From Dalmeny and Abercorn on the shores of the River Forth, south-westwards across the fertile tract of land through which ran the River Almond which traversed the main area, bounded by the Bathgate Hills to the north and the Pentland Hills to the south, into the moorland districts of Tarbrax and Cobbinshaw,

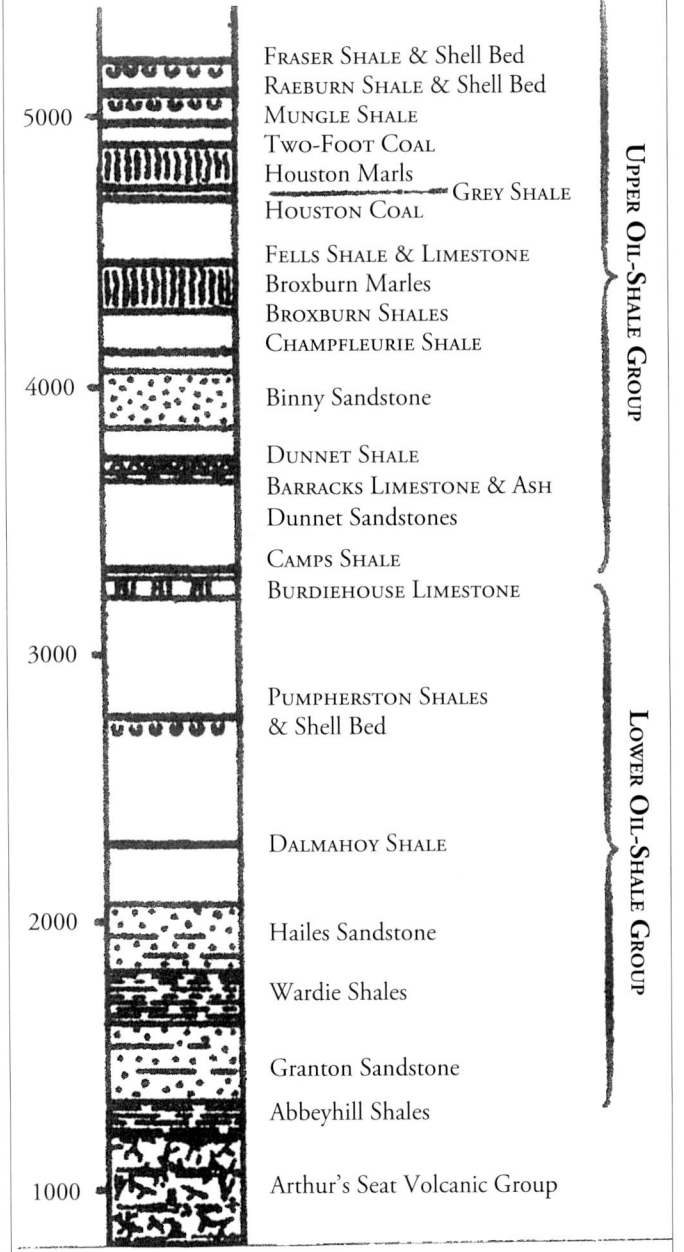

The general geological position of the Lothians oil shales in the upper and lower groups.

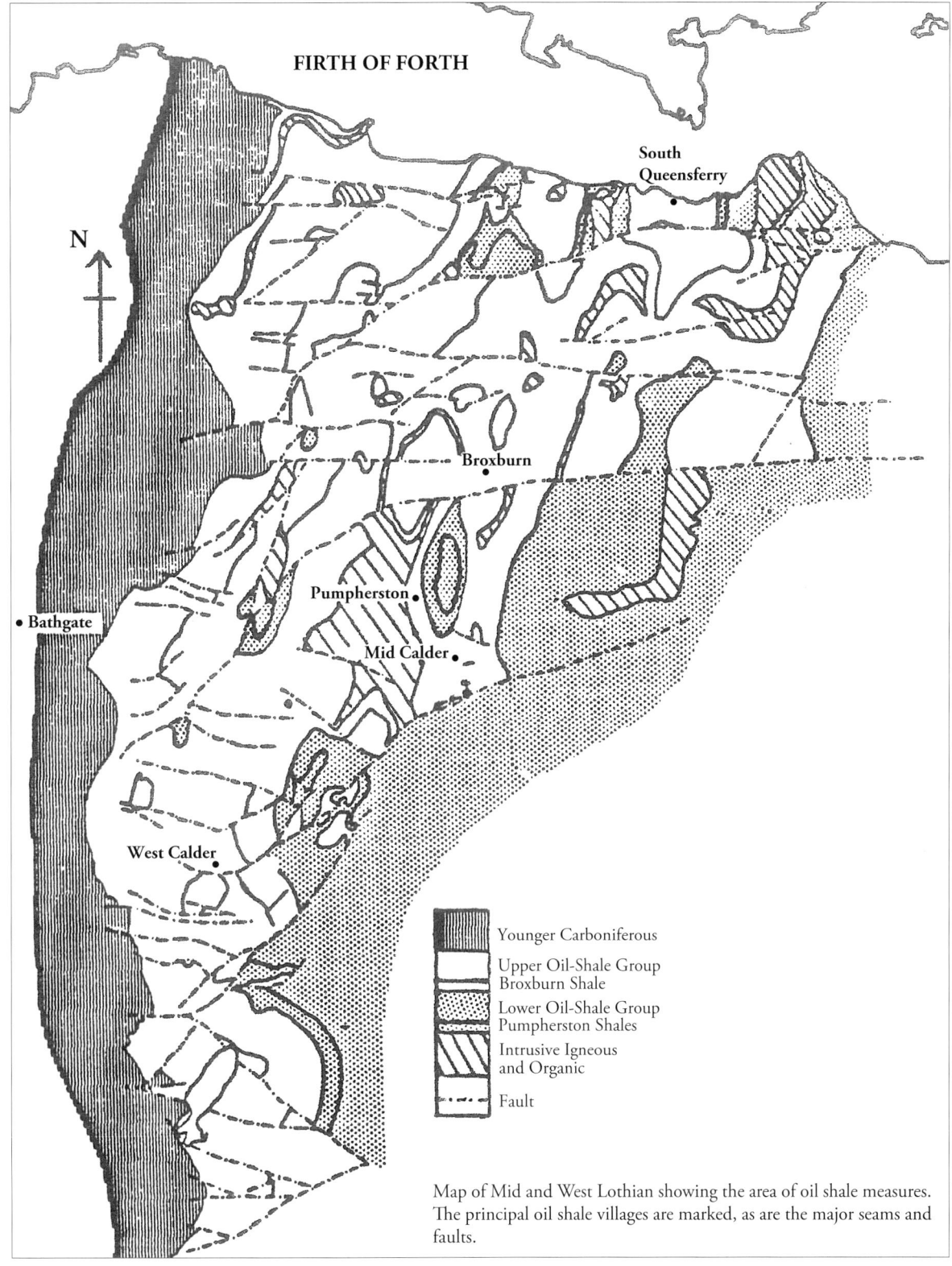

Map of Mid and West Lothian showing the area of oil shale measures. The principal oil shale villages are marked, as are the major seams and faults.

immediately on the Mid Lothian/Lanarkshire county march. The structure of the strata dipped overall from east to west and the seams which outcropped were generally of the younger Upper Oil Shale Group. The Lower Oil Shale Groups were found in the eastern side of the shale fields, these remained as the seams progressed westwards but only to be found at deeper levels. As discussed above, the rock folds produced an extremely complicated outcropping pattern and structure because of the complex east–west faulting, thus providing somewhat of a geological puzzle for geologists such as Geikie and Cadell to sort out. The land rose westwards throughout this area, from basically sea level on the shores of the Forth Estuary to a maximum elevation of around 900 feet at Cobbinshaw, with the highest ground running along the southern edge of the shale fields. On the higher ground at these western extremities there lay little else other than moorland and poor soil overlaying rich seams of coal and ironstone. To the north-east, the same oil shale seams extended under the River Forth to also make a brief appearance on the Fife coastal plain.

The shale seams varied greatly in thickness. At some locations they all but disappeared and merged into ordinary carbonaceous blaes, but at other locations they were found to be from 6 feet to as much as 15 feet thick. Volcanic intrusions (sheets of volcanic rock), up to 250 feet thick, were, in some places, found to have been forced between the sedimentary layers during the periods of volcanic activity previously referred to, the extreme heat from which had then effectively rendered the oil shales in the vicinity blind (that is, the oil had been driven out of the shale).

Cadell's work established that, out of fifteen seams recognised, there were seven main seams of oil shale, namely Raeburn, Mungle, Fells, Broxburn, Dunnet, Camps and Pumpherston. The seams varied in thickness across the field, the Dunnet seam being the most uniform in thickness over the intervening 25 miles between Cobbinshaw and Burntisland. The Broxburn and Pumpherston seams were closely-spaced and were generally worked together. Other identifiable seams included Chamfleurie, Grey and Fraser; yet another, the Dalmahoy seam, was believed to lie below the Pumpherston seam, but was never to be worked. Oil companies and miners had their own names for the seams worked, but they all were essentially part of the Cadell classification.

In addition, through his most valuable and important survey, Cadell established that there were five distinct main faults (vertical displacements of the strata) across the oil shale field, with a maximum displacement of up to 1,600 feet being recorded, together with a number of associated smaller fault lines. The main fault lines identified were the Murieston, Calder, Middleton Hall, Winchburgh and Ochiltree.

Chemical Composition of Oil Shale

The physical characteristic of the Lothians oil shale was a fine black or dark-brown clay-shale, but with some additional features which enabled it to be easily distinguished in the field. This oil shale was a fine-grained, laminated, indurated clay rock which was free from grit and which resembled hard, dark wood or dry leather. Its quality could be ascertained by the ease with which it could be cut with a sharp knife. When so cut, it tended to curl up and when crushed, produced a rich ochreous-coloured powder. Amongst miners it was known simply as 'shale'; geologists always referred to the stratified rock as carbonaceous shale, distinguished as a blaes by the bluish colour it assumed when it decomposed into clay. The two types of shale, oil shale and carbonaceous shale, were readily recognised in the field. Generally, the good-quality oil shale was darker in colour; it was also very tough, quite flexible and was very weather resistant. Inferior shales were considerably more friable under the same tests and they very soon disintegrated when exposed to weather. The highest quality shale was as described, and as the colour graduated through brown to grey, so the quality decreased. In internal structure, oilshale was finely laminated, as was very apparent in the 'spent shale' after retorting, when it was seen to be composed of extremely thin sheets like the leaves of a book.

Simply put, the oil shale of the Lothians consisted of what was, in essence, merely hardened clay combined with widely varying quantities and types of organic matter (sapropels). To the actual oil-yielding portions, the names of 'kerogen' and 'pyrobitumen' were initially and provisionally suggested by a noted gentleman of science, one Alexander Crum-Brown, Professor of Chemistry at Edinburgh (1869–1908), with the term 'kerogen' becoming the accepted description thereafter. This oil shale was discovered to undergo full or partial decomposition at given temperatures and under controlled conditions, giving off hydro-carbon gases of the paraffin, olefine and naphthene series, which, after condensation, produced a heavy crude oil accompanied by ammonia, the latter being derived from a portion of the nitrogen in the shale.

In reality, there was only limited knowledge of this remarkable material known as oil shale, with various theories being expounded over the early years as to the nature and origin of the kerogens contained in the shale. Some findings expressed the view that the origin of kerogen was organic matter, whilst another opined that it was the relic of a former oil field, derived from petroleum rather than being a source of petroleum. Generally, however, it is now accepted that the kerogens in shale were largely, if not entirely, vegetable in origin, derived from the decaying waste of plants growing on the swampy land surrounding Lake Cadell.

Analysis (proximate) of core samples of oil-bearing shale across the fields revealed the following:

- Moisture   between 1.40% and 4.16%
- Volatile matter   between 17.58% and 22.95%
- Fixed carbon   between 2.10% and 7.58%
- Ash*   between 69.58% and 76.00%
- Crude oil   from 17.19 to 30.44 gallons per ton
- Sulphate of ammonia from 32.55 to 85.51 lbs per ton
- Ferrous carbonate
- Sulphur
- Nitrogen
- Arsenic

  } minute traces only

   * The constituent composition of the ash was:
       Silica
       Alumina
       Ferric oxide
       Lime
       Magnesia
       Sulphuric acid
       Alkalies

The Fells shale seams offered the highest yield of crude oil but lowest yield of sulphate of ammonia, whilst conversely the Pumpherston No. 3 seam offered the lowest yield of crude oil but the highest yield of sulphate of ammonia.

In terms of overall yields for Scottish shale, evidence gathered over the years appeared to suggest that there was a tendency for the shale to deteriorate with depth. Shale mined from seams nearer the surface yielded more crude oil than the shales obtained at greater depths. Whilst this apparent trend was fairly constant with some shales such as the Broxburn seam, it was less apparent elsewhere and with different seams. The strata folding also appeared to indicate a similar trend where shale seams which had been put under great geological pressure revealed the same patterns of yield. However, controlled tests using the same shale from the same seam and mine, but retorted in two different types of retort, revealed that the quantities of crude oil and sulphate of ammonia varied significantly, and so many of the various theories then existing were rendered uncertain.

Petroleums (liquid oil) yielded around 90 per cent; shale, on the other hand, generally produced a uniform quantity of crude oil and across the board yielded some 75 per cent of refined products, however it contained far higher proportions of unsaturated hydrocarbons. This ratio was to put shale at a distinct disadvantage with the developing markets around the world for liquid crude oils.

3

THE SHALE FIELDS IN DETAIL

The shale fields were eventually identified by area and geological features as a consequence of the geological surveys carried out; thus, for the purposes of this book, that same geological identification has been followed and they are so described hereunder.

Southern Area: Cobbinshaw to Uphall

Cobbinshaw Shale Field

(See diagram on p. 156.) Cobbinshaw, lying on the Lanarkshire/Mid Lothian border, was the most westerly limit of the Lothian shale oil field and was somewhat unique by the fact that all the oil shale groups outcropped within this very small area. This shale field surrounded Cobbinshaw Loch. Two significant faults existed in the area, those being the continuation of the Wilsontown fault line in an east–west direction, which marked the southern boundary of this field, and the Murieston Fault which formed the northern boundary. No workable shale was ever found below the Fells seam in this field, although Broxburn, Dunnet and Camps shales were all present. Pumpherston shales were also found on the Harburnhead side of Cobbinshaw and at other bore sites, but no evidence was uncovered to suggest that these seams would be workable, given the exceptional condition of the overlaying Camps, Broxburn and Dunnet seams.

The pits/mines working these seams were: Cobbinshaw South Nos 1 & 2 and Cobbinshaw South Nos 3 & 5 mines, Viewfield No. 5, Tarbrax Nos 1 & 2 pit, Tarbrax (Viewfield) Nos 3 & 4 pit and Greenfield No. 1 pit worked the Fells seam; latterly Viewfield No. 4 worked the Fraser seam. The short-lived Cobbinshaw North pit (Nos 1, 2 & 3) also worked these seams.

Woolfords No. 5 pit and Viewfield pit produced coal for the oil works at Tarbrax. Later, the Pumpherston Oil Company, on taking over this oil works, also used this coal for Pumpherston refinery.

West Calder Shale Field

(See Map A, pp. 20–21.) It was in this field that shale was mined from the earliest days of the industry. It was the sheer quantity of shale available here that led Young to obtain six concessions for the extraction of oil shale and to set up his new oil works at Addiewell when the Torbane coal ran out, in order to exploit the potential offered by this area. Shale outcropped around West Calder and split shale had been widely used by farmers to line field drains over

The 'big' pit at Tarbrax, probably Tarbrax (Viewfield) Nos 3 & 4 pit.

THE SCOTTISH SHALE OIL INDUSTRY & MINERAL RAILWAY LINES

Map A: West Calder Shale Field (1922). Lower centre lies the large Addiewell Oil Works and village (rows). From the lower left and running immediately to the south of the oil works, the Caledonian Railway's Mid Calder to Cleland line of 1865 crosses the shale field. The connecting sidings from this line to the works are clearly visible at Addiewell Oil Works signal box. The Caledonian Railway West Calder Loop Line from Woodmuir Junction sweeps around the north side of Addiewell, crossing the NBR branch line from Bathgate to Addiewell on the level at Cuthull. Both companies are rail connected to the United Colliery Company's Foulshiels colliery (centre left) at this point.

To the north-east side of Addiewell Oil Works, the Addiewell & Polbeth mineral railway to Westwood and the Gavieside pits strikes off eastwards following Briech Water (the West Lothian/Midlothian county march at this point). Running south from the oil works and then turning north-east is the mineral railway to Burngrange No. 39 pit, which passes under the Caledonian Railway main line. The route of the former railway lines to the former Muirhall pits is clearly marked. At a later date, a connection off this branch was to access both the yet-to-be-sunk Fraser shale mine and Baads coal mine; also, at Burngrange, No. 39 closed and a new pit, Burngrange Nos 1 & 2 was sunk immediately south of the old pit.

North of Breich water in the centre of the map can be seen the remains of a curved railway embankment which carried the short-lived branch line from the CR West Calder Loop Line near Westwood Rows to Stewart's original paraffin oil works at Breich Dykes, closed in 1877. (Scale: six inches to one mile.)   *Courtesy of the Trustees of the National Library of Scotland*

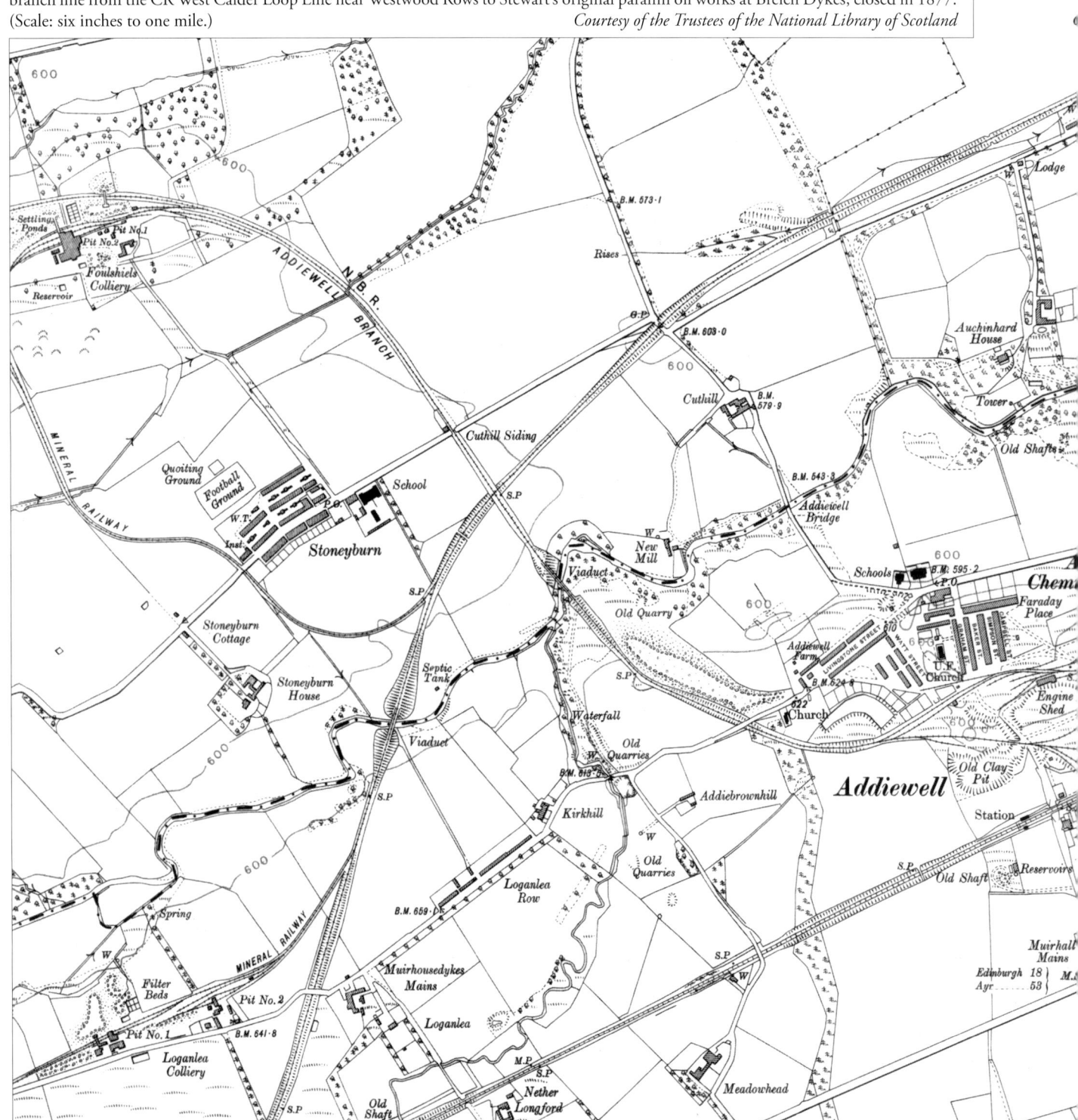

THE SHALE FIELDS IN DETAIL

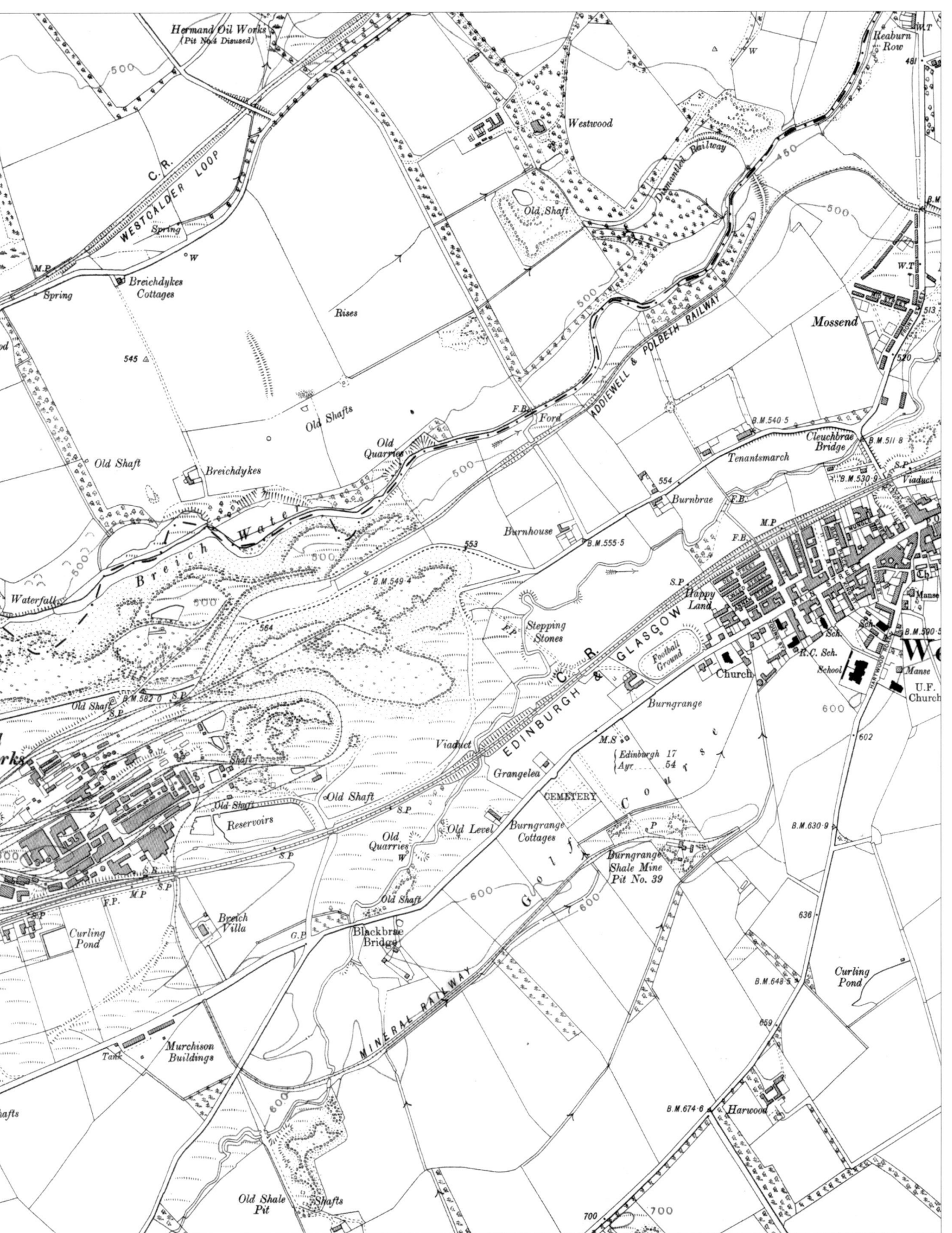

A view of the new Westwood pit, near West Calder.

the years (a fine example of such shale drains can still be found at Bridgend, near Linlithgow). This field covered an extensive area, bounded to the north by the east–west fault line south of Polbeth, to the east and south by the Murieston Fault and to the west by an outcrop of Cobbinshaw limestone.

Towards the west end of the village lay the centre of a broad anticline which crossed the shale field in a general north-to-south direction, with a deep basin occurring around Hartwood (or Harwood).* Evidence exists of many small inclined mines working the seams in the anticline close to the surface in the Addiewell/West Calder area, some of which only now have numbers to identify them, for example, No. 9, a small inclined mine known to have existed on the banks of West Calder Water just west of the village, close to the point where the A71 and A704 roads now diverge, on the north-east side of the bridge over the burn. The bifurcation of these roads is still, even today, referred to as 'No. 9' by locals. It is entirely likely that this small mine, which may originally have been Muirhall or Blackbraes No. 9, provided shale for the short-lived Burngrange (Blackbraes) Oil Works situated close by. The detail has, however, been lost in the mists of time! Other early pits were Muirhall Nos 1 & 2, No. 16, No. 18 and No. 19, the latter three being listed as coal workings. Baads No. 15, Baads No. 17 and Baads No. 22 were listed as shale mines (NT003615) working both Broxburn and Fells shale, but had closed by 1884. Addiewell Nos 2, 3 and 4 mines and Addiewell Nos 2, 5 and 7 pits worked both the Grey and Mungle shales close by, on the north-east side of the oil works.

The Fells seam, or 'thick shale of Addiewell' as it was known locally in earlier mining days, was the main seam worked, so-named since it was first mined by A. Fell at Gavieside pit, where the seam varied between 3 feet and 7 feet in thickness. This seam then gradually ran out, becoming too thin to mine in other areas. It outcropped alongside Hartwood Water south of the village and was found to be 6 feet thick at that point. The South Hermand field, consisting of both Broxburn and Fells shales, was worked by both inclined mines and pits belonging to the (old) Hermand Oil Company, but the field was soon abandoned owing to the great number of intersecting faults and the need for heavy pumping to control the continual ingress of water.

Fells shale yielded 26 to 40 gallons of crude oil, and 35lbs of sulphate of ammonia per ton. Addiewell was also the only area where the Grey shale seam was thick enough (around 4 feet) to be mined. Other seams being worked in this field were Broxburn, Dunnet and Raeburn. Muirhall Nos 1 & 2 worked the Grey shale.

The Raeburn shale was the highest producing seam worked until the Fraser seam was identified and opened up some time later, with this new seam being considered the richest oil producer in the West Calder district in the early days. This seam was worked by the Fraser pit at 68½ fathoms (411 feet).

---

* The name Hartwood, also known as Harwood, is common in and around West Calder, and both spellings can still be found. The estate lying to the south-west of the village was always Hartwood Estate, and the author was born in Hartwood Road. However, the situation then becomes confusing, for the farms surrounding Hartwood House on the estate are Harwood, Little Harwood, West Harwood and Mid Hartwood. The church close by was Hartwood Church, but on closure in 1960 and amalgamation with Polbeth Church (which was, incidentally, before the new church was built, the old winding house of Polbeth No. 26 pit), the new church became Polbeth/Harwood and still is today. Both spellings must, therefore, be considered to be correct and indeed Hartwood Water, referred to above, is also shown as Harwood Water on some OS maps.

THE SHALE FIELDS IN DETAIL

This was the only shale field where the Mungle shale was mined. Known also as 'Stewart's Shale', it occurred in 2ft seams adjacent to Breich Water at Addiewell. There were also thick seams of this same shale lying in the Hartwood Basin and the shale was considered to be the richest oil producer in the West Calder district. It was not found in any workable quantity elsewhere in the shale fields and the seam was not being worked at all by the 1920s. It is believed that the shale from this Mungle seam was used in the original Stewart's Oil Works which ceased production in 1877.

The Broxburn seam was certainly worked on the banks of Hartwood Water south of the village, and supplied the small oil works (retorts) at Shuttlehall (Old Hermand) (NT029625) in earlier days. The site of the retorts at Shuttlehall can still be identified by the small bing of spent blaes, now gradually being removed. Whilst this Broxburn shale was an inferior seam in terms of the thickness when compared to the other seams around West Calder (below 3 feet), it was worked by Burngrange No. 39 pit. However, just south-west of Danderhall, another inclined mine, Hermand No. 6, worked this seam at the point where it proved to be 6 feet thick.

The Dunnet seam was first worked to supply New Hermand Oil Works and was of sufficient thickness (up to 10 feet) to be worked all over the Lothians shale field. This seam was later to be worked by Hermand No. 4 mine lying on Hermand Estate to the south-east of the village, this having an inclined main shaft length of 2,000 feet. This same Dunnet seam was also worked at Burngrange Nos 1 & 2 pit on the west side of the village, at a depth of 72 fathoms (432 feet).

Young's also mined Cobbinshaw, or Hurlet, coal at Baads No. 42 mine (also sometimes called West Mains) (NT003613), lying to the west of the village, to supply Addiewell Works with the fuel for retorting, refining and other heating activities and this seam of coal was widely worked over a period of time between Baads Mill and Addiewell. Baads mine main shaft ran for a mere 144 feet in length at 1 in 3.8.

POLBETH/GAVIESIDE SHALE FIELD

(See Map B, p. 25.) It was in this shale field that, until at least 1925, the deepest pits in the whole industry were to be found. The pits, Gavieside No. 40, Polbeth No. 26 and Limefield No. 32, worked the Dunnet seam at a depth of 250 fathoms (1,500 feet) for the Young Company. The Polbeth shale field was a south-eastwards continuation of the Breich field, but with the strata lying in a deep basin at this point. A fault known as the Langside/Blackburn Fault marked the northern boundary of this field. To the north of this fault, lying between Gavieside and the River Almond, there was a further broad basin containing the Dunnet shale which was a northerly continuation of the aforesaid Polbeth/Langside Fault and the seam here, as worked by the Gavieside pits, varied between 6 and 16 feet thick, with, in Gavieside No. 40 pit, a general working thickness of 10 feet. Polbeth Nos 7, 8 and 10 worked this seam.

Baads coal mine at West Calder showing mine entrance. This coal mine was owned by, and provided coal for, Young's Paraffin, Light & Mineral Oil Company's Addiewell Oil Works. Note the quantity of new timber pit props waiting to be taken underground.

THE SCOTTISH SHALE OIL INDUSTRY & MINERAL RAILWAY LINES

Map B: Polbeth/Gavieside Shale Field (1909). From the bottom left, the Caledonian Railway West Calder Loop Line runs north-east serving (New) Hermand Oil Works before curving southwards around the Westwood site containing Westwood Paraffin Works and Westwood Nos 12 & 13 pit. It was on this site that the new and final shale oil works was to be built in 1941. On the lower right-hand side of the loop lies Gavieside, the site of the former oil works is marked by the bings. The Addiewell to Polbeth mineral railway passes under the West Calder Loop Line (with the connecting cord clearly shown) to serve Gavieside No. 4 pit, No. 27 pit and No. 40 pit before terminating at Limefield No. 30 pit. A reversing connection then gives access to Polbeth No. 26 pit (shown as closed), No. 21 pit and No. 20 pit. A new Polbeth No. 26 pit was sunk near the site of the former pit. In the right centre can be seen the remains of the bings created by the short-lived Grange Farm Oil Works served by a short branch line from the West Calder Loop and marked by a line of trees running south to north on the lower right of the site. (Scale: Six inches to one mile.)   *Courtesy of the Trustees of the National Library of Scotland*

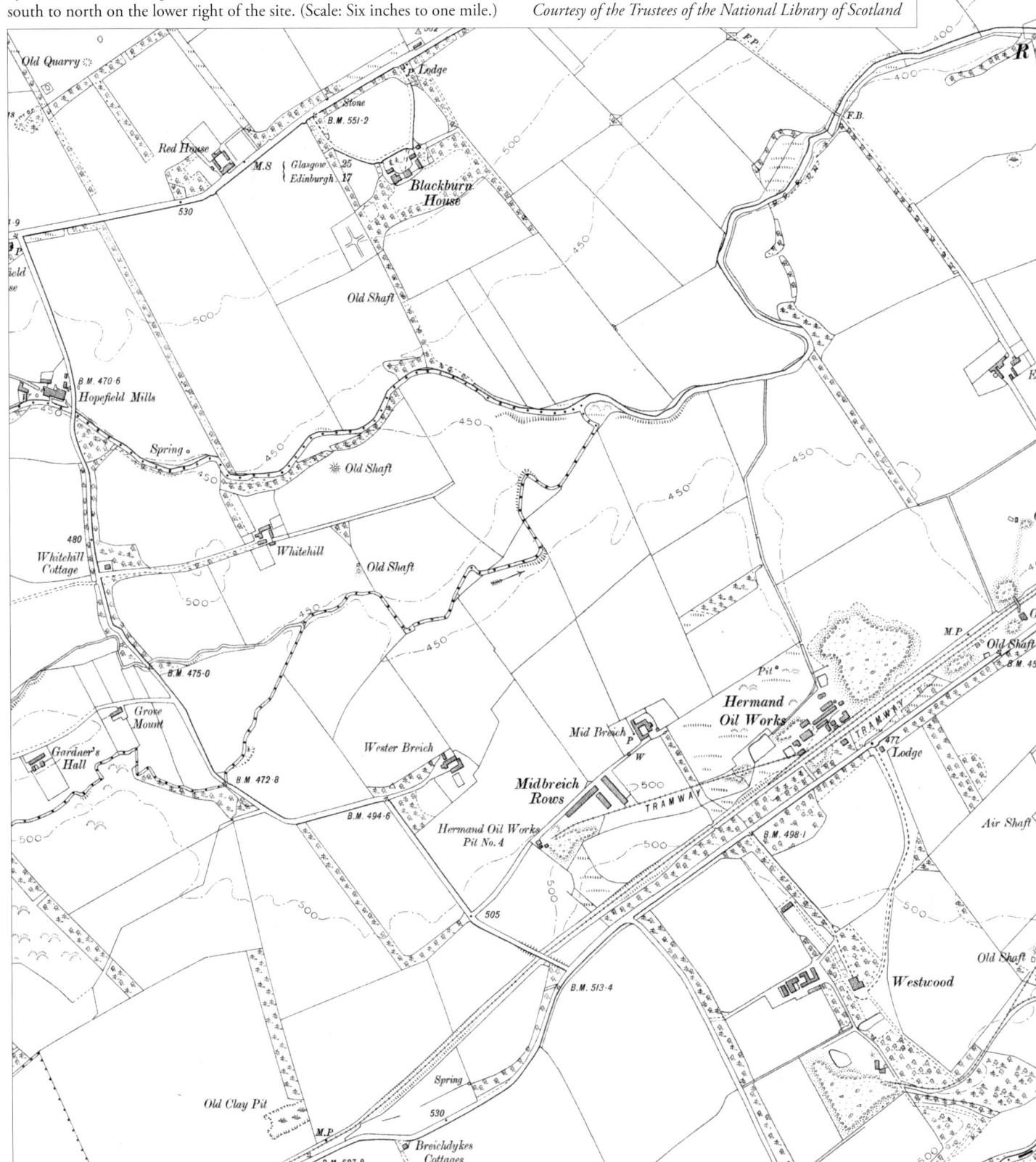

THE SHALE FIELDS IN DETAIL

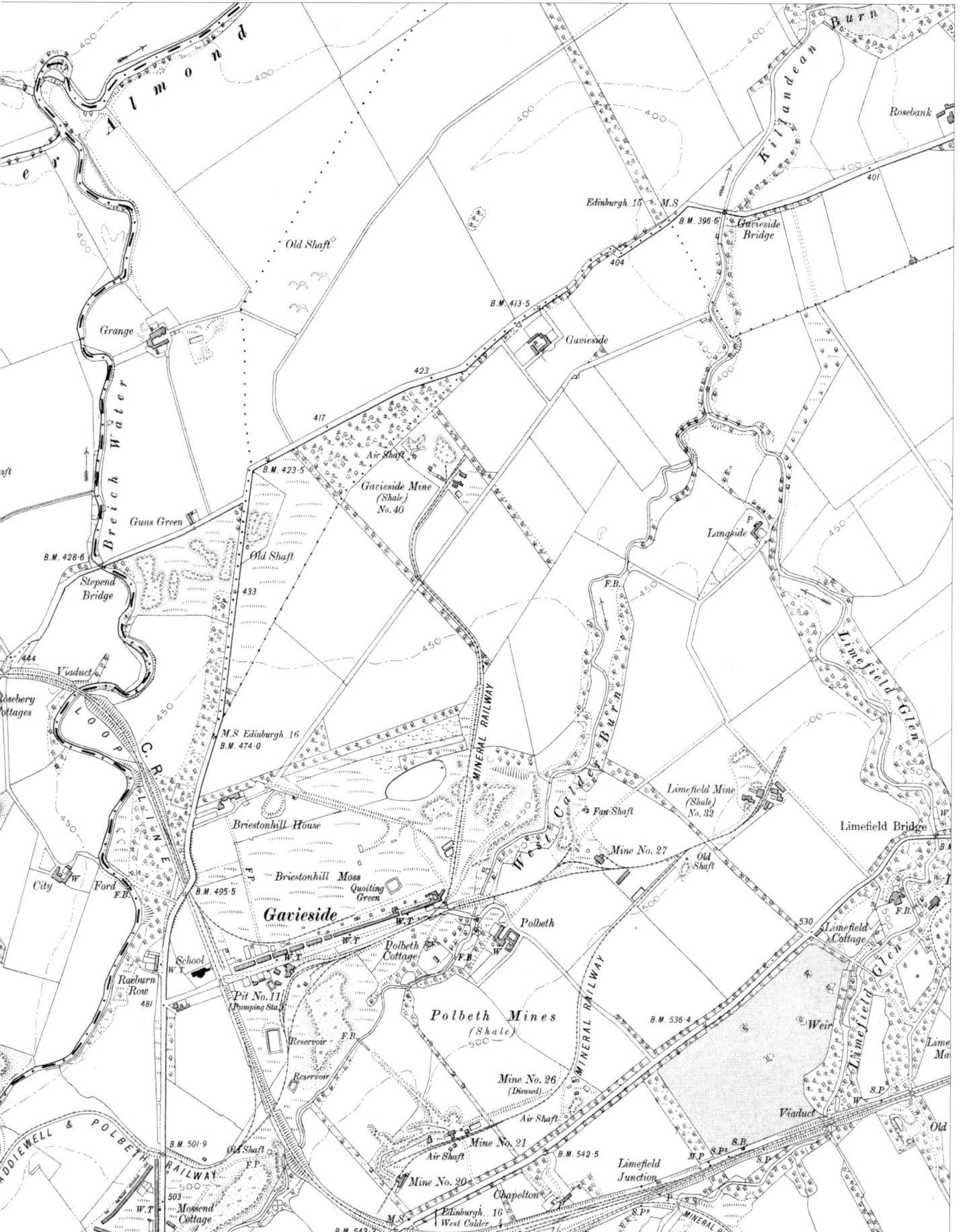

Polbeth No. 21 pit, also known as the Old Drove Road Pit, worked the Broxburn seam, and Polbeth (Gavieside) No. 11/Westwood No. 30 worked the Addiewell and Fells seams. Westwood No. 12 and Westwood No. 13 pits worked the Fells, Broxburn and Dunnet seams. The later (and last) Westwood pit worked the Broxburn seam (4 to 6 feet thick) at 70 fathoms (420 feet) and the Dunnet seam (9 to 10 feet thick) at 120 fathoms (720 feet).

Earlier, Houston coal had been worked along both sides of Breich Water.

Breich Shale Field

(See Map B, p. 24.) This field stretched from the aforementioned Langside/Blackburn Fault, south towards Mid-Breich, where it was separated from the West Calder field by a further east to west fault. The main axis of the Dechmont Arch ran in a southerly direction through this field causing severe inclination of the shale seams, with the Dunnet seam being almost vertical at one point. The angle of inclination gradually eased to the western edge, with the Broxburn seam lying at an inclination of around 60° and the Fells shale at 40°. In the trough on the south-east side of the main anticline lay these Broxburn and Fells shale seams, which were eventually worked out by the Pumpherston Oil Company. However, in the Breich field proper, the Fells shale was much thicker and the main Dunnet seam was 12 to 14 feet thick. Broxburn shales were also well developed, averaging around 4½ feet in thickness.

The Fells seam was worked by Easter Breich No. 1 and Easter Breich No. 3 mines, whilst Mid Breich mine worked both Broxburn and Dunnet shale and Breich Nos 1, 2 & 3 worked the Broxburn seam.

Cousland/ Seafield Shale Fields

These fields lay between the Breich and Deans areas, bounded by the Langside/Blackburn Fault to the south. Another fault, the great Middleton Hall Fault, ran through this area. The main seams worked here were the Dunnet group and the Broxburn group, the latter being considerably thicker and lying only 180 feet above the Dunnet. Most of the mining carried out here was concentrated in the area lying to the north of the Middleton Hall fault line. At Starlaw, further north, a small opening had been made into the Raeburn seam at one time, but mining this seam was never developed further.

South of the Middleton Hall Fault, the Dunnet and Broxburn seams were both worked on a dipping incline of 22° and test borings found the Fells seam was a mere 72 feet above the Broxburn seam at this location. Seafield No. 1 mine and Boghall Nos 3 & 4 worked the Fells seam, whilst Seafield No. 3 worked the Dunnet shale. Cousland No. 1 mine worked the Broxburn seam.

Livingston/Deans Shale Fields

These lay to the north of old Livingston village, but the two fields comprising this group were again separated by the Middleton Hall fault line which tended to run in a south-westerly direction in this area.

In the Livingston field, the Pumpherston seam outcropped in a sharp anticline just north-east of the old Livingston Station, however it was the Dunnet seams which were mined at Deans Nos 1 & 2 (Caputhall) mine.

In the Deans field, the main seams worked were the Dunnet seams, which curved around the southern edge of the Dechmont anticline. The main Dunnet shale seams were from 10 to 14 feet thick and were of excellent quality. Again, immediately above the Dunnet seams lay the Broxburn Shales seams, the latter not more than 30 feet thick in total, but only one of the Broxburn seams was to be worked in this field, at Deans No. 7 pit. Deans No. 3 (Starlaw) mine and Deans No. 4 mine worked the Dunnet shale whilst Deans No. 5 mine (Barracks) worked the Jubilee seam and Nos 6 & 7 worked the Fells and Broxburn seams.

The Mid Calder Shale Field

(See Map F, p. 166.) This small field lay between the River Almond and the Murieston Fault, extending a mere 4 miles from west to east. The Calder Fault, bisecting this area, had caused the most productive shale measures to lie to the northern side of the field, and these consisted of Under Dunnet, New Dunnet and Broxburn shales. These latter shales were of very good quality. These seams were worked by Mid Calder (Oakbank No. 1) mine lying on the eastern bank of Linhouse Water to the south-east of the village. Fells shale of high quality was also present, albeit only in the New Farm Trough. This Fells shale was worked by both Alderston No. 43 and Charlesfield No. 1; upon closure of both of these mines plus the Mid Calder mine, the Oakbank Oil Company sunk a new mine, New Farm Nos 3 & 4 mine, close by Dedridge Rows. The Fells shale was transported to Oakbank Oil Works by a unique overhead ropeway (see Chapter 17), but the New Farm mine was to be worked out and closed by 1919. Evidence of the original mining activity, in the form of headings running into the seams of shale outcropping alongside the gorge of Linhouse Water, can still be seen at Mid Calder.

Pumpherston Shale Field

This field was bounded by the triangle formed by the Middleton Hall Fault, the River Almond and Houston Wood, lying west of the village. The anticline known as the Pumpherston Arch, contained the Pumpherston group of shales (Jubilee, Maybrick, Curly, Plain and Wee), all confined within 92 feet of strata close to the surface. Numerous pits and inclined mines, including Roman Camp Nos 1, 2 & 3 and Pumpherston No. 5, were driven into these seams.

These seams were generally of mixed quality giving low yields of oil (16–28 gallons per ton) but producing a much higher yield of ammonium sulphate (30–70 lbs per ton), higher than any other Scottish shale. The anticline lay on an almost due-north/south axis and pitched away steeply, the angle of inclination being between 25° to 35° on the south side. At the southern end of the anticline the shales deteriorated rapidly. The Broxburn shales in this field, containing seams of 60 feet in thickness and lying in the south east corner, on a strata which was gently inclined, proved to be a small pocket of high quality shale when mined. A further basin was present on the north-west side of the Pumpherston Arch with three separate Broxburn seams each of between 5 and 6 feet thick of high quality shale.

The section of the Pumpherston shales at this location was found to be similar to the section found at Roman Camp, lying immediately to the east. These shales were also found in very good condition on the west side of the field near Livingston but lay in the deepest part of the trough, about 1,500 feet or more below the surface. Pumpherston shale was mined at Pumpherston, initially by opencast mining, later by inclined mines, finally with pits going down to over 160 fathoms (1,000 feet), as the seams dipped to angles close to vertical.

Camps shale was also worked at Pumpherston No. 4 mine and Roman Camp Nos 4 and 5 mines, with the seams at this point being between 7 and 9 feet thick and yielding a good to fair quality shale. The Dunnet shales were also worked in the Clapperton Trough on the east side of the Pumpherston anticline. Fells shale was also found in this field, but the seam was too thin to be worked economically.

Like other adjacent fields, the shales lying to the west of the Pumpherston field were found to have been rendered blind by a vast dolerite sill, a consequence of ancient volcanic activity.

Northern Area: Uphall to the River Forth

Broxburn Shale Field

(See Map C, pp. 28–29.) This shale field was, along with West Calder field, one of the earliest and most productive shale fields to be developed and worked (by Bell). It covered both the Uphall and Broxburn areas and was bounded to the south by the major Middleton Hall Fault, to the north the boundary was marked by the Ecclesmachan and Niddry Castle faults.

Initially it was the highest quality Broxburn shales which were to be worked here, these lying in the basin between Uphall and Broxburn. These seams were worked down to a depth of around 166 fathoms (1,000 feet), but after 1925 and the Broxburn shales had been exhausted, the Dunnet shale seam was then to be opened up. In the centre of this area lay the deep Middleton Hall Trough with, on the east side of the depression, the Broxburn anticline which was a continuation of the Pumpherston Arch, and on the other side the Binny anticline.

It was also in this field that evidence of earlier and significant volcanic activity was to be discovered. Due to immense temperatures associated with this activity, the Camps shale was found to be almost worthless. The area contained high quantities of dolerite, a volcanic (igneous) rock similar to basalt and typically found in the volcanic plugs which had, in earlier times, channelled the molten lava (basalt) to the surface.

The Broxburn shales had three principal seams with a total thickness of some 16 feet in a total strata thickness of only 45 feet, namely the Grey shale (6 feet), the Curly shale (5½ feet) and the Broxburn (Main) shale (5 feet). These shales yielded more than 40 gallons of oil per ton and 20 lbs per ton of ammonium sulphate. The seams were mined by Carldubs mine, Pyothill pit, Haycraigs mine, Albyn mine, Crossgreen mine and the Stewartfield mines. However, the Broxburn shales (Curly seam) had also been rendered somewhat useless by the intrusion of dolerite sills on the eastern side of the anticline; the dolerite was in extremely close proximity to the shale at the entrance to the Albyn mine. The lower band of dolerite found here proved to be of some interest, containing cavities filled with a yellowish-green mineral wax, suggesting that earlier volcanic activity had caused distillation of oil from the shale by the intense heat.

The Holmes Oil Company worked the Pumpherston, Dunnet and Barracks shales at Holmes Nos 1 & 2 pit. The Dunnet seam, a single seam some 8 to 10 feet thick, was worked to the north and east of Broxburn by two long inclined mines, Loaninghill No. 4 and Dunnet Sandholes. The seam dipped at an angle of about 1 in 2 here.

Houston coal, so named after Houston House, with two seams of some 5 feet in total, was also found in quantity in this area, this having been mined from earlier times. Its position was marked by a number of old outcrop pits running from the Middleton Hall Fault to Holmes and round to the west side of the Fivestanks Basin. Once more, however, intrusion of dolerite had in places rendered much of the coal in the seam blind and worthless. Around 1846, two pits in the vicinity of Houston House worked this seam at a depth of 22 fathoms (132 feet), supplying the local district with coal for many years.

Fivestanks shale, a composite seam which contained Fells, Mungle and Raeburn shales, was worked in the small Fivestanks Basin to the south-west of Broxburn. This Broxburn shale field was worked by some twenty-plus pits and mines over the life of the industry, possible the biggest concentration after the West Calder area.

Winchburgh and Humbie Shale Field

(See Map D, p. 148.) This shale field was bounded to the north by the great Ochiltree Fault, to the west by the Union Canal (between Winchburgh and Craigton), and to the south by the Winchburgh Fault which ran towards Kirkliston. No shale was mined in this district due to the underlying Dunnet seam having been infiltrated in numerous places by dark blaes and the considerable dolerite intrusion which had all but destroyed the small but potentially workable Broxburn and Fells seams.

Two important anticlines traversed the area from south to north, the first being a continuation of the Pumpherston/Broxburn Arch which ran out into the River Forth at Hopetoun House. The axis of the Arch ran from east of Winchburgh village, past Duntarvie Castle to Newton (Woodend). The other anticline was a continuation of the Kirkliston Arch bringing up the Burdiehouse Limestone near New Mains.

The massive civil engineering works associated with the construction of the Edinburgh & Glasgow Railway in this area between 1838 and 1842, such as the extensive 2 mile long Winchburgh Cutting and the 1,100 yard long Winchburgh Tunnel, had revealed no evidence whatsoever of oil-bearing shale. Borings indicated that a dolerite 'float' lay near the surface at least as far as Duntarvie Castle, and yet this shale field abutted with another, the Philpstoun field, which was to be the most productive shale field amongst all of the Lothian shale fields.

Ecclesmachan Field

Bounded by the Ochiltree Fault on the north and the irregular Ecclesmachan/Winchburgh faulting to the south, both Broxburn shales and Houston coal were worked here. The strata in this field consisted of a most complex pattern of folds, being a continuation of the other, similar, extreme foldings found in the Broxburn area. Two significant anticlines east of Hopetoun Rows brought up the Broxburn shales. Near Hillend lay the northern limit of the great Middleton Hall Trough.

Camps shale, although present, was unworkable, as were the Dunnet shales. The Broxburn shales were widely mined in this area by J. Ross & Company of the Philpstoun Oil Company. The seams, some 7 feet 10 inches thick, were worked by the Ochiltree pits in the Ochiltree Trough and also at an inclined mine at Glendevon, the main haulageway of which was some 230 feet deep. Hopetoun No. 2 pit also worked the steep syncline on this seam at 60 fathoms (420 feet). In addition, there was some opencast mining conducted at Glendevon for a time. Other pits working the shale in this field were Hopetoun (Glendevon) No. 5 mine, Hopetoun (Glendevon) No. 6 mine, Hopetoun (Threemiletown or Redhouse) No. 35 mine, Hopetoun (Fawnspark) No. 41 mine and Hopetoun No. 44 mine.

On the west side, the shale was again virtually destroyed as a result of the same volcanic activity which had thrown up Binny Craig and created the associated dolerite field which extended a mile or

THE SCOTTISH SHALE OIL INDUSTRY & MINERAL RAILWAY LINES

Map C: Broxburn Shale Field (1922). The area shown on this map contains the greatest single concentration of oil works in the shale fields. The Edinburgh & Bathgate Railway passes from bottom left to centre right. Following it in that direction, the first junction encountered is Bangour Junction where the private Bangour Railway diverged northwards and westwards, serving Dechmont station and terminating within the grounds of Bangour Asylum Hospital. Built and owned by the Bangour Lunatic Asylum Board, the railway was to be worked by the NBR over its lifetime, providing both passenger and freight services for the hospital.

The next location, centred on Uphall, was probably the most important for the railways serving the shale industry – although the site was more properly known as Uphall Station, lying about a mile south of Uphall village. At the eastern end of the station were the sidings serving Uphall Oil Works, and from these oil works a purely mineral line of railway then proceeded northwards passing Middleton Hall – later to be the HQ for Scottish Oils Ltd – then over Uphall West Main Street (Castlehill level crossing) and serving Forkneuk pit, before swinging to the east, passing through Ecclesmachan and terminating at Hopetoun Oil Works. From this line, near Hopetoun Works, another branch line diverged to the north-west to serve Hopetoun (Glendevon) No. 5 mine and, thereafter, Glendevon No. 6 mine, No. 35 mine at Redhouse and No. 41 mine at Fawnspark (all off this map). Another connection from Hopetoun Works gave access to Niddry shale mine to the south-east of the works.

From Uphall Junction (Camps Junction), the Camps branch line ran south and east serving, *en route*, Pumpherston No. 4 mine, Pumpherston Oil Works and Pumpherston No. 5 mine, with another connection accessing Roman Camp Oil Works. Passing over the River Almond, East Calder Goods station lies on the right-hand side of the line adjacent to the main road, whilst the line itself swings east to pass under the Caledonian Raw Camps Branch (entering at bottom right), to terminate at Raw Camps Quarry. A connecting spur provided an exchange facility between the NBR and the CR at this location. There is also a long siding from the CR branch serving Raw Camps limestone mine. The CR branch proper terminated at Camps Goods station (CR).

Proceeding eastwards from Uphall, a connection to the north served the site of Holmes Oil Works (closed) and, at Drumshoreland, another branch line diverged away northwards towards Broxburn, serving Broxburn Holygate Goods depot before crossing the Main Street to enter Broxburn Oil Works. Here, a mineral line crossed Greendykes Road (level crossing) and split, one leg running north then west to serve South Greendykes mine, Hayscraigs mine, Pyothill mine and Carledubs mine, whilst the other leg gave access to Albyn Oil Works and connected end-on with the branch from Broxburn Junction to Albyn Oil Works. NBR locomotives were barred from crossing the level crossing, so this was not a through route.

From the Drumshoreland to Holygate line, close by Powflats Farm, a further connection was laid in (Powflats Junction) for a branch line which curved away south-west to pass over the E&BR main line and served Roman Camp Oil Works.

Finally, further east on the E&BR main line, at Newliston Junction, a branch line diverged to the north-east passing under the E&GR main line 36-arch Almond viaduct to access Newliston No. 29 shale mine. (Scale : six inches to one mile, scaled to 40% to fit page.)

Courtesy of the Trustees of the National Library of Scotland

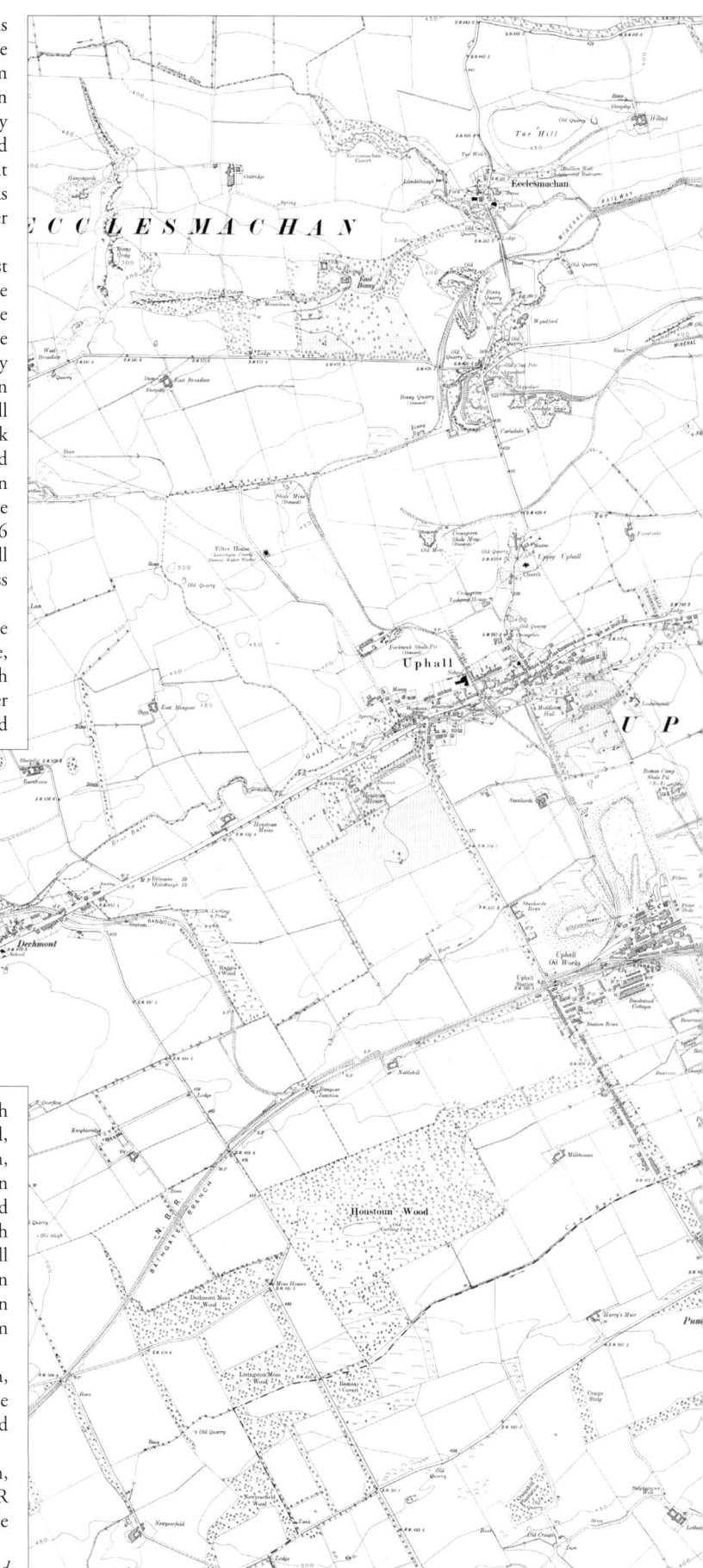

THE SHALE FIELDS IN DETAIL

more on either side. Once again, mineral wax, distilled by the heat generated by this volcanic activity, was widely to be found.

Fells shale was found, but in unworkable quantities. The Houston coal was worked to a considerable extent by the Linlithgow Oil Company for their furnaces and retorts.

CHAMPFLEURIE AND PHILPSTOUN SHALE FIELDS

These shale fields lay to the north of the Ochiltree Fault. The Broxburn and the Champfleurie shale seams were widely worked in this area. The latter shale was, in fact, more properly the Lower Grey Broxburn shale and was worked in the early days by the Linlithgow Oil Company (also sometimes known as the Champfleurie Oil Company) and later, with the demise of this company, by J. Ross & Company of Philpstoun. This shale was entirely local, being chiefly confined to, and worked in, the centre of the Gateside anticline, it lay in seams of 10 to 15 feet thick. However, these seams thinned rapidly both to the north and south of this area. There were numerous mines in this area over the years, Philpstoun No. 2, No. 3, No. 4 and No. 7, Champfleurie No. 1 and No. 4 and Ochiltree Nos 5 & 6, No. 7 and No. 8.

The Broxburn shales were extensively worked all round the Gateside Arch and were found in exceedingly thick seams close by the Nethermill Bridge. From the surface, some 600 feet of shale measures had been identified in a total depth of only 910 feet. Philpstoun No. 1 mine worked both the Broxburn seam and the Dunnet seam, and was somewhat unique in the industry since the mined shale was conveyed directly to the oil works via an underground continuous rope-worked tramway.

The Dunnet shale was also found in a valuable seam, some 10 to 12 feet in thickness north of Philpstoun, although south of Philpstoun it was unworkable. It outcropped at Whitequarries, but the seam thereafter inclined steeply westwards at 30° to 40°, the angle decreasing as it passed under Philpstoun House. The site was to be redeveloped in the late 1930s with a new inclined mine, Philpstoun (Whitequarries) No. 1, being sunk to tap the seams lying north of the Midhope Fault, whilst Philpstoun No. 6 mine was reconstructed and acted as the winding house for the new working, this being done to supply the new Oakbank Oil Company's Niddry Castle Oil Works with raw shale. Fells shale was also found but again in unworkable quantities. A new inclined mine was driven down the Dunnet seam at Whitequarries, which was also to supply large tonnages of shale to Niddry Castle Oil Works and both mines worked until the final days of the industry.

BLACKNESS SHALE FIELD

This field lay in the furthest north-west corner of the Lothian shale fields and consisted of an area of only 2 to 3 square miles around the village. It was closely linked with both the Philpstoun and Chamfleurie fields and whilst it was proved not to contain any worthwhile workable measures of oil shale, the area was nevertheless of extreme geological interest to both geologists and mining engineers in the shale oil industry. Borings proved quite conclusively that the stratum in this field mirrored that of the West Calder area and contained the same Riccarton limestone, Raeburn shale and Fraser shale, but with the latter having been largely destroyed by volcanic activity.

DUDDINGSTON SHALE FIELD

(See Map D, p. 148.) Lying north of the Ochiltree Fault and west of South Queensferry, the area contained two major anticlines. The Hopetoun Arch was a continuation of the Pumpherston Arch, whilst the other was in line with the Kirkliston/Humbie Ridge. The Duddingston Trough lay between these anticlines. The main shale worked was the Dunnet shales, found in two bands, each of some 5 feet in thickness and both were worked by Duddingston Nos 1 & 2 mines and the later Nos 3 & 4 mines.

Broxburn shales of high quality were also worked in the syncline at Totleywells mine where seams of up to 8 feet in thickness had been found. This shale yielded 30 gallons of oil per ton with 40–50 lbs of ammonium sulphate also being obtained. This shale field was worked by the Oakbank Oil Company and supplied shale to their works at Niddry Castle.

THE DALMENY/QUEENSFERRY SHALE FIELDS

These shale fields lay on the north-eastern corner of the larger shale region, close by the southern end of the Forth Railway Bridge and separated by the Ochiltree fault line.

The Queensferry field lay to the north of this fault line. Most of the various seams of this shale field were clearly exposed on the southern shoreline of the Forth Estuary. The main shale seams worked in this area were the Dunnet, Camps and Pumpherston shales. There was a thick seam (6 feet) of Dalmahoy shale underlying the Pumpherston seam, but this was never worked.

The Dalmeny area comprised of the seams lying on the south side of the Ochiltree Fault between Dundas Castle and Cramond. Here the Broxburn seam, the 6 foot seam and the Curly shale were worked by Rosshill Nos 1 & 2 mine, the Broxburn being the lower seam with a thickness of 7 feet. Duddingston Nos 3 & 4 also worked both the Dunnet shale and Pumpherston shale.

INGLISTON/NEWLISTON SHALE FIELD

This field lay south to south-east of Kirkliston, stretching almost to the Edinburgh city boundary. In the centre sat a broad anticline, the Kirkliston Arch. At Newliston No. 29 pit, Young's worked the Dunnet seam; at Ingliston, although the shale was different in appearance, it was believed that it was still the same Dunnet shale, and occurred in seams between 5 feet and 9 feet in thickness. This shale plus the Camps shale was also extensively worked by Young's at Ingliston No. 33 pit and Nos 36 and 37 pits, all traces of which now lie under the main runway 06/24 and surrounding area of Edinburgh Airport.

OTHER SCOTTISH SHALE WORKINGS

BURNTISLAND SHALE FIELD

The Dunnet seam ran north-eastwards under the River Forth, varying in thickness between five and nine feet and was worked by the Burntisland Oil Company in four inclined mines at Binnend, all of which lay on the north-east side of a fault line in Fife and close-by the oil works which had been established there. Further trial mines were sunk at the dip at Newbiggin, but the results with shale obtained there were disappointing, and no further development took place.

STRAITON SHALE FIELD

This small, self-contained field lay on the periphery of the West Lothian shale fields south-west of Edinburgh, around 1 mile north of the village of Loanhead. The field was bounded on the western side by the Pentland fault line and lay to the east of the Straiton anticline. Here, the Midlothian Oil Company set up mines, initially to work the Broxburn seam, but the Dunnet seam, between 5 and 6 feet thick in this area, was to prove the main oil producer. This latter seam was worked by pits at a depth of 196 fathoms (1,176 feet). The workings were taken over by the Clippens Oil Company (see below), but the operation was later to close because of litigation brought by the Edinburgh Water Commissioners regarding the stability risk posed by the mining operations to their main water supply pipes which bisected this field. Mortonhall Nos 9 & 10 mine worked the Dunnet seam, as did Straiton Nos 3 & 4 and Straiton Nos 7 & 8, whilst Pentland Nos 1 & 2 worked Dunnet, Broxburn and Fells shale.

This field was very much an isolated pocket of oil shale surrounded by rich limestone and coal seams, the limestone seams being around 30 feet thick at this location. Coal was to be widely mined at and around Loanhead, and was latterly mined at 400 fathoms (2,404 feet) by one of the new and most successful Scottish 'superpits', Bilston Glen, which opened in 1963.

West of Scotland (Various)

In Renfrewshire, in the west of Scotland, Lillie's coal shale, a canneloid material, was mined by the Clippens Oil Company and the Walkinshaw Oil Company at or near Johnstone. The oil produced from this material could not economically meet the changing quality demands; as a result, the Clippens Oil Company's oil producing process at Linwood, near Paisley, was eventually transferred to Straiton (Midlothian). The Walkinshaw Company also tried to establish a new operation in the Lothians but was unsuccessful.

Levenseat

Lying about three miles to the west of West Calder, Levenseat was not, in fact, a shale field in its own right, but rather was the point at the western edge of the West Lothian shale reserves where the seams were thinning out and disappearing. Levenseat neither lay on a shale field proper nor on the rich Lanarkshire coal reserves, although the latter were being worked a mere mile away. Rather, it was a sort of 'no man's land' where the principal mineral workings were to be limestone and ironstone, with very fine, high quality sand also being discovered. The sand was much used for horticultural purposes and was also transported by rail to Pettycur Bottle Works in Fife for glass-making. Overlying the significant limestone seams, however, there was a very thin seam of oil shale and this was included with mining lease for the limestone at a give-away price of 1/- per ton. This shale was therefore also worked, and went to supply the short-lived Handaxwood (Levenseat) Oil Works.

The Occurrence of Natural Oil and Gas

Natural Oil

The occurrence in the Broxburn Field of natural semi-solid oil in the form of a greenish waxy substance has already been mentioned. The presence of this material was considered to be the result of the considerable volcanic activity known to have occurred in this field during the Carboniferous Age, the intrusion of dolerite sills between the seams of oil shale and the high temperatures associated with this. The oil shale was rendered blind where this occurred and this semi-solid oil matter was the only evidence of the geological events which had occurred all those many years before.

At Westwood pit, in the Polbeth/Gavieside shale fields, natural oil was another unusual occurrence and was first observed in sandstone which was present, appearing as seepage through joints in the sandstone strata which formed the pavement of the workings. No evidence of volcanic activity or intrusion had ever been discovered in this neighbourhood and the oil was thus originally thought to be an entirely natural, albeit puzzling event.

However, in the adjacent Breich Field, Breich pit was to be another pit where natural oil was discovered, again in porous sandstones underlying the Lower Dunnet shale. When the shale was removed from above the sandstone, the relief of pressure caused the joints in the sandstone strata to move and flex, with an ensuing seepage of oil, accompanied in the early stages by salt water and inflammable gases. The oil, present in considerable quantities, was quite fluid at the working depths, but at surface temperature it set to a semi-solid greenish-brown mass. As in Westwood pit, there was no evidence or knowledge of earlier volcanic activity and again, the oil was considered to be an entirely natural, but unexplained phenomenon.

Further research carried out by the US Bureau of Mines, where samples from each source were later sent by Scottish Oils Ltd, concluded that the three oils were in fact all of an identical nature. Here was clear evidence. Whilst the volcanic activity at Broxburn was common knowledge and the Broxburn oil sample had been produced as a result of igneous intrusions, no such intrusions were known to have occurred at the other two sites. The only possible and logical conclusion to be reached was that since all the samples had been found on the same horizon as the Broxburn workings, it was most probable that they must all have been produced by the action of the significant Teschenite intrusion which was known to have occurred in the Broxburn/Pumpherston district; the oil thus released had then found a natural channel towards the final resting place in the deposits of porous sandstones to the west. This must have occurred prior to the subsequent faulting which then effectively separated the two areas, that is, the Broxburn field and the Polbeth/Gavieside/Breich fields.

The search for natural oil continued and in 1919 a series of test bores were drilled in the UK by S. Pearson & Sons Ltd, on behalf of the Mineral Oil Production Department of the Ministry of Munitions. One boring site was located in the West Lothian shale fields, at West Calder, just to the south of Burngrange Nos 1 & 2 pit. The drilling there encountered shows of gas and oil at depths from around 525 feet down to 3,390 feet, but nothing of any viable quantity was found to justify further exploration.

Natural Gas

Natural gas was also to be found in plentiful quantities in the Broxburn field, although it was never to be exploited commercially. It came from two main sources, the first being the firedamp given off by the coal and shale reserves underground seeping to the surface, the second being a permanent gas produced by the intrusions of molten igneous rock into the shale seams.

At an old bore hole (circa 1860) at Middleton Lodge, a constant stream of water seeped out some time after boring had ceased and for some time thereafter. At frequent intervals, around every four weeks or so, quantities of this natural gas would eject with considerable force and noise. This gas was lit and allowed to flare off in a great column of flame which then burned brightly for several days and could be seen from miles around.

An early view of Philpstoun Oil Works. J. Ross & Son's locomotive is shunting in the works yard with a rail tanker and two privately owned wagons. The locomotive, an 'Ogee' saddle tank, is thought to be *The Stag* (no number) built by Barclay & Co. in 1878 (works No. 249).

4

THE HISTORY OF THE INDUSTRY

The Early Years

It must, at this point of the book, be made absolutely clear that there had been a quite significant oil and chemical industry in Scotland well before James Young came on the scene. There were many earlier oil works located close-by east-coast ports where whale, seal and fish oils were processed to produce candles, grease and lamp fuels. Other oil works pressed seeds to extract vegetable oils for lamp oil and cooking purposes, and of course tallow works were set up adjacent to slaughter houses (abattoirs) to render the animal fats into oils, greases and tallow for candles. The burgeoning gas industry retorting coal to make coal gas also produced by-products which were refined to make tars, naphtha, benzoles and other solvents. This book is not interested with these activities, however, and thus it will hereafter concentrate on the use of shales in the production of oil.

An industrial process for the extraction of crude oil from shale was first patented in Britain in 1694. Discoveries were made by the noted German scientist Carl Ludwig von Reichenbach in 1830 and totally independently by Sir R. Christison of Edinburgh around the same time. Several patents were taken out for the making of oil from schists or shales in 1833, and Butler's specification (1833), Hompesch's specification (1841) and Du Buisson's specification (1845) were all registered. The first commercial plant was established in France in 1838. There the scientist Laurent had first obtained paraffines (from the Latin *parum affinus* or 'little affinity') – so named since simple hydrocarbons such as these did not readily take part in chemical reactions with other compounds. These paraffines were obtained by a distillation process of bituminous schist in 1830 and later, in 1835, Dumas obtained the same product from the distillation of coal tars. The product was, however, never anything other than a curiosity. There were known to be extensive oil shale measures in the Autun region of France and Laurent had suggested the working of this Autun shale in 1833. Products manufactured from this source were exhibited by Alexander François Selligue, a French engineer, in 1839, and manufacture of paraffin on a commercial basis in France was credited, in the main, to Selligue. A small shale oil production process was later operated in the Autun region for a number of years, using three benches of Scottish Pumpherston retorts, but the industry was gone by 1950.

In Britain, the true 'Father' of the Scottish shale oil industry was James Young. Born in Glasgow in 1811, he was the son of a cabinet maker and was himself apprenticed to this trade, he also studied chemistry by attending evening classes at Anderson's College in Glasgow (now the University of Strathclyde). After graduating and working at both Anderson's College and University College, London for some time, he branched out to work as an industrial chemist for the great Charles Tennant's (Glasgow) works in Manchester in 1843. In 1847, Professor Lyon Playfair, a fellow-student of Young's in Glasgow, had made a visit to his brother-in-law's (Oakes family) 'New Deeps' coal pit at Alfreton in Derbyshire, and his attention was drawn to the seepage of a thick, black, oily fluid which oozed through fissures in the roof. He spoke with Young, by then a highly respected chemist, and thus it was that Young in 1848, with capital supplied by the aforesaid Oakes family, set up a small business refining this oil that seeped from the colliery workings. In the course of his experiments he found that the crude oil, on being distilled, produced a pale yellow oil upon which floated crystals of paraffin. Young made a great study of this natural 'crude oil' and came to the conclusion that it had been distilled out of the cannel (or parrot) coal by the intense heat of much earlier volcanic activities. This belief then led him to evaluate the use of coal and cannel coal as an

A picture of the great man who started it all, Dr James 'Paraffin' Young.

A very old view of Uphall village before the shale oil industry made its mark.

oil-producing medium through pyrolysis. Two years later he had a share in a patent for extracting oil from cannel coal.

Young had a ready market for this oil produced, with the cotton mills of Manchester by now requiring considerable quantities of lubricating and other oils. The oil at Alfreton eventually dried up and Young immediately turned his attention elsewhere.

He had discovered that canneloid coal, known locally as 'Torbanite' coal, was being widely mined at the Boghead, Torbane, Torbanehill and Hopeton pits lying immediately west of Bathgate, by James Russell & Company. Thus Young, in partnership with Messrs Edward Meldrum, another former classmate from Glasgow, and E.W. Binney, a Manchester solicitor and amateur geologist who had acted for Young in Alfreton, returned to Scotland to set up their (and the world's first) oil refinery at Bathgate in 1851, under a patent taken out in 1850 (Patent No. 13292), 'Improvements in the Treatment of Certain Bituminous Mineral Substances and Obtaining Products Therefrom'. Young also took out an equivalent patent in the USA. The partnership also sunk their own pits, numbering seven, in the Torbane cannel coal fields (Torbane No. 2 pit through to Torbane No. 8 pit) but all were quickly exhausted. The Young extraction process by 'slow distillation', which was established at Bathgate, produced crude oil, paraffin oil, paraffin wax, naphtha, gas, coke and – some years later, at the new Addiewell Works – ammonium sulphate, with the latter then in growing demand as a fertiliser.

To give an indication of just how oil-rich this Torbanite coal was: because it was being mined in significant quantities around the Armadale area, it had to be stored in great heaps (bings) pending transportation to the oil works. In June 1872 a huge bing of stored coal lying close by Torbanehill House was, despite the presence of watchmen, set alight – the canneloid coal, unlike oil shale, igniting very easily. Such was the scale of the fire that fighting it was an impossible task and it burned for some days, sending up great vast quantities of black, oily smoke which could be seen for miles. Crude oil from the cannelloid coal, released by the intense heat, flowed like water and men were employed to dig trenches and deep pits to trap and contain this oil. Train-loads of empty barrels were sent from Boghead and Bathville Oil Works, filled with the oil captured in these pits and returned. Arson was suspected but no arrests were ever made. Over the years, twenty-nine separate pits at Torbane and Torbanehill were mining this material.

The Bathgate (Boghead) Oil Works, situated south of the coal miners' 'rows' at Durhamtown (now Whiteside), aroused much local and not so local interest. The new company took great pains to keep the complete process a strictly guarded secret, so much so that the works became known as the 'Secret Works' because of the high wall which surrounded the site and the misleading names given to many of the processes. These Bathgate Works, much enlarged in later days, carried on as a complete establishment until 1884, but in 1887 the complex was dismantled with much of the equipment being transferred to Addiewell and Uphall. Part of Bathgate Works survived as a sulphuric acid-making plant for another few years.

The original Bathgate site occupied some 25 acres and the works were immediately (and fortunately) connected to a main railway line (the Wilsontown, Morningside & Coltness Railway Company's 1851 extension to Bathgate) and thus all inwards crude oil and outwards finished products were able to go by rail. Initially the works prepared four main products from the crude oil, namely naphtha, paraffin, lubricating oil and paraffin wax. In the early days, the works were producing some 900 gallons per week and the paraffin wax commanded very high prices. All the products were in high demand and were soon returning high profits. Boghead coal yielded an extraordinary 120 to 130 gallons of crude oil per ton and by 1860, fully 10,000 tons of Boghead coal were being retorted there each year. By 1869, there were 200 (horizontal) retorts on site, with much of the incondensable gas produced being supplied to Bathgate town for street lighting purposes.

The world had lacked a cheap and reliable source of lighting. Since the early days of the original discovery of fire, apart from firelight, it had been largely dependent upon vegetable oils, animal tallow (fat), seal oil and whale oil as some of the better sources of artificial lighting. It is known that the Romans grew rape oil seed and produced oil for lighting by pressing the seeds. Church candles, a traditional and important product, were made with beeswax which was extremely expensive. By 1861, Bathgate was producing around 5 tons of paraffin wax per week and this was sold on to the candle-making industry. Candles made with this paraffin wax were found to burn with a clear, bright light and soon were in great demand. Indeed, as the industry grew, several works built associated 'candle houses' to take advantage of the ready market; Addiewell Works in particular was to become famous for the intricate designs, decorations and quality of its candle products.

Young had also discovered at an earlier stage that one intermediate boiling fraction of the mineral oil was totally suitable as a burning oil, to which, in 1856, he had begun to refer to as 'paraffin'. The problem of smell was quickly addressed and suitable oil lamps (the Clissold Lamp) were soon being manufactured by Young's own lamp company at the Clissold Lamp Works in Birmingham. Bathgate was the first place in Britain to benefit from the new 'brilliant light' and soon a full one-quarter of all lamp oil used in London was being produced in the Bathgate Works. Young used the brand name 'Paraffine Oil' for this product. Young was thereafter to be known by the nickname, James 'Paraffin' Young – the name by which he is still remembered today.

The Growth Years

By the mid 1850s, with a virtual monopoly, the oil-based products at Bathgate were returning a revenue of around some £57,000 each year and other companies were looking on with no little anxiety. Binney's interest in the company was waning and finally, in 1864, Young bought out both partners for £32,000. However, the Boghead coal reserves used at the Bathgate Works were confined to a very small basin of about 2,500 acres only, the seams were narrow and of no more than 28 inches in maximum thickness, narrowing out until they disappeared entirely, so the supply of Torbanite coal was rapidly running out. Meanwhile, others such as Robert Bell, a Wishaw Iron Master, had been exploring the possibilities of using oil shale as a means of circumventing Young's Patent, which was based on retorting canneloid coal.

The modern development of the shale oil industry can be dated from the development of Bell's Stewartfield Works at Broxburn in 1862. This pre-dated the expiration of Young's patent by two years. In 1858, Bell had been granted the right to work coal and ironstone found on the estates around Broxburn owned by the Earl of Buchan. In 1861, a supplementary agreement to the original gave him the rights to work *'the oil mineral or shale in said ground'* – the use of this material was free from Young's patent, but the methods of refining the crude oil so produced were not. Stewartfield Works

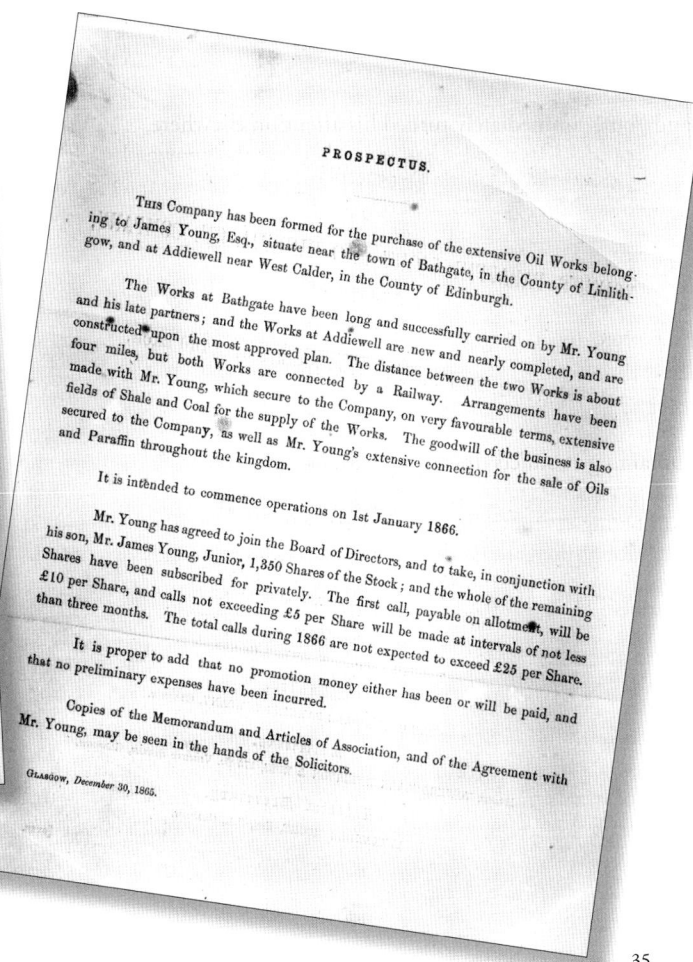

This quite historic document is the cover and page 2 of the prospectus for the creation of Young's Paraffin, Light & Mineral Oil Company.

A view of the candle house at Pumpherston Oil Works.

were therefore to be a crude oil producer in the first years. Thus in Scotland, the honours for the first extraction of oil from shale must lie with Bell.

The success of the Bathgate Works also led to paraffin oil being manufactured on a small scale by many concerns in the chemical trade with access to supplies of canneloid coal and under payment of royalties to Young for use of his patent, but there were also many infringements of the patent and Young had to resort to much litigation in the enforcements of his rights.

Young, however, had not been letting the grass grow under his feet and had turned his attention to West Calder, where he had been carrying out his own explorations. By the time the Torbanite coal was finally running out, oil-bearing shale had been discovered in quantity around West Calder, specifically at Muirhall, Addiewell and Breichmill, and in particular alongside a burn (stream) called Hartwood Water, lying just to the south of the small village. Here, on the banks of the stream, oil shale outcropped and traces of oil continually floated upon the surface of the water. An 1855 survey on behalf of the Geological Society had identified these rich seams of shale oil around West Calder. Young, with great foresight, purchased six shale concessions in the area and went on to establish an oil works at Addiewell, just a mile west of the village, which was to be the largest of its type in the world at that time. Here he established both a system of oil extraction and refining through distillation, using the raw shale mined locally around the village.

Young also purchased Limefield House and Estate, lying about one mile to the east of West Calder, for his home; it was to here that the famous missionary and explorer of the time, Dr David Livingstone, another one-time classmate and close friend of Young, was a sometime visitor. This friendship was to be a life-long one, with Young providing much financial support for Livingstone's work in Africa and also financing an expedition to find Livingstone at a later date. Livingstone laid the foundation stone for the new works at Addiewell in 1865. Later, Young retired to Kelly Estate near Wemyss Bay in the west of Scotland, where he died in 1883 and was interred in the graveyard at Inverkip, overlooking the Firth of Clyde.

Before the new Addiewell Works were actually completed, Young sold out and formed the whole undertaking into a new limited company, registered on the 4th January 1866 with a capital of £600,000, the purchase price being around £400,000. Young was then the largest shareholder and acted as General Manager with his son, James Young, Junior, acting with him. The new company became Young's Paraffin, Light & Mineral Oil Company.

In the same year, 1866, Young's Paraffin, Light & Mineral Oil Company opened their new works at Addiewell, about one mile west of West Calder, on a 70 acre site, which was the largest in the world at that time. Here, the same processes as were carried out at Young's Bathgate Works were perpetuated, but on a far larger scale and using vertical retorts, and a fifth valuable by-product was soon to be discovered, that of sulphate of ammonia. This by-product was soon to be in high demand as a fertiliser and was later exported in large quantities to the West Indies for the cultivation of sugar-cane and to Europe for sugar beet. Later still, the main markets were to be the USA and Japan with, of course, an extensive home market. The real value of this commodity was that it could not be obtained from the refining of liquid petroleum and could neither be manufactured nor marketed by petroleum companies. It was unaffected by the severe price fluctuations prevailing in the oil markets, although this was to change later as we will see. Addiewell's renowned candle house was the largest in the industry.

Young's patent had expired in 1864 and this immediately opened the floodgates for a massive increase in industrial workings. In that same year, some thirty-eight new works were to spring up; between 1855 and 1865, some 120 oil companies were established, although not all involved in shale oil, and a list of those works which can be identified is contained at Appendix A. This list records the known shale oil works and refineries, plus other oil works established in those early days, but records are somewhat sketchy in this regard.

One example of the new works was Young's former partner, Edward Meldrum, who re-appeared as Meldrum & Company at a new crude works at Starlaw near Bathgate in the late 1860s. By 1871 his company had become Meldrum, McLagan & Simpson, later converting to become the Uphall Mineral Oil Co. Ltd, with retorting and refining works at Uphall Station. In 1884, this company amalgamated with Young's Paraffin, Light & Mineral Oil Company, thus giving Young control of significant additional resources with which to supplement his new works at Addiewell. Many of the new companies were extremely small concerns which soon were to close or be absorbed by other, bigger concerns, and this process of absorption, through bankruptcy or closure, gradually reduced the number of crude oil works and refineries to a total of seven by 1912.

Addiewell was supplied with raw shale from a number of pits adjacent to, or within a 2 mile radius of, the Addiewell Works and it is recorded that a total of five works' locomotives (the ubiquitous pugs) were kept fully employed for this purpose. An annual output of 8,000,000 gallons of crude was intended but, immediately the works at Addiewell were commissioned, American competition brought the price of crude oil and associated products tumbling and it was to be several years before Addiewell reached its target production.

The yield per ton of the oil-bearing shale seams worked in West Lothian varied between 30 to 40 gallons at best. The higher yields of crude oil were obtained at the expense of solid paraffin and the quality of the heavier oils, whilst the oil yields from the inferior shales were smaller but produced a higher quantity of sulphate of ammonia

This serious competition to the Scottish shale oil industry arose when in 1859 Col. Edwin Drake, later to form the Seneca Oil Company of the USA, struck liquid oil (petroleum) at Drake Well, Venango County, Pennsylvania and in the space of three short years was producing 3 million barrels (100 million gallons) per year. Imported oil was to have a severe and continuing effect on the Scottish oil industry. When Young's patent was in place, burning oils had been selling at 2/6d to 3/6d per gallon (12½p to 17½p), by 1875 they were selling at 10d (about 4p) per gallon and by 1894 the

Uphall passenger station with Uphall Junction signal box in the background but overshadowed by the chimneys of Young's Uphall Oil Works, in NBR days.   *Bill Lynn collection*

Deans Oil Works circa 1912 showing the benches of Pumpherston retorts and the condensers.

USA was offering kerosene (paraffin) for 4d (about 2p) per gallon, with the barrel thrown in for free!

The heyday of the industry was to prove to be from around the turn of the century until after the First World War, from 1895 until 1925, with over 10,000 employees at the height of the industry. In 1873 some 524,000 tons of shale were mined, by 1913 this had grown to 3,280,000 tons in the year, but by 1961 shale production had declined to only some 468,000 tons. Whilst the earliest demands had been for paraffin, paraffin wax, naphtha, burning, lamp and lubricating oils, plus sulphate of ammonia, with the advent of the internal combustion engine there arose the growing demand for motor spirit, or petrol as it became better known in the UK. This in turn led to further firsts for the area, the Scottish shale oil industry achieved distinction as the first source of automobile fuel, with the first filling station being opened in Newton Village (near South Queensferry) by J. Ross & Co. of Philpstoun Oil Works, who were initially the sole producers. A second filling station, or rather roadside mounted petrol pumps, quickly followed in the High Street in Linlithgow. The other oil companies soon followed suit!

The average production costs for obtaining finished products from one ton of Scottish oil shale in 1924 were:

Mining	48.55% of total cost
Retorting	25.94%
Refining	13.87%
Sulphate of Ammonia production	11.64%

The industry continued to flourish by dint of amalgamations and closure of the smaller concerns, and in the Second World War it was to play a major role in producing fuel and other oils for the military. Workers in the industry were considered to be in a reserved occupation and vital to the war effort.

The Years of Decline

In 1912 there was a national strike by coal miners which led to lockouts in the shale oil industry. National coal strikes occurred again in 1920, 1921 and 1926, and there was the same knock-on effect with lockouts again being imposed.

In 1921 there was a lockout at Broxburn. The shale miners were not on strike, they were simply denied access to their place of work as a result of the coal miners' strike of that year. The supply of coal, the essential commodity for boilers, gas plants and retorts, had dried up and production stopped. Stewartfield No. 1 mine was shut for nineteen weeks, but at both the Dunnet and Roman Camp mines, the lockout lasted for fourteen months. During this time the oil works at Winchburgh, Philpstoun and Pumpherston, with alternative supplies of company-produced coal, worked as normal. Even after the lockout was withdrawn, production was never again to reach pre-1921 levels. The following years were to see a major emigration of workers, out to America, Canada and Australia.

Further lockouts occurred in 1925 and led to the complete closure of the works at Tarbrax and Dalmeny and the mines which serviced them. During this time, competition for jobs was fierce and wages were reduced. In the years that followed, despite this great depression, working conditions improved, with a 7-hour day for miners being introduced; with corresponding lower hours for the oil workers, however, wages remained depressed due, in no little way, to the cheap oil being imported at this time.

In 1918, under the Financial Act of that year, the Government relieved indigenous light hydrocarbon oils of part of the excise duty, a move which was extremely beneficial to the industry at the time. This preference was withdrawn in 1921 and it was not until 1928 that the Government re-introduced a preferential excise tariff for shale oil. In that year, Winston Churchill exempted the industry from the 4d excise duty on each gallon of petrol and wages were increased by 10 per cent. Ammonium sulphate was to prove to be the continuing shining star in the shale oil firmament with demand for the fertiliser remaining high. However, overall, the industry was still uncompetitive and it was not helped when the lucrative ammonium sulphate product was later to be challenged by advancing technology, thanks to a new process (the Haber process) which produced ammonia straight from atmospheric nitrogen.

The workforce dwindled from around 5,000 to 3,500. In 1930 2,023,110 tons of shale was mined, producing 160,030 tons of crude oil.

Supplies of imported petrol continued to flood in with a disastrous effect on locally produced spirit, as a result of which several companies suffered heavy losses. Philpstoun Works closed in 1932 along with the associated mines; the remaining oil works quickly established a system of three weeks working and one week idle for all employees, in order to avoid wholesale redundancies. Indeed, some 900 men were re-employed to ensure the success of this scheme. At that same time, Oakbank Oil Works and Seafield Oil Works and associated mines also closed. 1932 was also to see the remaining oil works merge to form Scottish Oils Ltd. These measures, whilst helping employment, caused further financial hardship and the hard times continued until 1938.

In 1934 – after the Government extended the 4d exemption, first announced in 1928, for a further ten years and extended this to include diesel oil – Scottish Oils Ltd confidently expanded the industry to meet the additional demands and new pits were opened at Burngrange (West Calder) and Drumshoreland. Money was invested in Pumpherston oil refinery which had produced 14.5 million gallons of motor spirit up to that time. Uphall refinery had been running down from 1921 and was finally closed in 1936. In 1938, the Government gave a further commitment to maintain

Tarbrax Oil Works.

Westwood Oil Works, West Calder. This was the newest of all the Scottish oil works, opening in 1941. To the right is the retort bench with the conveyor belt feeding crushed shale to the retorts in the centre background.

the duty preference at a sum equivalent to 8d per gallon over the following twelve years. Once more, Scottish Oils Ltd had good reason to remain confident and a completely new oil works was built, opening in 1941, at Westwood to the north-east of West Calder. Two benches of retorts were constructed with a retorting capacity of 1,200 tons of shale per day. They could, and did, produce 150,000 gallons of crude oil, 30,000 gallons of naphtha and 100 tons of ammonium sulphate each week. The raw shale to feed the retorts came mainly from one of the refurbished Westwood pits which had worked both the Dunnet seam and the Broxburn seam at about the 12 fathom (720 foot) level, and a newer mine (No. 4) at Hermand, just south of West Calder, which worked the Dunnet shale.

Five crude oil works remained operational by 1938: Addiewell, Deans, Roman Camp, Hopetoun and Niddry Castle, with Deans by that time being the largest crude works in the industry. Deans and Hopetoun oil works were closed in 1946. Pumpherston was by then refining all the crude oil from the five remaining oil works although the two sulphuric acid works remained at Bathgate and Broxburn.

The onset of the Second World War increased the economic security of the industry enormously and wages started to rise. Paid holidays were also introduced for the first time. However, it was during the Second World War that the shale industry began to lose its hold on the area. Shale mining and engineering work in the oil works were declared reserved occupations, with the men so employed being exempted from active war service. Nevertheless, many men enlisted voluntarily and, at the conclusion of hostilities in 1945, many of those who had survived never returned to work in the industry, leading to a gradual contraction of operations.

Now a problem arose because of the labour monopoly that shale had held in the area over the years. This actually inhibited any alternative industrial or commercial developments. Indeed, in about 1947, Ferranti, the developing electronics giant, was set to extend operations and submitted a planning application to establish a new factory in West Calder. Scottish Oils Ltd objected on the grounds that they as a company could gainfully employ all the available labour, including school-leavers, in the area for the foreseeable future, and Ferranti was forced to look elsewhere. The truth was that the writing was on the wall for shale oil and increasing contraction of the industry towards final closure saw unemployment inexorably rise.

By 1954, 1.4 million tons of shale were being mined annually, producing 97,000 tons of oil and employing 3,500 workers. Roman Camp Works and some others were by now under threat of closure, and local MPs and the Shale Workers Union lobbied Parliament to remove all duty on shale products, which was at that time 2/6d per gallon. The loss to the Treasury would have exceeded £1 million per year and regrettably the minister responsible, Emmanuel (Mannie) Shinwell (who, with the support of the same shale oil workers, had become the MP for West Lothian in earlier days), turned a deaf ear to their plea and the industry was doomed. More closures went ahead in that year and by 1957 only 2,740 men were employed in the industry. Niddry Castle Oil Works and associated mines closed in 1960; production that year was a mere 669,000 tons of shale, falling to just 500,000 tons in the last full year of the industry. With the ending of the duty preference in 1962, the closure of the remaining oil works – and, indeed, the whole industry – quickly followed, causing much hardship in West Lothian as unemployment figures soared to a level higher than anywhere else in the UK.

When introduced in 1908, the preferential duty exemption was 6d per imperial gallon, by 1950 it had risen to 1/6d per gallon, by 1952 was 2/6d per gallon and from 1961 to 1963, when it ended, was 2/9d per gallon.

The Government, now stung into action to mitigate the problem they had had a hand in creating, instructed the British Motor Corporation (BMC), later to become Leyland Motors, the giant

The detergent plant at Pumpherston Refinery under construction.

car-making company, to establish the planned new factory for its Truck and Tractor Division at Bathgate, instead of in the traditional Midlands heartlands, and this was done. Whilst this immediately provided much needed employment for many, it proved, in the longer term, to have been a bad decision for other reasons, but that is a story for another book.

However, in the early 1940s, Scottish Oils chemists had made an exciting discovery, one which led to alkyl sulphate detergent being obtained from shale oil olefins, in short, it was the discovery of the detergent. After the war a new plant for production of this detergent was opened at Pumpherston under the name of Young's Detergents and the detergent (liquid) was marketed under the trade name Iranopol. This proved to be an unacceptable name with the public through association with the Middle East and it was soon changed to By-Prox, the association with Scottish Oils and BP becoming obvious and widely recognisable. This was the first of the new detergents which were to revolutionise domestic washing-up, right up to the present day. By the 1970s the plant employed 270 workers and was producing some 40,000 tonnes of detergent products annually. Somewhat ironically, right up to final closure in 1993, Pumpherston was producing a special detergent which was capable of dispersing crude oil spillages in the new Scottish oil industry, North Sea offshore oil production.

There was another ancillary product of the shale industry which also was situated at Pumpherston – this was the brick making plant, making bricks from the spent shale. The bricks were marketed as 'SOL' bricks (Scottish Oils Limited) and production carried on well after the refinery at Pumpherston closed. There was, however, a small problem with the bricks produced, as it was found when they were used in an exterior brick finish that any 'harling' (roughcasting) then applied tended to come away after a period of time.

Renaissance (but not Shale)

Scottish Oils had become a subsidiary of the Anglo-Persian Oil Company in 1919, in 1954 this company was renamed the British Petroleum Company Ltd (BP). In 1970 BP was the first company to strike oil in the North Sea, in the Forties Field, and a new era in Scotland's oil history was now to unfold. By 1991, BP was producing around 500,000 barrels of crude per day and employing in excess of 9,000 people, with 5,000 being employed in the actual off-shore activity. Much of the oil was refined at BP's Grangemouth Refinery near Falkirk, by now one of the few crude oil refineries left in the whole of the UK. In 1998, BP merged with the American Amoco Oil Company to form the third largest oil company in the world, BP-Amoco. However, that also is a story for another book.

Once more, oil shale is in the limelight as a possible means of augmenting the diminishing world-wide resources of liquid oil and natural gas. In 2010 Estonia was reported as one of the world's chief

Packing boxes of By-Prox detergent at Pumpherston.

5

THE MINING OF OIL SHALE

SHALE MINING,* UNLIKE COAL mining, was never operated as a commercial activity, but rather was an essential overhead incurred by each and every each oil company in order to extract as much shale as possible to keep their retorts in full operation. Thus, whilst much of the mining might now appear to have been short term and uneconomic, the truth was that mining was undertaken to tap the seams as far as it was practicably possible, but not necessarily financially viable, and to work the seams up to the respective workable boundaries. As a result, many mines had a very short life, and over the years of the industry a great many mines and pits were sunk having varying life spans. Shale mining was simply nothing more than an essential activity, but it was the one which attracted the highest costs (48.55 per cent) in the whole oil production chain. Within these total mining costs, wages accounted for 45.48 per cent of all expenditure.

Mining Techniques

The critical first steps in identifying the presence of workable shale seams were based on visual examination of surrounding rock exposures and the outcropping of blaes and shale beds by geologists and mining engineers. Through proper identification of the strata and its inclinations (perpendicular to horizontal), bore holes could then be sunk down to the seam or seams to test distance from the surface and their thickness and quality. Adequate test boring was essential to identify possible slips, faults or other troubles so that accurate calculation as to the position of the shale could be documented, since, as discussed in Chapter 3, folding and inclination of shale seams was somewhat more the rule than the exception throughout the Lothian shale fields. Test borings revealed, in many cases, the existence of rich and thick seams of shale which then simply disappeared within a very short distance.

Since the shale measures occurred, in the main, either in synclinical basins or where seams outcropped or appeared very close to the surface in the anticlines, the shale could be worked by either opencast mining or inclined mining. The latter was the mining technique where the headings were driven down through the seam, with horizontal levels being established to work the shale or access other nearby seams. Thus inclined mines were in the majority. Opencast mining of shale was carried out, but only on a very small scale within the Lothian shale fields. Where shale occurred in the troughs of synclinical basins at greater depths, vertical pits were sunk to extract the shale.

In the early days of the industry, many small inclined mines were established working into seams which were very close to the surface and generally these mines were small concerns and did not extend to any great depths. Mines of only 300 yards in depth with cross level workings of a mere 300 feet on either side have been recorded. Such mines had a very short life as the shale measures were quickly worked out and deeper, more productive workings became the norm. The main access roads of the inclined mines were, in the later days, generally constructed 12 feet wide, 6 to 20 feet high and could, and often did, run into the seam at an angle as steep 1 in 2. The entrances were bricked for proper support. As a point of interest, the entrance (now completely bricked up) to such an inclined mine can still be seen up on the moorlands at the site of Cobbinshaw No. 5.

Inclined mining was the method chosen to access the five seams of Pumpherston shales which dipped from 29° to 38°; later, however, vertical shaft working was developed to work the deep seams and was to be the method of working adopted to access the rich Broxburn and Dunnet seams in other fields, even where the seams lay very deep underground as in the Polbeth/Gavieside field for example. The advent of steam-driven horizontal pumps made inclined mining a much more popular method of working since the problems of pumping water back up an inclined shaft were then satisfactorily overcome.

A team of two or three men worked together at each face. Faceworkers were generally referred to in the all-encompassing terms as 'miners'. In practice, in shale mining the miner was the 'contractor' or 'placeman' who negotiated a daily rate for driving the levels with the management and he was responsible for the safe and proper working at the face. From this agreed sum, he paid the wages of his 'drawer(s)' and also paid for all explosives used. He also supplied and maintained all required tools, known within the industry as the 'graith'. He worked the face whilst the drawer, or drawers, filled the hutches, drew them to the inclined shaft or nearest haulage road and brought empty hutches back in to the face. When a drawer had three or four loaded hutches ready to be taken up the shaft, he waited for a turn or 'ben' to have the hutches taken along to the pit or mine bottom and raised to the surface. In a good seam, with a level 12 feet wide and 8 feet high, an extraction rate of 12 feet per week was entirely possible, giving a output tonnage of 3 tons per day per man, although this was often exceeded. Indeed, under the longwall method of extraction in Breich pit, 5½ tons per man per shift was the norm. It was, however, a known practice for a 'contractor' to have several teams of miners and drawers on whose behalf he undertook all negotiations and from whom he extracted his share of the proceeds.

In support of the mining activity, whilst the two- or three-man team formed the production unit, in order to maintain workfaces and haulage roads in each working panel (a panel was six active workfaces with up to eighteen men), a further team of around eight men (the on-cost), including the safety expertise of an oversman or fireman, were employed.

---

* To aid clarification, where the term pit and mine is used in this book, whilst generally these terms were used in the context of a colliery (coalerie) or coal pit, in Scotland a mine (or surface mine) was always an inclined mine driven into the strata from the surface, whilst a pit had a vertical shaft to access the deeper seams of shale.

The pit bottom at Westwood pit with a cage containing miners ready to ascend to the surface. The man in the foreground is the 'pit bottomer', his left hand on the signal lever with which he signals to the winding engineman on the surface. Drill boring bits rest against the wall waiting to be taken up to the surface for resharpening. *Courtesy of Grangemouth Heritage Trust*

To ensure that an accurate record was maintained of all persons who were underground, each miner took a 'check' or 'token' from a special check-board located in an office, usually the time office, near to the pithead. This check normally took the form of a numbered brass tag and each man had his own unique tag. It was a mandatory requirement for each man to take one before going underground, and return it to the board at the end of the shift. In this way, management could ensure, (a) how many men were underground at any time, and (b) that every miner had returned to the surface at the end of the shift; in cases of an underground emergency, it would be clear whether any men, and how many, remained in danger.

The mining of shale, even to the last days of the industry, was carried out by the same time-honoured method which involved little if any mechanisation. Two well-known methods of working shale seams were adopted in the working of Lothians oil shale, namely 'longwall' and 'stoop and room'. Each had its own advantages. The shale seams were generally thick, but the Mining Regulations for the industry generally forbade the working of seams above a height of nine feet, this being the maximum height of 'roof' permitted underground (although this was in fact exceeded in practice). For the thinner seams, the longwall system of working the seam was adopted, with the thicker seams being worked on the stoop and room method. In the Lothian Basin, the seams of shale were numerous but varied considerably in thickness, and the workings of the average thickness of seam did not go much below 4 feet but could be as great as 11 feet, despite regulations.

Longwall Working

This method can be best described as the extraction of the whole of the available shale seam in one continuous operation, by means of a long 'face' moving forward with the working. This was the preferred method in coal extraction, stoop and pillar working being considered both wasteful and dangerous, and this also became the principle method of shale extraction adopted for about the first thirty years of mining, for two reasons:
- being the method used in coal-mining, it was well understood by the miners who were experienced in the method, and
- longwall mining was best suited to the extraction of the thinner seams of shale.

The advancing face was generally between 100 and 200 yards long, divided into fourteen to sixteen workplaces or 'stints', each worked by two or three men – a faceman and one or two drawers. The shale was undercut from floor level to a depth of about 3 feet 6 inches (by pickaxe in early days) and then two parallel rows of holes were drilled, by hand, along the face. These holes were then packed with explosive, Polar Ammon gelignite being used in the lower holes and gunpowder in the upper row. After shot-firing, the broken shale was loaded into hutches by the drawers and the loaded hutches were taken out to the main haulage roads for conveyance to the mine bottom.

The roof at the face was initially supported by timber pit props, but later, as the face progressed, the space created behind the recovered shale was supported by more substantial timber (generally old sleepers) packs, filled with the rubbish (waste) from the seam. The roadways to the face were protected by these packs which could withstand the overhead pressures for a lengthy period of time, and the size varied dependent upon the available material and waste in the working, but also on the height of the seam. Generally they were spaced between 6 to 12 feet along the face and were the equivalent of the 'stoops' described hereafter. Spaces were left in between the wooden walls of the packs; these were known as 'cundies' and above which the roof was left unsupported in order to allow the strata to subside naturally, thus taking pressure off the roof at the face and the actual walls of the packs themselves. Longwall mining was widely used in the extraction of the Fells shale, but both methods could be found in the same pit; however, it then rather fell out of favour and was to be seldom used anywhere between 1890 and 1940.

In 1941, the longwall method was again introduced on an experimental basis in Breich pit to work the Broxburn seam. Two faces were so worked to evaluate modern technology, such as coal-cutting machines and conveyor belts and based on a three-shift production cycle. Basically, a 50hp coal cutter, fitted with a 3ft 6in wide heavy chain cutter of 11 lines and 39 picks, was used in the undercutting process. The shot placement and firing remained the same, but with the broken shale being loaded onto conveyor belting which ran out to the haulage roads. H-section steel props were used to support the roof with strip steel being incorporated into the roof packs.

The three-shift cycle was as follows:
- late (back) shift moved the conveyors and chocks and extended the rubble roof packs as required;
- night shift bored shot holes, undercut the face, extended the haulage roads and fired the lower row of shots;
- day shift removed the shale brought down on the night shift, fired the top row of shots, cleared this shale and set up supports for the conveyor track.

In truth, there were inherent problems from the outset with this system of mining, and the difficulties were soon to manifest themselves. It was proving difficult to organise the available labour in Breich pit into cohesive units for the proposed cycle of mining, owing, in the main, to a marked lack of team spirit. The wear and tear on both coal-cutting machines and conveyor system was much heavier than first anticipated with consequential higher than allowed for maintenance costs, and there was a substantial loss of shale through shattering during the undercutting process – a feature not encountered in traditional mining methods. A second trial of longwall mining on similar lines had been introduced at the Fraser pit, near West Calder, but by July 1946 both trials were abandoned not only for the reasons stated, but also because roof weaknesses had been identified at both locations which made longwall extraction an unsuitable and high risk method of mining.

Longwall Working in Steps

In Oakbank pit the Dunnet seam was in two parts, separated by an intermediate bed of 3½ft thick calmy blaes. To address, as far as possible, the event of roof falls which frequently occurred along the face, the shale faces were worked in a stepped manner. The lower seam of shale was extracted by normal boring and blasting, generally for a distance of 9 to 12 feet deep, with the seam of calmy blaes overhead being supported by timber props and timbered packs 3 to 4 feet square along the sides of the roadways. When the bottom seam of shale had been extracted, the props and packs were removed except for those immediately in front of the lower face. The calmy blaes, which parted easily from the upper portion of the shale seam, was then allowed to fall naturally, breaking at the point of propping, thus protecting the lower working face. After the calmy blaes had been 'brushed' and packed into the roadside packs, the top seam of shale was then similarly dealt with, the roof being again supported by timber props.

Stoop and Room Working

Where a shale seam was relatively free from blaes, flakes and ribs of unproductive rocks, and the workable shale seams varied between 4 feet to 10 feet (7 feet as an average), the stoop and room working method of mining was the most effective. It was therefore the method more or less generally adopted throughout the entire shale district after 1890. It consisted of two operations, the 'whole' or first working and the 'broken' or second working.

The whole working involved a series of excavations in the seam dividing it into rectangular blocks or pillars (stoops). Generally, a main roadway, 12 to 13 feet wide, was driven deep into the seam until the mine boundary or a fault was encountered, with a roof height of 9 feet. Sections were then opened out on either side of the main roadway. These excavations were called 'rooms', one set being driven at right angles to the dip of the shale and at regular distances from one another (levels) and another set being driven to the rise of these levels and at right angles to them (ends). The top of the seam was known as the 'roof' and the lower, the 'pavement'. The nature of the roof and pavement determined the width to which the rooms were driven. In the event of a bad roof, if rooms were driven at intervals 12 feet wide, there was a higher likelihood of roof falls and thus more props (timbers) would be required. If the rooms were curtailed to intervals 10 feet wide, less timber was required thus there was a cost benefit as well as the obvious safety benefit. The stoops formed by the driving of right-angled rooms were left *in situ* for the purpose of supporting the superincumbent strata, and since increased depths of the workings posed increased pressure on the stoops, the requisite size of the stoops was extremely important. At 60 fathoms (360 feet) depth, the stoops were 60 feet square, but at 200 fathoms (1,200 feet), the stoops required to be about 200 feet square. In highly inclined workings, the stoops had to made longer to the rise than at right angles to the dip (rectangular) because the inclination of the seam decreased the actual supporting area of the stoop and with the inclined stoops, because of the greater weight of the overlying rocks, the roof had a tendency to slip.

There was, in fact, no hard and fast rule for the size of stoops, this being left to the sheer professionalism of the mining engineer.

Miners at the face. Both men are using naked flame lamps and the miner on the right is smoking a clay pipe. He is working a hand rachet boring drill, the body of which is supported against a pit prop put in for that purpose (a boring tree). The miner on the left is using his pick to clear holes in preparation for inserting the gunpowder cartridges which are lying on the pavement beside him. On the actual shale face, the fireman who made the last safety inspection of the workplace has chalked his initials and the date.
Courtesy of Grangemouth Heritage Trust

THE MINING OF OIL SHALE

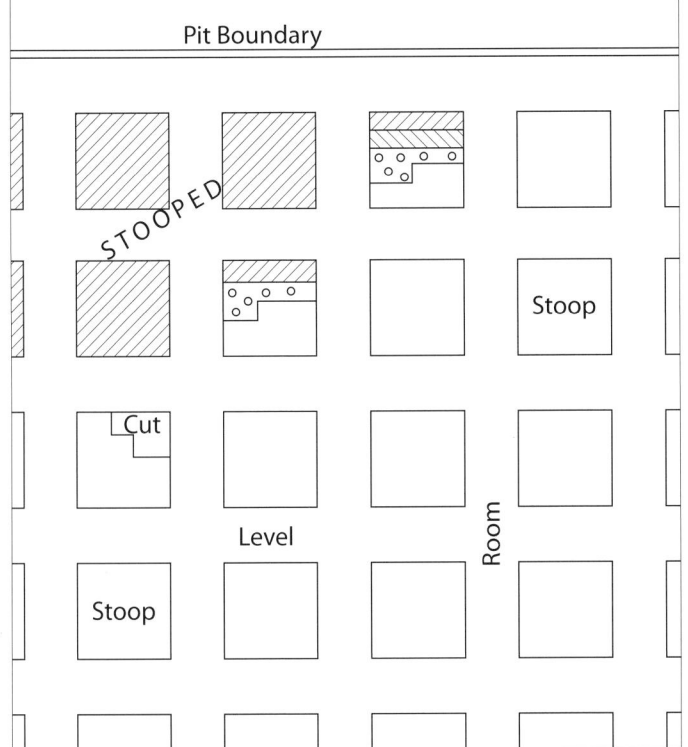

Drawing showing the method of stoop and room and retreat mining. It shows how stoops were removed in slices in shale mining as it was the norm to allow the roof to collapse behind stoops when recovered.

One of the most significant dangers associated with mining (any form of mineral mining) was 'creep' (the slippage referred to above) since, when it started, it was unlikely that anything could be done to check its extremely destructive effects. First indications of creep were generally the lifting or heaving of the pavements in the rooms, causing timbers to bend and break. The primary cause of creep arose when the stoops which had been formed and left to support the roof became unable to withstand the pressure they had to bear. When the whole working had reached the boundary of the shale field (or limits of economical working), if the stoops were then left *in situ* too long, they began to show effects of the extra weight they had to support. It was thus necessary to ensure that the system of stooping employed was adequate to minimise the pressures on the stoops left supporting the roof.

On balance, shale miners (and coal miners) preferred wooden pit props (always referred to as 'trees') as a means of roof support, since the wood gave fairly accurate and advanced audible and visual warnings of any form of roof movement, by creaking, bending and, in the worst case scenarios, actually splintering

MODIFIED STOOP AND ROOM WORKING

With severe labour shortages during the Second World War, conveyor belts were introduced on a wider application to reduce the demands on manpower. This then led to quite a radical change in the traditional method of stoop and room mining. In mines equipped with conveyor belts, the main haulage roadway was driven to the full dip and was supported by permanent roof supports. The hutch railway and the pumping column ran along this roadway and endless rope haulage was provided. Parallel to this main haulageway, a companion roadway was driven some 35 yards or so on the higher level which carried the main conveyor belt to the hutch loading point. This was continuously fed from the conveyor belts running from the work faces at the boundary of the mine, on level roads at right angles to the main haulageway. These level roads also had a companion road. At right angles to the level roads, a series of 'inbye ribs', merely long stoops, approximately 110 yards long and 16 yards wide, were established which when fully developed allowed total extraction by retreat mining to commence, working backwards from the boundary towards the main roadway. The shale was conveyed from these workplaces on shaker belts, being transferred to the 'level' belts and finally by the main conveyor belt to the hutch loading point.

BROKEN OR RETREAT WORKING

When the furthest extent of the seam had been reached, the removal of the stoops was then commenced in a retreat operation, and what was known locally in shale mining as 'broken working' was in fact a recognised mining technique called 'retreat mining'. Where the roof was strong, the stoops were split by driving roads 12 to 15 feet wide (splits) through them thus forming a workface. Where the roof was weak, the stoops were removed in a 'slicing' operation, forming what were 'mini pillars' around 12 feet square and with the roof area of the last slice being supported by timbers. These timbers could be recovered and re-used as the recovery of the shale stoops progressed, but what then happened was that the unsupported roofs eventually fell and a great disturbance of the strata overhead followed. This eventually created subsidence at surface level, a problem that the villages in the whole shale area suffered over the years. During retreat mining, the roof was often opened out, sometimes up to 12 feet above the pavement to maximise the tonnage of shale being recovered, but the extraction was done in a manner so as to maintain a roof fracture-line of roughly 45° to the level. There was a greater risk of roof falls in broken working as opposed to whole working and thus this was probably the most dangerous activity in shale mining.

OPENCAST WORKING

Apart from some opencast working of outcropping seams early on in the history of the industry, opencast working as a means of shale extraction disappeared for a time. In the latter days it was mainly carried out by J. Ross & Company Philpstoun Oil Works. From April 1946, the Middle Dunnet seam at Whitequarries was quarried from the surface. This seam outcropped over an area of 21 acres with fault lines lying at either end. After work commenced, difficulties were posed by the belated discovery of a volcanic pre-glacial river bed which effectively divided the strata in this site. The thickness of the shale seam varied between 11 and 15 feet and the seam lay at a depth of 150 feet at the western end and 80 feet at the eastern boundary. Both workable areas were overlaid by about 8 feet of surface deposits consisting chiefly of sand and gravel plus an overburden of 50 feet of carbonaceous shale.

Opencast mining was also carried out at Livingston where the opencast recovery method employed was the usual opencast practice of step-mining, using tractor scrapers and dragline crane-lifts to remove debris and expose the face over a width of around 80 feet. The shale face was then worked by boring shot holes of a depth between 16 and 18 feet by means of compressed-air hand-held percussive drills, packing the explosives and firing them. Between 1947 and 1950, the site produced slightly in excess of 3,000 tons of shale weekly with a workforce (including lorry drivers) of only twenty-six persons.

Another underground view showing a miner at work operating the hand ratchet drill supported against the boring tree. Beside him are the various lengths of drill bits used in the hole drilling process. Again, note the naked flame headlamp. *Courtesy of Grangemouth Heritage Trust*

THE MINING OF OIL SHALE

A shale miner at work at the face, using an air-powered drill. Lying on the ground beside him are his explosives (cardboard tubes of gunpowder) and a length of fuse.

MINING EQUIPMENT

WINDING SYSTEMS

The early winding engines for both inclined mines and vertical shaft working were powered by steam and used parallel winding drums. An electric winding engine was introduced at Cobbinshaw No. 1 pit in 1904. This consisted of a 200hp DC motor connected directly to the winding drum shaft. This engine was subsequently transferred to Viewfield pit at Tarbrax. It was later replaced by the superior Ilgner system, which utilised a 3,000V induction motor variable DC voltage generator, an exciter and heavy flywheel, and which was capable of winding an output of 80 tons per hour from 75 fathoms. In 1912, a similar system was introduced at Breich pit. All winding engines in use after 1935 were converted to use an AC motor of between 170 and 225hp, and by the 1950s only two pits still used steam powered winding, Westwood and Hopetoun No. 35.

In 1950, Philpstoun No. 6 mine and Hermand No. 4 mine employed electric winding on both inclined shafts, whilst the vertical shafts at Burngrange Nos 1 & 2, Breich pit and Fraser pit were similarly equipped. The winding drums were concave in shape to ensure that the rope ran on to and off the drum in a controlled fashion. The winding engineman had one of the most responsible jobs in any mine or pit since he held the safety of all who had to descend or ascend from the pit bottom in his hands. In inclined mines, haulage of shale was provided by 2ft gauge rail tracks arranged in pairs (one bringing loaded shale hutches up to the surface, the other for empties going back down underground) and in the main using the continuous wire rope and pulley system.

In locations where the seams were far underground, vertical pits had to be sunk. The pits in the Polbeth/Gavieside area were for many years the deepest pits in the industry, although the Pumpherston pits also went down to deep levels post 1925. When sinking a pit shaft, this generally took the form of either an elliptical or a circular shaft (although square shafts were common in older pits) lined with 9-inch brickwork. Two shafts were provided in later pits, an upcast and a downcast, to help regulate the ventilation in the workings. Cages for winding shale (and men) were generally capable of taking either two hutches if single-decked or four if double-decked. The early cages were merely a steel floor with four supports to which the winding rope was attached and had little or no other protection for miners whilst ascending or descending; later cages had sides and were gated for safety. Steel cage guides were fixed to the side of the shaft (generally steel rope) and buffer ropes were employed on the inward side of each cage to ensure that cages did not collide when passing each other in the shaft.

Where a pit had two shafts, each was numbered separately, thus, for instance, Burngrange Nos 1 & 2 was in fact one and the same pit, but provided with two shafts (upcast and downcast).

The winding ropes (1¼ inch diameter wire ropes) passed over the winding wheels (the whorls or 'horrals') on the head frames and were attached to the four corners of the top frame of the cages. The whorls were grooved with grooves three times the diameter of the winding ropes to ensure that the ropes could not jump out.

Miner at work with an electrically powered drill in Burngrange pit, West Calder.

A miner working on an undercut, holing the face with his pick. This is a 'longwall' face where the shots are arranged in a lower level and an upper level, and he is preparing the lower line for shot firing.
Courtesy of Grangemouth Heritage Trust

Underground Haulage Systems

Narrow (2ft gauge) railway track was used underground in the main haulage roadways connecting the pit bottom to the levels at the working faces. This track was of light construction and could be easily moved as the workings progressed. The majority of the shale seams were inclined and at the faces gravity was cleverly used to assist the supply and removal of hutches to and from the main haulage roads.

Four principal methods of counter-balanced operation were common in shale workings:
- *Cousie Braes.* Involved two or more hutches running on adjacent parallel tracks and linked by a chain passing over a pulley near the face. Descending loaded hutches then pulled up empty hutches.
- *Cuddie Braes.* In principal, similar to cousie braes but it used a smaller loaded hutch (the cuddie*) running permanently on a parallel narrow gauge track alongside the haulage road to haul up empties.
- *Cut Chain Braes.* Again based on the same basic principal of balance and counter-balance but with the refinement that hutches could be let down or hauled up part of the way from one level to another.
- *Carriage Haulage.* Like the cuddie brae, but with specially designed platform carriages upon which the full hutches were placed and carried to main haulage roads.

When loaded hutches were brought out to the mine haulage roads, they were assembled into trainloads and worked forward to the pit bottom by a variety of means. Ponies were quite widely used in the early days and lasted until 1948 in some pits and mines. The ponies were gradually superseded by either the introduction of small diesel or battery powered narrow-gauge 12hp locomotives (as used in Burngrange Nos 1 & 2 pit), which could handle up to thirty loaded hutches on gradients as severe as 1 in 120, or by direct or continuous rope haulage. Haulage engines were all steam powered prior to the introduction of electricity in the first decade of the twentieth century. Steam was used up to closure in Polbeth No. 26 pit, where they operated counter-balanced hutches assisted by steam haulage, a system which was widely used and preferred where gradients exceeded 22°.

---

* In Scotland, a 'cuddie' is a pony or small horse, hence the analogy.

THE MINING OF OIL SHALE

Ventilation

A supply of fresh air was effectively maintained by the use of two shafts, an updraft and downdraft, the latter being the principal ventilation shaft. In early workings, most pits were of the single shaft arrangement with this one shaft being divided into downcast and upcast sections by nothing more than wooden battens. In such pits the air was assisted in its circulation by the use of a cube furnace at the bottom of the upcast shaft, which caused the heated stale air to rise up the shaft thus causing fresh air to enter by the downcast shaft, but naturally, with the presence of timber, this posed a real fire hazard underground and the use of this method was abandoned after the disastrous fire which occurred in Starlaw shale pit (near Bathgate) on the 9th April 1870.

Extraction fans were then introduced to force the air flow and initially this was achieved by a Guibal steam-driven fan. This was a large fan, about 30ft in diameter, powered by a static steam engine and named after its Belgian inventor. This drew the stale air up the shaft and through a specially designed chimney to the open air. By 1927 there were four main types of ventilation fans in general use, namely Capel, Guibal, Wassel and Sirocco, all working on the same principle and all driven by electric power. Fresh air flow was directed through the workings by means of air-tight doors and cloth screens.

Because of the large quantities of gunpowder being fired each shift, there were complaints from miners that the fumes that were left made working conditions difficult and extremely unpleasant. The ventilation systems employed were generally adequate to remove these fumes and any residual gases, but at the furthest working faces and in steeply inclined workings this could take some time.

Lighting

Shale workings were considered to be 'safe' workings with a low risk of gas. Electric lighting was trialled in one pit, Westwood, in 1948, where diffused incandescent lighting was installed along the main haulage roads, but not at the face. Generally, lighting underground was confined to the immediate pit bottom and, elsewhere in the workings, to the miner's headlamp. In the earliest days, most miners used tallow lamps (tally lamps) which consisted of a small kettle-

A view of the short-lived experiment with coal cutters at a longwall face in the Fraser pit. Use of these machines was discontinued after it was found that the harder shale (compared to coal) led to higher than anticipated maintenance costs of the cutting equipment. Note the use of steel channelling supports.

Miners at the face. Here the miners are clearing space to work after shots have been fired. Again, both men are using naked flame lamps.
Courtesy of Grangemouth Heritage Trust

THE MINING OF OIL SHALE

A miner and his drawer at work at the face in Haycraigs mine, Broxburn.   *BP Archives*

shaped reservoir which was filled with seal oil or tallow (animal fat) wax, which was softened and pushed into the reservoir. A wick led down a short funnel into the reservoir and the heat of the burning wick kept the wax soft. These lamps were, of course, just naked flames and were generally hooked into the miners' cloth caps. Later, miners were to use open acetylene gas (calcium carbide) lamps attached to their helmets which were again naked flame lamps. Postwar, battery electric and flame safety lamps, similar to those used in coal mining, were provided. However, old habits died hard, and even after the Burngrange disaster, the inherent gas-free conditions meant that both open and safety lamps were still to be seen in use, even in certain sections of the same Burngrange pit.

Calcium carbide (CaC_2), a grayish stone-like substance which the miners used in their lamps, reacted when exposed to water, producing acetylene (ethyne) gas (HC_2H), a colourless gas which was inflammable and, when lit, provided illumination for working underground. Like gunpowder, the carbide was purchased in the local Co-operative ironmongery department.

The safety lamps also provided a source of light at the workfaces, but this was poor in the extreme and working in such poor lighting conditions could bring about an eye problem called nystagmus (a sideways rapid oscillation of the eye) or, as it was known in Scotland, 'Glennie Blink', which could lead to blindness. Although this was widely known in coal workings, no evidence of any occurrences amongst shale miners has been found.

All electric lamps and safety lamps were held in a special lamp house from where they were distributed at the beginning of a shift and returned at the end of the shift. All safety lamps were cleaned, filled and lit by the lampman before being provided to the miners. Electric safety lamps were kept on chargers to charge the batteries when not in use. Earlier 'tally' lamps and acetylene lamps were the personal responsibility of each underground worker and were not held by the lampmen.

Power Supplies

Initially, steam was the sole power supply and was widely used right across the industry for many purposes. All winding haulage was steam driven and coal, much of it intentionally mined from the thin seams often found amongst the shale seams, was then used in the furnaces for steam generation. The Scottish Oils colliery at Baads near West Calder mined the Hurlet seam for this purpose.

In 1902 the Oakbank Oil Company opened their completely new oil works at Niddry Castle, Winchburgh, and from the very first this works was completely electrified, the electric power being generated

An evocative picture of a drawer loading a hutch with shale by hand. This was hard work and it can be seen that he is loading as much as possible into the hutch since payment was paid by tonnage. Note the angle of the pit props supporting the roof. *Courtesy of Grangemouth Heritage Trust*

in-house. Duddingston Nos 1 & 2 mine, and later Duddingston Nos 3 & 4 mine and Totleywells mine, all of which served Niddry Castle, were also connected to this new power supply. By 1938 all shale mines and pits were generally electrically powered, served by an internal grid supply owned by Scottish Oils Ltd.

Drilling

The actual extraction of the shale was achieved for many years by mainly hand labour. Up to around 1860, shot holes were made by the miner's pick, but after 1860 the use of a hand-operated ratchet boring machine, more commonly known as a 'rickety', was introduced.

In ratchet boring, a pit prop, known as a 'boring tree', was fixed in position close to the face being worked and to which the body of the drill was then fixed at the barrel end, the drill end being supported by the drill bit against the shale face. The boring consisted of a backwards and forwards movement of the ratchet handle until the screw was extended to its full length. This bit was then removed and a longer drill bit put in its place and the action repeated until the hole had reached the desired length.

Between 1900 and about 1910, some companies introduced hand-held, percussive boring machines. Whilst these were successful in certain conditions, overall they proved unsuitable and incurred disproportionately high maintenance costs. They proved only to be of any use in ideal conditions and were unsuitable for bad shales and steep inclined mining. Coal cutters were also trialled (as discussed above) but the difficulty in transporting them to the scattered workings led to their abandonment before the Second World War.

One air-powered boring machine was used, however, out of sheer necessity and this was the Burnside Borer which had certainly been employed in Westwood pit as early as the 1920s. This was a specialised piece of equipment, the use of which was mandatory and imposed by law. Mine operators were required to test for flooding where adjacent abandoned workings were known to be present and this was done by the boring bit drilling through a water-tight rubber seal which prevented a possible deadly inrush in the event of the drill hitting water, but permitting the water to flow in a controlled manner onto the pavement.

A new design of hand-held electric drill was re-introduced at Burngrange Nos 1 & 2 pit in the early 1930s. This consisted of a drill with a 600rpm, 1½hp motor and proved to be more successful.

In 1940/41, tungsten carbon-tipped drill bits were introduced with success at Burngrange and by 1948 the Siemens Schuckart high-speed electric drill was making an appearance in some pits.

Loading shale to hutches from a 'shaker conveyor' underground in Breich pit.

These were found to be capable of drilling shot holes 4 to 6 feet deep in a very short time and proved to be entirely successful. By 1950 more than 100 electric drills of this type were in regular use across the industry.

Explosives

In the early days, black powder (gunpowder) was the only explosive used, since it had a better spreading action when fired. The black powder used contained 75 parts saltpetre, 15 parts carbon and 10 parts sulphur. It was locally produced at Camilty Powder Mill near Harburn, a mill which had been started in 1878 by West Calder Co-operative Society, later taken over by the Midlothian Gunpowder Company in 1893, and finally became part of the ICI empire, closing down in 1931. The black powder was packed into strong paper cartridges of varying lengths/weights and sold to miners through the ironmongery department of the local Co-operatives (licenced to hold and store explosives under the Explosives Act of 1853). The black powder was ignited by the lighting of a 'strum' safety fuse. This was rope-like fuse with a black powder core which was inserted into the cartridge. The loose end was then lit, either by naked flame or by a chemical Bickfort ignitor, with the length of fuse used being dependant on the time required to access a place of safety after ignition.

In wet conditions, or when hard rock seams were encountered, gelignite was used as an explosive but, since it was a high explosive, it tended to shatter the shale. Black powder remained the preferred explosive since it did not damage the shale to any great extent. Polar Ammon gelignite was used in the longwall extraction method adopted for mining the Broxburn seams, although not generally used elsewhere.

In blasting, the holes which had been bored out were cleaned and the black powder cartridge inserted. These cartridges were of varying length depending on how much explosive they contained and were an easy fit in the hole, the forcible ramming of explosives being prohibited by the Coal Mines Regulation Act of 1887. A length of fuse was then inserted in the end and the cartridge gently pushed to the back of the hole by a 'stemmer'. The hole was then filled by a mixture of broken material and dust in a paste form. Inadequate or careless packing could result in a blown-out shot, where the hole acted like a cannon. This was of no service whatsoever and was merely a waste of explosive, as well as being very dangerous. The fuse was then lit either by the flame on the miner's helmet or by a chemical igniter and men had then time to reach a place of safety.

As a safety measure, under the provisions of Special Rule 1A, as agreed by the various mining managers across the industry, not more than two shots were permitted to be fired in any working place at any one time, unless in upsets of more than 1 in 3½, where four shots were allowed but with a single person lighting only two shots each. That was the rule, but it is recorded that many miners fired up to six shots at a time; indeed, it is reputed that in Hermand mine the placeman fired up to ten shots at a time by cutting the strum longer. There is a report later in this volume of the circumstances which were apparently pertaining in the Pentland pit at Straiton, where a miner was using straw to fire his shots, with tragic consequences.

In the event of a misfire, the miner had to allow 30 minutes to

The winding gear at Philpstoun No. 6 mine after the site at Whitequarries was re-opened in the late 1930s. Philpstoun No. 6 also acted as the winding house for the new Philpstoun (Whitequarries) No. 1 mine, opened in 1939 (see illustration on facing page).

elapse before going back to the face. It was the general practice to fire shots immediately before a break for food so that the men could then remain clear of the face until the fumes and dust had dissipated.

Averaged out across the shale fields, total costs of explosive (including fuses) per ton of shale was calculated to be no more than 5½d (just over 2p).

Whilst control of explosives was tightened up considerably over the years, in the heyday of the industry the miner would often take any unused explosives home to be used another day – and this was thrown under beds or in cupboards. Indeed, the author's grandmother often used to relate the following story, maybe funny now, but certainly was not at the time.

We resided at No. 15, Hartwood Road, West Calder. It was the lower flat of a two storey building. At a point close to the start of World War I, the neighbours above were, inevitably, a mining family, and into three bedrooms and a recessed bed space in the living room, the father (a miner), mother, five sons (all miners), one daughter and three lodgers (all miners) managed to live in some (close) harmony. One night, or early morning, I was awakened by a loud banging at the door. It was the woman from upstairs, to tell me that the house must be on fire since the upstairs accommodation was filled with smoke. A quick search revealed nothing untoward and off went Mrs L.

Later that day, I asked if any reason for the smoke had been discovered and was informed that the source of the conflagration had been found to be the net curtains in the lodgers' room, they had blown across a candle which had not been extinguished and had caught alight. The fire, such as it was, was quickly put out and all was well, at least until Mrs L. dropped her bombshell. 'You know, Mrs Knox, it was just as well we found the fire because there was about 20lbs of gunpowder lying on top of the wardrobe.'

This small mining community living above them had collectively been purchasing explosives, as was the norm, taking with them just enough for a daily supply (normally 5lbs per man) and just throwing the rest on top of the wardrobe 'for another day'.

Progressively, under the provisions of the 1853 Explosives Act, the oil companies applied for licences to hold explosives in a secure place at each pit or mine. This change took place over a period of some time, and usually the licence covered the provision of small, lockfast storage boxes at each location. Sometime later still, and probably before the Second World War, it became compulsory for each pit or mine to have a 'magazine', a building located at a safe distance from the pit head, where all explosives were securely held and allocated to miners at the beginning of each shift. Unused explosives had to be handed back to the magazine at the end of the shift, thus ensuring all explosives were accounted for and doing away with the 'gunpowder under the bed' culture.

A view of the new Philpstoun (Whitequarries) No. 1 inclined shaft driven into the Broxburn and Dunnet shale seams lying to the north of the Midhope Fault. This new mining facility was built on the site of the former Philpstoun Nos 1 and 6 mines closed earlier in the decade.

Water Control

The earliest mines and pits were shallow and thus water never presented any significant problems, but in later workings where water ingress was proving troublesome, 'ram' or 'bucket' pumps were used which could lift the water to a height of between 180 and 270 feet to the surface. This was the adopted method until 1876. In that year, the Oakbank Oil Company installed the first large-scale 'Grasshopper' pumps with a 62 ins × 10 ins stroke cylinder in their pits and in 1882 a similar design of pump was installed in Gavieside No. 11 pit.

Later, direct action steam pumps were installed underground for use in 'dripping' workings (that is, very wet pits). At the turn of the twentieth century, electricity brought the triple-ram and multi-stage turbine pumps and by 1950 turbine pumps were the norm across the industry.

Pits and mines which were especially prone to water ingress were Duddingston Nos 1 & 2, Philpstoun No. 6, Pyothill mine (where a Reidler pump was installed), Stewartfield No. 1 and Viewfield No. 5 (where an Oddie Barclay pump was installed). Westwood pit also had the reputation of being a 'dripping pit'. The rate of water ingress varied, but in the worst case scenarios this could exceed 600 gallons per minute.

Inside a winding house at an unidentified mine, showing electric winding engine and engineman.

6

MINING: SAFETY, HEALTH AND ACCIDENTS

Amongst people either employed in the industry, or who were brought up and merely lived in the many shale villages, there is a lasting perception now, in the twenty-first century, that the mining of shale was a hazardous and indeed dangerous activity with a history of serious accidents and injury. Was this in fact, really the case?

Safety

Shale mining overall, was, when compared to coal mining, conducted in a relatively cleaner, safer, roomier, better ventilated working environment, and one where gas – methane, 'firedamp' or 'chokedamp' – was not often found in a dangerous quantity. Apart from one major incident, explosions proved not to be a highly significant factor in the shale mining operation. Shale dust was found to be a retardant to flame propagation and, unlike coal dust, was neither a potential fire nor explosion hazard. Besides, shale dust was not particularly prevalent in the mines and pits.

Yes, in the twenty-first century, with the focus on current Health & Safety legislation, there is little or no doubt that the shale mining process as it stood in the nineteenth and early twentieth centuries would be largely unacceptable; it would surely be the subject of far more stringent working practices and conditions, but that is now. In its time, however, shale mining was never considered to be unacceptably unsafe.

The mining of shale was regulated by the same legislation governing coal mining such as:
- Mines Act 1850
- Coal Mines Regulation Act 1887
- Coal Mines Act 1911
- Mining Industry Act 1920
- Mines & Quarries Act 1954.

The Mines Act of 1850 led to the appointment of Mines Inspectors although it was to be the 1870s before they had any real powers to enforce any rulings made.

Generally, the emerging legislation over the years addressed such issues as the management of mines, conduct of employees, layout of underground workings, fire precautions, fire and rescue procedures and provision for first aid, the storage, security and issue of explosives, and so on. All serious accidents or fatalities occurring both on the surface and underground were reportable (after 1866), as were all instances of ignition of firedamp (explosions). The health and welfare of the humble pit pony was also addressed in the 1911 Act. Safety in the industry was gradually tightened over the years and was ensured by careful planning and direct supervision. Shale mining was always to be seen as a significantly safer industry than coal mining for the following reasons:
- workings were generally larger and thin seams were not worked;
- the roof was generally more stable with less risk of crumbling;
- coal dust in itself was highly explosive when mixed with air, whilst shale dust inhibited combustion;
- the risks of gases collecting were considered to be much lower.

Despite the low levels of gas in shale workings, just like the coal industry each working face had to be inspected by a mines deputy (oversman) or fireman prior to work commencing at the beginning of every shift, and by the faceman on return to the workplace each time a face had been left unattended. Under the 1911 Act, this required a full safety inspection and the use of a safety lamp to detect the presence of any gases. The prevalent gases underground were:
- *Whitedamp*. Otherwise carbon monoxide (CO), which was a toxic and explosive gas.
- *Blackdamp*. Also known as chokedamp, was a mixture of nitrogen (N), carbon dioxide (CO_2), argon (Ar) and water (H_2O), and was a particularly pernicious asphyxiant.
- *Firedamp*. Mainly methane (CH_4), a highly flammable gas and explosive at a concentration of between 4% and 6% in normal atmospheric conditions, and violently explosive at a concentration of 10%.
- *Stinkdamp*. Hydrogen sulphide (H_2S) was both a toxic and inflammable gas, smelling of rotten eggs.
- *Afterdamp*. A residue of non-inflammable gases left after an explosion caused by firedamp, with high carbon monoxide and carbon dioxide levels creating an oxygen deficiency situation. It is reckoned that afterdamp caused more deaths than the actual firedamp explosions. However, with the low levels of gas in shale pits, this was to be more of a risk in coal workings.

The term 'damp' is believed to have come from the old German word *dampf* meaning vapour and was the common term for all gas in the mining industry. Certainly within shale mines, both firedamp and blackdamp were known, but not in the same quantities as experienced in coal workings.

The safety lamps in use were exactly the same as those used in coal working and were based on the lamps variously designed by Dr Clanney in 1813, and by both George Stephenson and Sir Humphrey Davy in 1816. These individual safety lamps all had worthwhile features but employed the same principles. Later safety lamps were to be a culmination of the principles established by these inventors. The most famous was perhaps the Davy Lamp, where the naked flame was surrounded by a metal screen, actually wire gauze, and it worked on the principle that for an inflammable gas to ignite, it required to be heated to its 'kindling' temperature. This gauze screen, by virtue of the greater area of metal surrounding the flame in the lamp, dissipated the heat being generated by the flame by conduction, thus keeping the temperature well below the ignition point of methane (firedamp). The lamp also provided a crude but effective means of testing for the presence of gas, since if methane was present the flame would burn with a higher, blue-tinged flame. If the lamp was placed on the pavement and white or blackdamp was present, or the air otherwise oxygen-poor, the lamp would extinguish. If the lamp did go out, it could not be re-lit underground, but had to be returned to the surface lamp room to be opened and re-lit. The safety lamps were fuelled with colzalene, a flammable spirit with a low ignition point (very similar to lighter fuel).

There was another version of safety lamp, supplied only to firemen and pit deputies, which could be re-lit *in situ* underground if necessary, this being achieved by use of a ratchet (which could be operated from outside the lamp) and a flint, the latter being purely internal to the lamp, the spark igniting the colzalene, very much akin to the principle of the standard cigarette lighter

Dr Clanney later designed an improved version of the Davy lamp, replacing Davy's wire gauze by a glass cylinder enclosing the flame with a wire gauze cone atop (a hooded flame). The safety lamps in use thereafter varied in type, but not in principle. In the latter days the safety lamps in use were typified by the Type MG40, manufactured by The Protector Lamp & Lighting Company of Eccles, Lancashire, which used a double wire gauze cone atop a glass cylinder.

Perversely, Scottish miners always referred to safety lamps as 'Glennie' lamps irrespective of design, probably a corruption of Dr Clanney's name, whilst miners in the north-east of England stuck to their 'Geordies'.

The 1911 Act referred to above also banned the use of naked flame, or any object which could cause ignition to be carried underground, but although that became the law, there were to be pits which were considered to be sufficiently safe (gas-free) so as to permit naked lights. Because of the relatively gas-free working conditions underground, these 'naked light' shale mines and pits were unlike coal pits where cigarettes, matches etc. were considered contraband and it was illegal for miners to take them underground. Such items appeared with unfailing regularity in shale pits in the earlier days. There is included in this book (p. 48) a photograph of a miner working at the face whilst smoking a clay pipe. Naked lights (latterly acetylene lamps) were thus a common feature of shale mining right up to the latter days of the industry.

Under Sections 64 and 65 of this Act it was a statutory requirement that the faceman in charge must test the workplace for gas using a flame safety lamp each time the workplace had to be entered. To this requirement, Scottish Oils Ltd also added a requirement (Special Rule No. 38) that the faceman should always wear a safety electric cap lamp and have a safety lamp hanging in a prominent position in the workplace. He was then responsible for ensuring that the workplace was gas-free and this was tested every time the workplace was entered, before permitting the drawers, who continued to use naked flame lamps, to enter and start work. This was a requirement over and above statutory requirements.

Protective clothing was minimal and mostly self-provided by the men. Safety boots were generally worn and West Calder possessed a well-known shoemaking family, 'Forrest the Bootmaker', famed for the quality of both their miners' boots with full metal toecap and heel protection (incidentally, also their 'shepherds' boots which were made on a special 'hill' last).

Otherwise, safety equipment was sparse and even as late as 1937 there were only 625 hard hats in regular use across the industry, soft caps (bunnets) being the favoured head gear particularly in the earlier years. Face masks had been provided in Pumpherston No. 6 mine, but proved unpopular and were not widely adopted. After the Second World War, the situation did improve and by this time all remaining pits and mines had been equipped with baths and First Aid stations.

Health

In the earliest days of the industry it was soon recognised that some workers in contact with oil and paraffin wax were developing epithelial tumours on hands, arms, legs and face, which led to early identification of the hazard of occupational cancers, but rapid response in the shape of an improvement in hygiene methods, washing facilities and general workforce surveillance soon countered this trend and virtually eradicated this problem by the beginning of the twentieth century. Sadly, this was not before many workers were to be badly affected. Otherwise, the shale oil industry was recognised as a fairly healthy environment.

It was the general opinion, revealed by a survey of miners, that shale dust was neither prevalent nor a perceived problem underground but, rather, the fumes from the large quantity of gunpowder being used in shot-firing were the principle source of complaint amongst underground workers when working at or near the face.

In 1981, the publication of an account of four shale miners who were portraying symptoms of pneumoconiosis and progressive massive fibrosis (PMF) led to some comprehensive research being undertaken to determine just what risks had been associated with the Scottish shale industry. The research was initiated by the Institute of Occupational Medicine and was undertaken by highly qualified medical practitioners with expertise in this particular area, including Dr W. Rhind-Brown, a close personal friend of the author. Whilst this was undertaken some twenty years after the end of the Scottish shale industry, the research was commissioned to determine potential risk to miners in the USA, by comparison with Scotland, where the development of shale extraction was then under serious consideration. The study group selected (volunteers) were all members of the Scottish Oils Provident Fund and a random sample group of 200 men was chosen. For comparison purposes, 'control populations' were also selected from men within the industry who were not directly exposed to either shale dust or oil, and also from coal miners from a colliery in North-east England. It was found that the highest exposure to shale dust arose, as to be expected, at the working faces in the mines, at the crusher at the retorts, and on the tips at the bings. The findings, after very carefully controlled statistical comparisons, concluded:

- There was no evidence of continuing excessive skin disorders or skin diseases in the industry and shale workers had no higher prevalence of probable skin disorders than did either of the control groups.
- Pneumoconiosis, well recognised in coal miners, was not described in shale miners during the life of the industry and with a particularly low relevance in the group where this might be expected to occur, for example miners and retort men. Shale dust in fact contained no free silica, unlike coal dust which had significant traces. Any pulmonary disorders were generally found to be associated with heavy smoking and not mining activity.
- Shale miners were spared the degree of hazards that coal miners were exposed to.

An interesting aside relating to health, which came to light when researching this book, was the presence of vermin underground. The author had known from his own underground visits to coal pits where pit ponies were still employed that mice were prevalent, it being thought that they came down in the ponies' foodstuff, and that where there were mice, rats were not to be found. This matter caused great debate amongst shale miners, some of whom had encountered mice, some rats, but never the two together. Eventually it became apparent that the only workings where rats were to be found were, in fact, inclined mines and never in the pits – suggesting that the rats could merely run down the inclined mine workings but would have had to have used the cages in pits, where, of course, they would have been more likely to be seen and immediately killed. Mice were virtually unknown in pits where ponies were no longer employed. There is no evidence to support any suggestion that the presence of vermin underground impacted on the health of shale miners, although with rats the risk of Weil's Disease (leptospirosis) was always present and indeed had posed significant health risks for miners in coal workings.

In other ways, however, the mines and pits were unnatural and unsanitary places and it has been described elsewhere how both the pit ponies (where used) and men ate, drank, urinated and defecated wherever they could. The darkness was absolute, broken only by naked lights or electric safety lamps which, in truth, gave precious little effective overall illumination. Indeed, the darkness underground was palpable, as the author, when only fourteen years of age, found to his cost on an underground visit to a coal pit (Woodend, near Armadale) with only an engineer for company during the pit summer holiday period. His headlamp gave out when in a 3-foot seam and his heart nearly followed! Underground workings were, without doubt, often foul-smelling, dark, hot, wet and claustrophobic places.

ACCIDENTS

Mining was, under any circumstances, a relatively hazardous occupation and accidents did inevitably occur, but information regarding the incidence of serious injuries or fatalities occurring before 1866 is scarce and now largely non-existent. The first recorded accident occurring in shale working was a fatality in Gavieside pit in 1865 when a 'bottomer' jumped out of an ascending cage before it stopped and was crushed. The first recorded mining accident was at Broxburn No. 2 pit, where a miner was caught in a fall of shale at the face. In 1866 it became mandatory for all serious accidents and fatalities to be reported to HM Inspector of Mines, who maintained records and statistics. With this better reporting it has been possible to identify the accidents with greater accuracy, but even with the good records existing there is no guarantee that every incident was, in fact, reported. Indeed, the records of HM Inspector of Mines in Scotland have in part been compiled from accident reports submitted, from newspaper cuttings and from information being supplied even yet from families. The real question is, just how dangerous was the shale mining activity?

Over the 110 years of the industry, the main causes identified in accidents reported were:
- *Roof Falls.* Without any doubt this was the single greatest hazard. In the years from 1866 up to the closure of the industry, some 395 accidents of this category were reported with 104 fatalities and five serious injuries.
- *Moving Equipment and Crushing.* Derailments, failure of equipment and particularly failure of personnel to take due care for their own safety accounted for a total of thirty-nine moving equipment fatalities and twenty-one crushing fatalities.
- *Explosives (Shot-firing).* Given that on average over half a million shots were fired every year between 1866 and 1963, it is perhaps remarkable that only thirty-three fatal accidents were reported, with carelessness or inattention in the handling of explosives being the major cause.
- *Explosions.* Between 1866 and 1963, a total of 2,126 accidents due to explosions were recorded with a total of sixty-four fatalities and fifty-three serious injuries. In this total are included the fifteen men who perished in the single worst accident in the history of the shale industry at Burngrange Nos 1 & 2 pit in 1947 (described below). Further explosions were reported in 1950, 1952, 1954 and 1955, and all were due to the presence of firedamp gas. This was, under any circumstances, not an excessively bad safety record when viewed in the context that from 1931 until 1951, some 8,500 men were employed underground, and from 1952 until closure this figure was still in excess of 5,500 men.
- *Other Causes.* Slipping, tripping and falling (including material falling on persons) accounted for thirteen fatalities and two serious injuries, with a further three cases of suffocation by the ingress of whitedamp being reported. Some six miners were killed by falling down the shaft and eight men were killed by an underground fire which spread to the winding shafts. One man was crushed as he was caught between the side of the shaft and the cage.

Some of the early accidents make harrowing reading and, amongst the very worst, the following are worthy of special mention.

STARLAW PIT BY BATHGATE (1870)

This was an old pit and one which had only a single shaft. To assist ventilation, this shaft was divided by wooden boarding right up the centerline from pit bottom to surface forming the equivalent of an upcast and downcast facility for air circulation. To assist the free circulation of air, a cube furnace was positioned at the foot of the upcast shaft, the hot, stale air rising, causing fresh air to be drawn in via the downcast side of the shaft. Each half of the single shaft had its own cage with winding gear. On the mid-morning in question, the wooden structure at the foot of the upcast shaft caught fire and the flames quickly spread upwards and also began to break through to the downcast side. The alarm was quickly raised and the winding engineman kept the cages operating in both shafts despite the fire. The cages held but four men under normal circumstances, but, in an effort to reach safety, eight men were cramming in at each run. The winding rope in the upcast shaft burned through, rendering the shaft useless but still the cage went up and down in the other shaft, by now passing through flames on the way. The situation became somewhat chaotic and two miners who had just ascended to safety were, by erroneous information given to the winding engineman, straightaway lowered back through the flames again, down to the pit bottom. Fortunately, before they could gather their senses, the cage immediately began to ascend once more and they eventually were able to scramble out to safety. Three further fruitless ascent attempts were made before another two men, both badly burned, came up with the fourth ascent. One eventually succumbed to his injuries. All but seven men had been rescued and these seven were waiting at the pit bottom for the final cage This cage was never to arrive as the fire broke through and rendered the shaft impassable. It proved to be impossible to lower the remaining cage for a final time and the seven unfortunate men perished in a dreadful manner, leaving five widows and fifteen children fatherless.

Legislation was later enacted to require that all mines/pits were equipped with two shafts, nor only to facilitate ventilation but to ensure that an escape route was maintained should one shaft become obstructed. It was, sadly, all too late for the unfortunate miners of Starlaw!

Gavieside No. 2 Pit, West Calder (1871)

On the 16th January 1871, two miners entered the cage to descend to their place of work. On the descent, the winding rope broke and the cage dropped to the pit bottom, hurling both men to their death. The two involved were a 34-year-old miner and his eleven-year-old son.

Hayscraigs Mine, Broxburn (1882)

In August 1882 there was an explosion of firedamp which led to the deaths of four men. Two of their number, an oversman and a fireman, both men the keepers of underground safety, inexplicably went into a face which all knew was unventilated, with naked lights on their caps. The resulting explosion caused injuries which proved in a short time to be fatal, but all the men were able to say exactly what had taken place before succumbing to their injuries.

Gavieside No. 11 Pit, West Calder (1885)

This pit worked at a depth of 100 fathoms (600 feet). On the morning of 16th March 1885 the miners on the early shift were going down the pit in batches of six. Between 06.00 and 07.00 the cage had descended seven times, on the eighth descent the cage apparently left the slides and became jammed in the shaft, about 20 fathoms (120 feet) down. The circumstances went unnoticed by both the winding engineman and the pithead foreman, with the result that the engine kept lowering the winding rope, which then gathered in coils on top of the cage. At some point, possibly because of the added weight, the cage then came free and immediately, and without any warning, dropped some 30 fathoms (180 feet) before being brought to an abrupt stop. When the jerk was felt in the winding house, the engine was immediately stopped, but four of the six miners had been thrown out of the cage and fell to their deaths some 300 feet below. The four were all young men, indeed merely boys, a thirteen-year-old, a fifteen-year-old, a nineteen-year-old and a twenty-year-old.

Burngrange Nos 1 & 2 (The Burngrange Disaster) (1947)

Burngrange Nos 1 & 2 pit at West Calder was a more modern 1930s development, replacing an earlier pit, Burngrange No. 39, and working the extensive and thick seams of Dunnet shales at 78 fathoms (468 feet). This was to be the scene of the very worst, and most tragic, accident ever to occur in the Scottish shale oil industry – one which was outstandingly unusual, and indeed unique, in the

Burngrange Nos 1 & 2 shale pit, West Calder. A view of the pit head. The nearest shaft with winding gear is No. 1 downcast shaft whilst the further away shaft is No. 2 upcast with fan house close by. This was the pit where the worst accident in the history of the Scottish shale industry occurred in January 1947.

annals of shale mining. As discussed above, shale pits generally, because of the nature of the material being mined, were not 'gassy' pits in the way that many coal mines were; Burngrange was no different and naked flame lamps were thus common underground.

A total of 205 men were employed at Burngrange, twenty-nine surface workers and 176 underground workers. On the late afternoon on the 10th January 1947, the afternoon (back shift), consisting of seventy-six men underground, were at work. In No. 2 district, a miner (faceman) and his two drawers were working two adjoining stooping levels, the miner and one drawer working in a rise split, outbye from the level and the other drawer working at No. 14 level face. The miner and his drawer re-entered the rise split at No. 2 dook at around 8.00pm after a meal break, the miner wearing the approved electric safety lamp on his helmet and carrying his safety lamp as required by Mining Regulations (see page 33). He was, however, quite unintentionally to then breach Special Rule 38 by failing to prevent his drawer, who had a naked light on his cap, entry into the workplace until it had been declared safe. The miner, on re-entering the rise split, was very surprised to see the wooden pit props shattered and part of the roof down (the effects of weighting) since there had been no hint of any problem as they left to take their meal break and, in the heat of the moment, called his drawers to come and see. At that point, the naked flame of the acetylene lamp on the drawer's cap ignited methane gas, a small pocket of firedamp, just slightly above the ignition concentration and a minor explosion occurred, strong enough to blow the miner and his drawer out of the split and cause the second drawer to be thrown down on to the pavement where he suffered fatal head injuries. A second and more extreme explosion followed, when a larger accumulation of gas released by the roof fall ignited and fire quickly followed.

Some fourteen men were at work at the inbye side of No. 3 dook at No. 3 level of the pit and were totally unaware that there had been an explosion. From evidence given at the Inquiry, it transpired that two of the drawers from that part of the working had come out to the main haulage road at No. 13 level with loaded hutches and were speaking with a third drawer from another workplace when the explosion occurred. However, in the confined conditions underground, the noise was a mere thud and wafts of displaced air extinguished the acetylene cap lamps they were all wearing. On re-lighting the lamps, whilst they pondered what might have caused it, the two drawers then merely took empty hutches back into the working and resumed work, and perished. Had they raised the alarm, or even told the more experienced miners of what had transpired at that point, then it is perhaps possible that there would have been but one casualty. On the surface, when the alarm was raised at around 8.20pm advising of the accident and calling for medical aid for an injured drawer following an explosion, there was an element of disbelief that there had in fact been an explosion and a fire, and some valuable time was lost before the Mines Rescue Service was called and the full rescue plan swung into action with

Burngrange Nos 1 & 2 pit after the 1947 disaster, showing miners, mines rescue personnel and members of the National Fire Service at the pit bottom during the rescue attempts to reach fourteen trapped miners. This photograph is quite unique in that members of the NFS did not generally go underground, but such was the close-knit community that, when faced with a tragedy, these men went way beyond their call of duty to help friends and neighbours.

ABOVE: Underground in Burngrange Nos 1 & 2 pit after the disaster in 1947. Here sand bags have been used to seal off the road between No. 10 and No. 11 levels, with a member of the National Fire Service present.

BELOW: The Burngrange Mines Rescue Brigade are seen underground in the aftermath of the explosion and fire in 1947. The Brigade were to the fore in the rescue attempts and Brigadesman Jim McArthur, (right seated) was to be awarded the King's Commendation for his own personal bravery in the number of heroic attempts he made to reach the fourteen trapped miners.

Rescue Brigades attending from all parts of the shale field. Despite the initial desperate (and heroic) attempts by two men to reach the men trapped underground, they were driven back several times as the fire took hold. The men were the oversman, David Brown (the author's second cousin) and a miner (fireman), Jim McArthur (a family friend), both of whom were trained members of the Mines Rescue Brigade at the pit. David Brown had several times gone in along the haulage road with wet towels wrapped around his head in an effort to reach the trapped men but was driven back each time until finally he asked for two sets of Proto-Breathing Apparatus from the National Fire Service members who had arrived on the scene underground. Despite the best efforts of Brown and McArthur, they were still unable to reach the men, the way being obstructed by several fires and fallen debris.

Oil shale is very hard to set alight, but the heat generated by the explosion and thereafter contained by the cramped underground conditions acted almost as a retort and released vast quantities of hydrocarbon gases (the crude oil) from the oil shale *in situ*, which then ignited. One can only imagine the appalling atmospheric conditions and extreme heat which confronted the would-be rescuers.

The fires raged for four days before being contained, and at one point arrangements had been put in hand to completely seal up the affected area. In the event, this was not done and No. 3 level, where the men had been working, was finally reached by Captain James Readdie and his team from the Threemiletown (No. 35 pit) Mines Rescue Brigade. All the miners were found dead, all but two of the bodies totally unmarked, having being asphyxiated as the fire drew all the oxygen out of the pit air. Of the two that showed signs of having suffered slight burns, it was assumed that these two men had made an attempt to leave the workplace and had suffered burns in consequence. However, the empty hutches which had been taken back into the workings were found to be fully loaded and it was obvious that the men had not succumbed quickly, expert opinion suggesting that to load the hutches would have taken at least twenty minutes. David Brown was awarded the Edward Medal (later to be replaced by the George Cross) for his bravery and Jim McArthur, the King's Commendation for Brave Conduct.

In all the many acts of heroism at the time, there was one quite extraordinary and extremely unusual event which went largely unrecorded outwith the village. It was relatively unknown throughout the mining industry for members of the National Fire Service, as it then was, to go underground, this generally being left to the professional Mines Rescue Teams; but at Burngrange on that late afternoon, the West Calder Brigade – a retained brigade and volunteers all, and a part of the other NFS brigades in attendance – to a man volunteered to go underground to fight the fire and assist the Mines Rescue Teams in the rescue effort, giving a clear indication of just how close-knit the shale mining community was. These firemen had friends and neighbours amongst the trapped men and to their everlasting credit they thus responded in a manner which was way beyond their call of duty, by going underground. The courage these NFS firemen displayed was without doubt one of the epics of what was a heroic service. This selfless action certainly has a valid place in this book.

At the Public Inquiry, held in the Seafield Institute on 15th September 1947, under the chairmanship of A.M. Bryan, HM Chief Inspector of Mines, amongst other issues it was concluded that whilst there had been some initial delay in sending for the rescue services, this had no bearing on the final outcome and that had the alarm been raised immediately, nothing would have changed. He considered that, even with the enhanced safety requirements over and above the statutory requirements in place on the night in question regarding lamps, these requirements had proved to be insufficient and thus he recommended that Section 32 of the 1911 Act be thereafter strictly enforced, which called for only locked safety lamps to be used in a pit where an explosion had caused injury/death in the preceding twelve months. However, because the 1911 Act was somewhat ambiguous about the storage and use of explosives in shale mines, he did suggest that some further special instructions be employed in shale in this regard.

Burngrange Nos 1 & 2 pit after the 1947 disaster showing wooden pillar supports being set in place between No. 10 and No. 11 levels. Note the canary in the cage as the means of testing for gas.

Burngrange Nos 1 & 2 pit after the 1947 disaster showing the tunneling through the shale which had been burned *in situ* by the fire. Again, note the presence of the canary for precautions against the presence of gas.

Summary

To answer the question posed at the beginning of this chapter: overall then, despite the perceptions, and indeed the many recorded memories within former shale communities that shale mining was an extremely dangerous activity, with inherent risks to both health and safety, the opposite was in fact the case. When compared with coal mining, shale mining was a relatively 'safe' industry or as safe as underground mining could be. Whilst it must be accepted that there were many, many more coal mines and pits and many more men employed therein, when the accident records are examined and compared, as was necessary for the compilation of this book, for sheer number of fatalities and the instances of multiple fatalities, the coal mining industry, even on a *pro-rata* basis, stands out head and shoulders above the shale industry. Consider, for instance, the worst mining disaster in Scotland, at High Blantyre Pit, Hamilton, on the 22nd October 1877, where 233 miners perished in an underground explosion, and the worst-ever pit disaster in the UK, at the Universal Pit, Senghenydd, Glamorganshire, in South Wales where, on the 14th October 1913, 440 men were lost in an underground explosion when a large quantity of coal dust was ignited by a firedamp explosion. The explosion at Burngrange pit in January 1947 with fifteen fatalities was alone to be the single worst disaster in the 110 years of shale mining in the Lothian shale fields. Even the aggregated number of recorded fatalities over the life of the industry do not even begin to approach the numbers recorded in the coal industry.

Indeed, when the accident records of the author's own chosen industry, the railway, are examined, particularly in the earlier years and during the period when the shale industry was very active, shale mining compared most favourably in risk terms. As an example, in the single year of 1890, and within the five Scottish railway companies, eighty-four men were killed whilst at work and 229 were seriously injured.

As with all underground mining, yes, there were significant risks, accidents did happen, and fatalities and serious injury did occur. However, it has to be said that some were self-inflicted by sheer carelessness or by failure to comply with standing safety instructions. Consider, for instance, the miner in the Pentland pit at Straiton who tried to fire his shots, not by using fuses, but by lighting them with burning straw. It cost him, and his drawer, their lives.

It goes without saying that any accident was one too many and all the fatalities which occurred were harrowing and unfortunate. Reading through the individual annual accident records brought home the real impact and real victims when a bread-winner was lost. Wives were not only widowed; in many cases large families, and even in one case sixteen children, were left fatherless. What was even worse, if that were possible, was a practice common until the 1930s: the bodies of miners and workers killed at work were often just taken home in the condition found and it was left to the widow or family to arrange burial.

A splendid view of the Westwood (Breich pit) to Seafield Oil Works haulageway operated on the endless rope principle. The building is the Almond pump-house.

7

THE PITS AND MINES

Some of the early pits around the shale fields, particularly those serving Young's Oil Company, appear to have had numbers only. Many of these mines were very small concerns and were quickly worked out or later superseded by larger and more productive pits and mines. Research is difficult, as on the early maps and records all the test boring sites were also identified by number. There were, undoubtedly, several small inclined mines driven into outcropped seams – known as 'in-gau'in' ee's' (in going eyes) and worked by a few men – which were unrecorded and had a limited life.

It was not until 1872 that the government of the day introduced legislation to require mine owners to advise and provide details and plans of mines and pits being closed, under the provisions of the Coal Mines Regulations Act, 1872. This legislation covered all mining activities, not being solely confined to coal, despite the title. The plans were required to show the workings relative to surface features and to record information concerning orientation, scales, contours, boundaries and faults. These Abandonment Plans formed the last record of workings, thus ensuring the safety of any future operations.

Unfortunately, many underground workings from the shale industry were already long abandoned by the time this legislation was introduced, and the knowledge and understanding of such unrecorded workings have been, in many cases, lost forever. It is thus now difficult to list with any certainty all the pits or mines where shale was extracted to serve the various retorts and refineries, but one thing is clear, there were many. Certainly, over the life of the industry, well nigh 100 and more pits and mines must have been in operation, the lifespan of many being five years or less.

Listed below are the pits and mines that can be identified, mostly from Abandonment Plans. Between 1901 and 1920, only forty-five were recorded as still working – indeed, forty-five was about the number being worked at any one time. By 1926 only twenty-four remained active and post 1950 only thirteen are recorded as being in production, although these figures may not, in fact, be wholly accurate. It is interesting to note, however, that during the life of the industry, West Lothian pits and mines were to supply 77 per cent of all mined shale, with the Midlothian workings accounting for the balance.

Those named (as opposed to being identified by merely a number) were as shown in the following table.

Deans No. 5 mine. Note the secondary upcast (and emergency) inclined shaft in the foreground. *Courtesy of Grangemouth Heritage Trust*

West and Mid Lothian

Pit	District	Main Seams Worked	Date of Abandonment Plan
Philpstoun No. 1 mine	Abercorn	Broxburn & Dunnet	1931
Philpstoun (Whitequarries) No. 1	Abercorn	Broxburn upper, Main Dunnet	1962
Philpstoun No. 2 mine	Abercorn	Broxburn	1890
Philpstoun No. 3 mine	Abercorn	Broxburn	1892
Philpstoun No. 4 mine	Abercorn	Broxburn	1899
Philpstoun Grey No. 4 mine	Abercorn	Grey	1912
Philpstoun No. 5 mine	Abercorn	Broxburn	1911
Philpstoun No. 6 mine	Abercorn	Dunnet	1932 reopened/closed 1962
Philpstoun No. 7 mine	Abercorn	Dunnet/Champfleurie	1931
Duddingston Nos 1 & 2 mine	Abercorn	Dunnet, Camps	1941
Duddingston Nos 3 & 4 mine	Abercorn	Pumpherston shales	1956
Champfleurie No. 1 mine	Linlithgow	Broxburn	1902
Ochiltree No. 2 mine	Linlithgow	Broxburn	1912
Ochiltree No. 3 mine	Linlithgow	Broxburn	1912
Ochiltree No. 5 mine	Linlithgow	Broxburn	1889
Ochiltree No. 6 mine	Linlithgow	Champfleurie	1916
Ochiltree No. 1 coal mine	Linlithgow	Houston coal	1903
Addiewell Nos 2 & 3 mines	West Calder	Grey Shale	1865
Addiewell Nos 2, 5 & 6 pits	West Calder	Fells	1870
Addiewell No. 1 mine/No. 4 pit	West Calder	Raeburn	1866
Hartwood mine	West Calder	Fraser	1875
Gavieside Nos 1 & 2 pit	West Calder	Fells	1879
Gavieside No. 3 pit	West Calder	Raeburn	1879
Gavieside No. 27/Polbeth No. 27	West Calder	Under Dunnet	1900
Gavieside No. 40 pit	West Calder	Dunnet & Under Dunnet, Barracks	1927
Muirhall Nos 1 & 2 pit	West Calder	Grey	1870
Polbeth Nos 7, 7½ & 10 pit	West Calder	Thick	1885
Polbeth Nos 8/20 pit	West Calder	Fells	?
Polbeth No. 21 pit (Old Drove Road)	West Calder	Fells	1907
Polbeth No. 20 pit	West Calder	Fells	1906
Polbeth No. 26 pit	West Calder	Dunnet	1947
Polbeth No. 31 pit	West Calder	Dunnet	1892
Limefield No. 32 mine	West Calder	Dunnet	1923
Polbeth No. 11/Westwood No. 30	West Calder	Addiewell, Fraser & Thick	1927
Hermand No. 4 mine	West Calder	Dunnet	1961
Hermand No. 5 mine	West Calder	Broxburn	1894
Hermand No. 6 mine	West Calder	Fells	1893
Hermand Nos 5 & 6 mine	West Calder	Broxburn, Fells	1894
Cobbinshaw Nos 1 & 2 mine	West Calder	Fells	1926
Cobbinshaw South mine	West Calder	Raeburn	1874
Cobbinshaw No. 5 mine	West Calder	Fells	1926
Cobbinshaw South No. 3 mine	West Calder	Fraser	1887
Cobbinshaw South No. 28 mine	West Calder	Fraser	1887
North Cobbinshaw mine	West Calder	Fells	1873
Tarbrax Viewfield No. 4 pit	West Calder	Fraser	1920
Tarbrax Viewfield No. 5 pit	West Calder	Fraser	1926
Tarbrax No. 1 pit	West Calder	Fells	1903
Tarbrax No. 2 pit	West Calder	Fells	1903
Tarbrax No. 3 & 4 pit	West Calder	Fraser	1920
Lawhead No. 1 pit	West Calder	Fells	1906
Greenfield No. 1 pit	West Calder	Raeburn	1886
Greenfield No. 3 pit	West Calder	Raeburn, Fraser	1903
Burngrange No. 39 pit	West Calder	Broxburn	1912
Burngrange Nos 1 & 2 pit	West Calder	Dunnet	1958
Fraser pit	West Calder	Fraser	1951
Baads No. 9 (Blackbraes No. 9)	West Calder	Fells	1868

THE PITS AND MINES

Pit	District	Main Seams Worked	Date of Abandonment Plan
Baads No. 15 mine	West Calder	Fells/Broxburn	1883
Baads No. 17 mine	West Calder	Dam shale	1873
Baads No. 22 mine	West Calder	Fells	1897
Charlesfield Nos 1 & 2 pit	Mid Calder	Pumpherston	1873
Charlesfield No. 3	Mid Calder	Raeburn	?
Alderstone No. 43	Mid Calder	Pumpherston (Jubilee)	1921
Oakbank Nos 1 & 2 pit	Mid Calder	Big shale, Dunnet	1909
Oakbank Calderwood mine	Mid Clader	Broxburn	1870
New Farm Nos 3 & 4	Mid Calder	Lower Dunnet, Fells, Broxburn	1919
Pumpherston pit	Mid Calder	Big shale	1873
Pumpherston No. 1 mine	Mid Calder	Pumpherston shales	1901
Pumpherston No. 2 mine	Mid Calder	Pumpherston shales	1901
Pumpherston No. 3 mine	Mid Calder	Pumpherston shales	1901
Pumpherston No. 4 mine	Mid Calder	Camps	1927
Pumpherston No. 5 mine	Mid Calder	Dunnet	1925
Pumpherston No. 6 mine	Mid Calder	Broxburn	1912
Ingliston No. 33 pit	Kirkliston	Dunnet	1894
Ingliston Nos 36 & 37 pit	Kirkliston	Camps	1926
Newliston No. 29 mine	Kirkliston	Dunnet	1937
Crossgreen No. 1 mine	Uphall	Broxburn, Grey, Curly	1910
Crossgreen No. 2 mine	Uphall	Broxburn	1899
Crossgreen No. 3 mine	Uphall	Broxburn	1899
Carledubs	Uphall	Grey, Broxburn	1922
Pyothill No. 5	Uphall	Camps & Broxburn	1919
Hopetoun No. 1	Uphall	Broxburn	1881
Hopetoun No. 2	Uphall	Broxburn	1882
Hopetoun No. 3 Niddry	Uphall	Broxburn	1884
Hopetoun No. 4	Uphall	Broxburn Main, Upper Grey	1890 reopened/closed 1944
Loaninghill No. 4	Uphall	Camps & Dunnet	1930
Dunnet Sandhole	Uphall	Dunnet	1927
Stewartfield No. 1 mine	Uphall	Broxburn	1897
Stewartfield No. 2 mine	Uphall	Broxburn	1914
Stewartfield No. 1 pit	Uphall	Broxburn	1897
Stewartfield No. 2 pit	Uphall	Broxburn	?
Stewartfield No. 3 pit	Uphall	Broxburn	1914
Stewartfield No. 4 pit	Uphall	Broxburn, Curly	1914
Albyn	Uphall	Broxburn & Camps	1905
Hut	Uphall	Broxburn & Camps	1912
Drumshoreland Muir	Uphall	Lower Dunnet	1901
Holmes Nos 1 & 2 mine	Uphall	Pumpherston, Maybrick, Plain	1900
Forkneuk No. 1	Uphall	Broxburn	1909
Fivestanks mine	Uphall	Raeburn	1919
Stankyards No. 1 pit	Uphall	Broxburn, Grey	1873
Stankyards No. 45 mine	Uphall	Broxburn, Grey	1912
Roman Camp No. 1 mine	Uphall	Jubilee, Curly, Maybrick	1918
Roman Camp No. 2 North mine	Uphall	Pumpherston shales	1893 reopened/closed 1930
Roman Camp No. 2 South mine	Uphall	Pumpherston shales	1930
Roman Camp No. 3 mine	Uphall	Jubilee, Pumpherston	1901
Roman Camp No. 4 pit	Uphall	Broxburn, Grey, Fells	1926
Roman Camp No. 5 mine	Uphall	Camps	1934
Roman Camp No. 6 mine	Uphall	Camps	1956
Roman Camp No. 7 mine	Uphall	Dunnet	1951
Greendykes North mine	Uphall	Broxburn, Grey, Curly	1874
Greendykes South mine	Ecclesmachan	Broxburn	1915
Haycraigs	Ecclesmachan	Camps & Broxburn	1919
Hopetoun Glendevon No. 5 pit	Ecclesmachan	Broxburn Main, Upper, Grey	1946
Hopetoun/Glendevon No. 6 mine	Ecclesmachan	Broxburn Main, Grey	1957
Hopetoun (Redhouse) No. 35 pit	Ecclesmachan	Broxburn Main	1958

Pit	District	Main Seams Worked	Date of Abandonment Plan
Hopetoun Fawns Park No. 41	Ecclesmachan	Broxburn Main, Upper Grey	1927
Hopetoun No. 44 mine	Ecclesmachan	Broxburn	1950
Forkneuk No. 9 pit	Ecclesmachan	Broxburn, Upper Grey	1909
Forkneuk No. 10 pit	Ecclesmachan	Broxburn, Upper Grey	1909
Forkneuk No. 38 mine	Ecclesmachan	Broxburn, Upper Grey	1909
Breich Nos 1 & 2 pit	Livingston	Broxburn	1927 reopened/closed 1961
Breich Nos 2 & 4 pit	Livingston	Broxburn	?
Breich No. 3 mine	Livingston	Broxburn, Fells	1887
Easter Breich Nos 1 & 2 pit	Livingston	Fells	1892
Seafield No. 1 mine	Livingston	Fells	1926
Seafield No. 3 mine	Livingston	Upper Dunnet & Dunnet	1917
Westwood Nos 1 & 2 pit	Livingston	Dunnet	?
Westwood No. 12 pit	Livingston	Fells, Dunnet & Broxburn	1887
Westwood No. 13 pit	Livingston	Fells, Broxburn & Dunnet	1896
Westwood No. 30 pit	Livingston	Thick	1893
Westwood New pit	Livingston	Broxburn & Dunnet	1961
Totleywells No. 1 mine	Dalmeny	Broxburn	1960
Rosshill Nos 1 & 2	Dalmeny	Dunnet	1921
Dalmeny No. 1 pit	Dalmeny		
Dalmeny No. 2 pit	Dalmeny	Broxburn	1894
Dalmeny No. 3 pit	Dalmeny	Broxburn	1914

The winding house at Breich pit.

THE PITS AND MINES

Westwood pit-head with No. 2 upcast shaft to the left with air duct in the fan house and No. 1 downcast shaft with staging for receiving loaded hutches from the pit bottom on the right. The building in the centre is the steam-driven generator house.

Pit	District	Main Seams Worked	Date of Abandonment Plan
Dalmeny (railway) Nos 1 & 2	Dalmeny	Never worked any seam	1914
Mortonhall Nos 9 & 10	Edinburgh	Dunnet	1900
Pentland Nos 1 & 2	Edinburgh	Dunnet, Broxburn & Fells	1901
Boghall Nos 3 & 4	Bathgate	Fells	1871
Starlaw/Deans No. 3	Bathgate	Dunnet, Upper Dunnet	1920
Cousland No. 1 mine	Bathgate	Broxburn	1912
Cousland No. 2	Bathgate	Dunnet, Upper Dunnet, Pattison	1913
Deans Nos 1 & 2 mine	Bathgate	Pattison, Stanley, Dunnet	1904
Deans (Caputhall)	Bathgate	Fells	1886
Deans No. 3 mine	Bathgate	Under Dunnet, Dunnet, Barracks	1920
Deans No. 4 mine	Bathgate	Under Dunnet, Dunnet, Barracks	1930
Starlaw/Deans No. 5 (Barracks)	Bathgate	Barracks, Jubilee, Dunnet	1931
Deans Nos 6 & 7	Bathgate	Fells, Broxburn	1931
Deans (Broxburn) Shale	Bathgate	Broxburn	1891
Starlaw pit	Bathgate	?	post 1870

THE SCOTTISH SHALE OIL INDUSTRY & MINERAL RAILWAY LINES

FIFE

PIT	OWNER	MAIN SEAMS WORKED	COMMENTS
Binnend (Winnyhall) No. 1		Dunnet	All closed when the Burntisland Oil Company went into receivership in 1880.
Binnend (Winnyhall) No. 2		Dunnet	
Binnend (Whinnyhall) No. 3		Dunnet	
Binnend (Grangehill) No. 4		Dunnet	
Newbigging Nos 1 & 2 mine		Dunnet seam opened but never worked	
Kilrenny No. 1	J. Youill		Closed by 1873. Shale was mined in Kilrenny No. 1 and retorted at Rowatt & Youill's Kilrenny Oil Works (closed in 1879)
Lady Erskine	J. Youill		

RENFREWSHIRE/AYRSHIRE

PIT	OWNER	MAIN SEAMS WORKED	COMMENTS
Abercorn mine	Merry & Cunningham	Lillies	The shale was retorted locally in oil works set up in association with the pits. Most, if not all, pits were closed by the late 1800s.
Barbush mine	Dixon's	Lillies	
Clippens mine	Merry & Cunningham	Lillies	
Douglas mine	Merry & Cunningham	Lillies	
Fulton (Johnstone) mine	J. Liddle	Lillies	
Linwood mine	J. Anderson	Lillies	
Milliken mine	Merry & Cunningham	Lillies	
Arnicklodge Nos 1 & 2 mine		Lillies	

DENNY (STIRLINGSHIRE)

PIT	OWNER	COMMENTS
Banknock	William Wilson	Shale was retorted in the Coneypark Oil Works.
Coney Park	J. Barr	

THE LAST PITS AND MINES RECORDED AS BEING OPERATIONAL BY 1950

SEAM	MINE OR PIT	OTHER SEAMS WORKED
Dunnet Shale	Burngrange Nos 1 & 2 pit	
	Hermand No. 4 mine	
	Roman Camp No. 6 mine	Camps shale
	Roman Camp No. 7 mine	
	Westwood pit	Broxburn shale
	Whitequarries mine*	Broxburn shale
	Philpstoun No. 6 mine*	Broxburn shale
Broxburn Shale	Westwood pit	Dunnet shale
	Hopetoun No. 6 mine	
	Hopetoun No. 35 mine	
	Whitequarries mine*	Dunnet shale
	Philpstoun No. 6 mine*	Dunnet shale
	Totleywells mine	Pumpherston shale
Pumpherston Shale	Duddingston Nos 3 & 4 mine	
	Totleywells mine	Broxburn shale
Fraser Shale	Fraser pit	
Camps Shale	Mid-Breich pit	
	Roman Camp No. 6 mine	Dunnet shale
Hurlet or Cobbinshaw Coal	Only Baads No. 42 mine passed into NCB hands in 1948, by then employing 300 men and producing 420 tons of coal per day for use at Addiewell Oil Works. The mine closed in 1962.	

* Whitequarries mine and Philpstoun No. 6 mine were linked in production.

8

EXTRACTION AND REFINING OF SHALE OIL

Crushing the Shale

The raw shale, as mined, was generally too bulky to undergo retorting since the larger pieces were likely to cause congestion inside the retort, and could neither be heated evenly nor fully. To counter this problem and to ensure uniformity in the retorting process, the shale was broken into approximately 4-inch cube pieces. Initially, this was done by hand but this was laborious, resource-intensive and costly work, and so shale crushing machinery was developed. This consisted of steel rollers, some 5 feet in length and 2 feet 10 inches in diameter, and fitted with 336 specially hardened steel teeth. The rollers were set with an approximate ½-inch clearance between the opposing teeth, and revolved at 4½ rpm. The shale passing through was crushed into suitably small pieces and so successful and efficient this method proved to be that, even after some years in constant use and crushing some 1.5m tons of shale, the wear on the teeth was negligible, being ½ inch or less.

The Shale Retort

Over the life of the industry, a variety of retort types were used right across the shale fields. The early retorts were modelled on the coal-gas retorts which were in common use by the mid-1800s, but with a basic difference in design in that they were required to provide low-temperature distillation to avoid unnecessary fracturing of the hydrocarbons. It has been suggested that by 1850 some fifty-two oil works were in production, mainly using horizontal retorts.

At Bathgate, the birthplace of the industry, Young used a horizontal common gas retort. This D-shaped horizontal gas retort was fed by a rotating screw and designed for use in retorting the 'Torbanite' coal. It was an inexpensive design to construct and was heated in a coal-fired brick oven. The residence time for the shale was between 16 and 24 hours. Steam injection, to become a feature of later designs, was not used. The yield was around 35

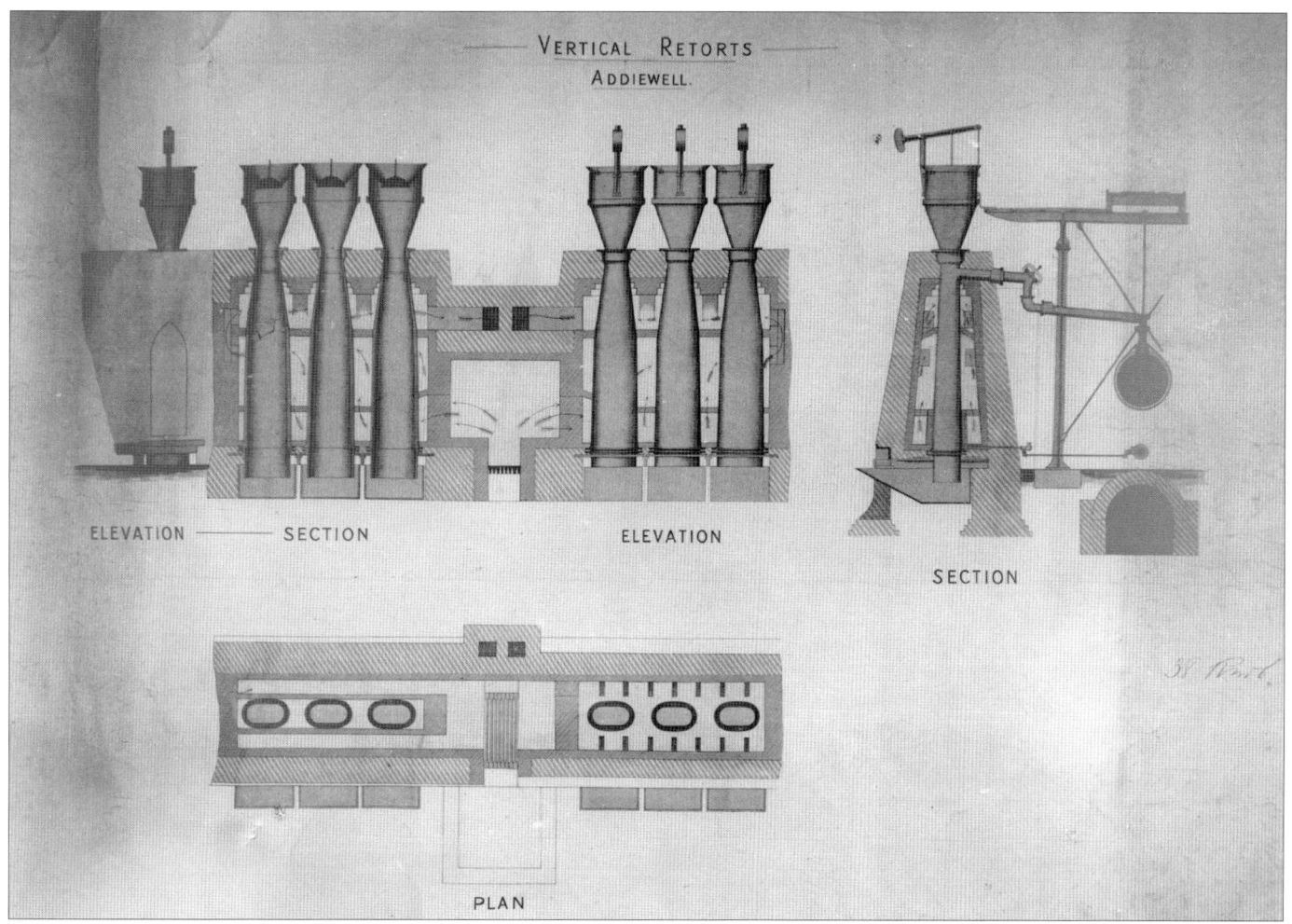

A plan of the vertical retorts at Addiewell Oil Works.

gallons per ton of shale, although significantly higher yields were obtained when the cannel coal was involved in the process. There was a downside to this type of retort, however, in that it was difficult to maintain a constant heat and higher temperatures tended to fracture the hydrocarbons.

During the 1860s and '70s, after Young's patent had expired, many different types of retorts were patented. There were at various times retorts of various designs in use such as horizontal, inclined, vertical, rotating, single and multiple offtakes and upward and downward distillation, to name but a few. The vertical retorts, however, all followed the same basic design of being a vertical pipe with a furnace set into the base. The upper ends were open with a funnel shaped top which could be stopped up (fitted with lids sealed by sand). The raw shale was charged at the top and the shale descended by gravity, being exposed to heat at all times. At the lower end, the spent shale was drawn off without any interruption in the process. Under influence of the heating process, the temperature of which increased as the shale descended, the shale decomposed and

Shale being tipped into the crusher from railway wagons at Westwood Oil Works.

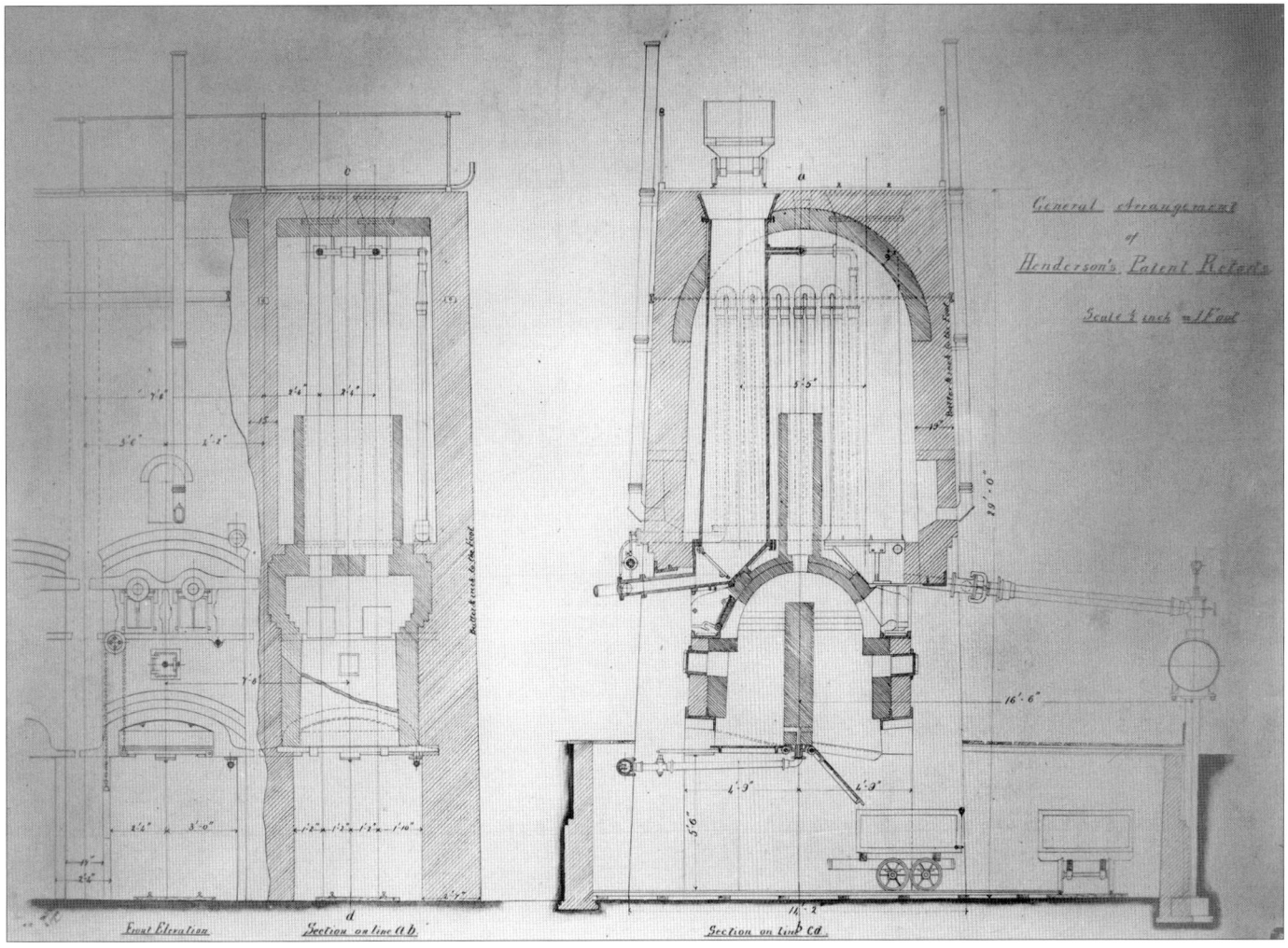

A plan of the Henderson vertical retort.

the oil was drawn off as a hydrocarbon vapour and collected for condensation as crude oil. This basic design of retort was improved by one A.C. Kirk whose contribution to the retorting process was to design a retort with an elliptical cross-section and which was larger at the bottom than the top. This gave a larger heating surface area and dispensed with the need for mechanical agitation. However, it was less than perfect and had disappeared from use by around 1885.

The benefits of using vertical retorts soon became obvious: they operated by gravity thus reducing manpower, they permitted almost continuous operation thus retorting higher tonnages of shale and reducing heat wastage. Indeed, by employing banks (benches) of retorts, the principles of mass-production were established.

It was the Kirk innovation which prompted the manager of the Oakbank Works, Mr N.M. Henderson, to develop a revised form of retort in 1873, the Henderson retort (Patent No. 1327) which offered the complete 'burning' of the raw shale which was, itself, then used as an additional heat source, saving around 50 per cent in fuel costs and allowing extraction of oil vapours at the lower end of the retort. This resulted in a superior quality of oil, capable of yielding higher percentages of marketable finished products. The Broxburn Oil Company adopted the Henderson retort in 1879 with great success, the shareholders capital being repaid in the first four years of operation. By 1880, Young's were converting to this retort and they were also in use at Linlithgow (Bridgend Works) and Burntisland.

However, retort development did not stand still and, in 1881, G.T. Beillby patented a completely new retort which heated the shale in two stages, the first a cooler stage used to maximise oil extraction without excessive cracking, and the second stage where higher temperatures were employed to maximise the extraction of sulphate of ammonia. The Beillby retort yielded similar quantities of crude oil as the other types but offered a threefold increase in the quantity of sulphate of ammonia produced. However, despite the very conceptual advances, the retort was in fact a commercial failure.

Undeterred, Beillby joined forces with William Young of the Clippens Oil Company and they jointly patented the new Beillby and Young, or Pentland Composite retort as it was better known (Patent No. 1377), in March 1882. This embodied the concepts of the original Beillby retort but was an improved version, offering greater reliability and ease of operation. This retort was capable of yielding higher tonnages of sulphate of ammonia. The design was very successful and offered, for the first time, a retort that was capable of continuous operation, 24/7. It could be charged every 6 hours, with an 18-hour dwell time for the shale, and could deal with a throughput of 28cwt of raw shale every 24 hours. The advantages were so significant that Young's at Addiewell, despite the fact that they had only renewed their retorts in 1883 using the Henderson retort, began to supersede the latter type with the new Pentland retorts as quickly as money would allow. By 1889, over half of the

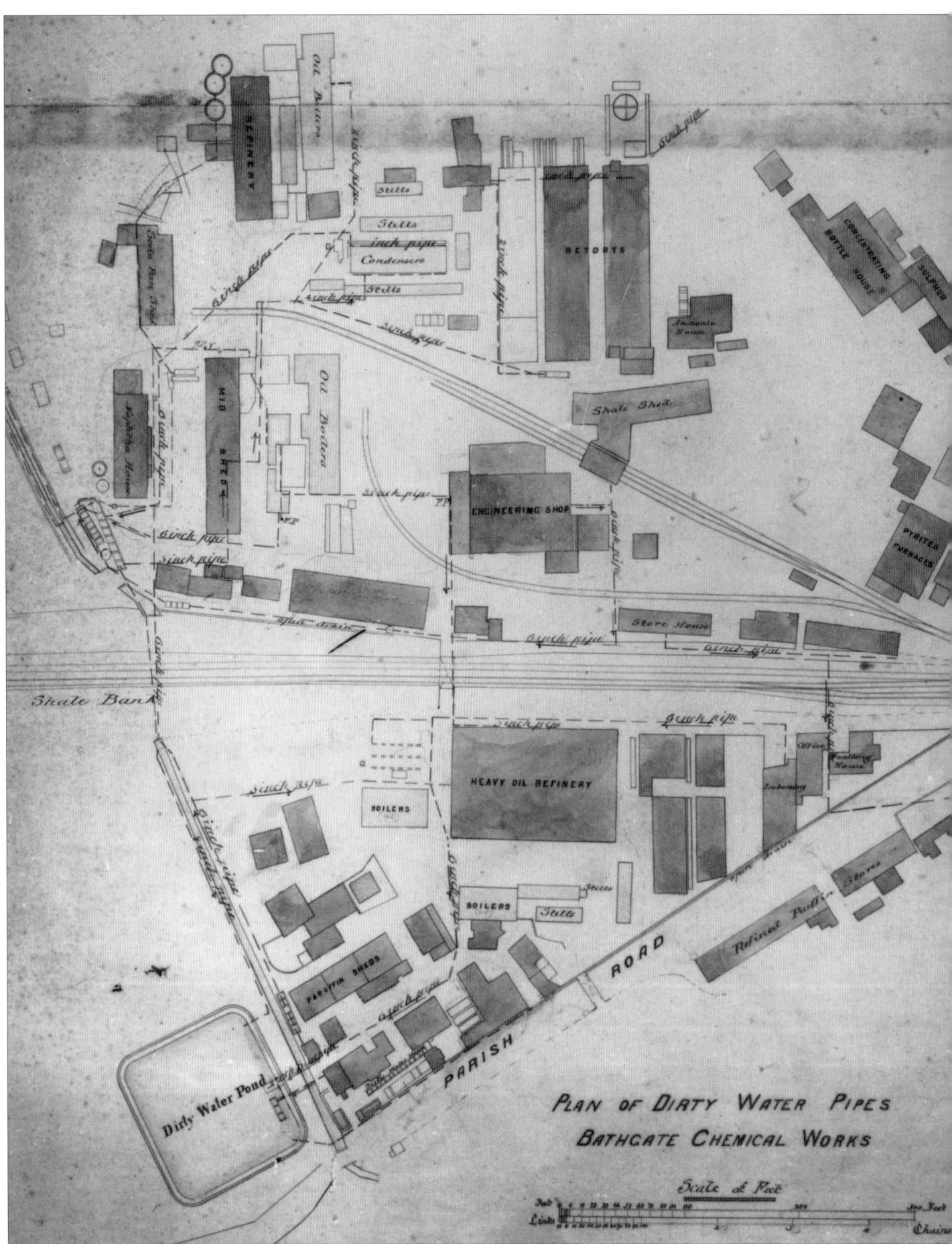

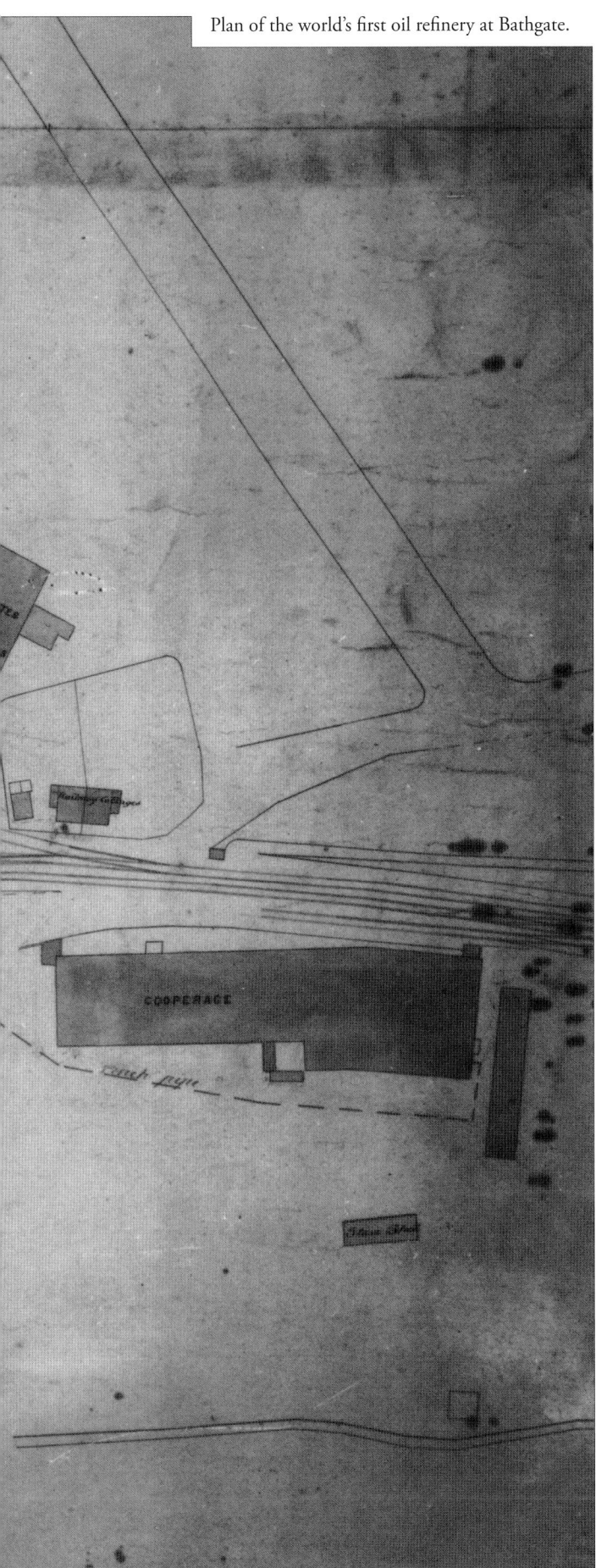

Plan of the world's first oil refinery at Bathgate.

5,000 retorts in use, some 3,636, were of the Pentland type, with the Henderson retort accounting for 896 and the three other types accounting for only 396 of the remainder.

The Henderson retort, yet further improved, regained its popularity in 1889 as a continuous-working gas-heated retort, similar to the Pentland design, but with a significantly higher yield of sulphate of ammonia which at this time was attracting £22 per ton.

In 1894 the Bryson or Pumpherston retort appeared (patents Nos 8371 of 1894 and 7113 of 1895), retaining all the valuable principles of the Pentland retort but with added improvements in mechanical design and offering increased fuel economy. It consisted of an externally heated continuous vertical retort with a cast iron upper section and a lower section of firebrick construction. The throughput was raised from around 27 to 28cwt of raw shale per day to a staggering 4 to 5 tons per day. This Pumpherston, or 'Scottish', retort was quickly to become the industry standard and was to be the only type left by 1938.

Development did not cease, however, and Mr Henderson set about eliminating perceived deficiencies in the Pentland retort by introducing the new, modified Broxburn retort using a heat range from 900°F through to 1300°F and which had the capacity to retort 3.2 tons of shale with a residence time of only 4½ hours. These retorts were erected in banks of 88 retorts per bank and each had a throughput of 160 tons of shale every 24 hours. This retort became noted for reliability.

In 1937/38, experiments with air injection concluded that by injecting air at the bottom level the retort – the retorts used in the experiment were Young's, Pumpherston and Broxburn retorts – the process more than doubled the throughput of the retort and by 1938 air injection became the norm in the retorting process.

A Philpstoun retort was patented by Mr A.H. Chrichton. Based on the Pentland design, it failed to attract widespread attention although it was to be the mainstay of Philpstoun Oil Works until closure in 1938. Other designs were to come and go without any significant impact on the industry from that time. However, in 1939, anticipated increased demand due to the forthcoming hostilities led Scottish Oils to create a centralised retorting facility capable of producing large quantities of crude oil; the new oil works at Westwood, near West Calder was constructed between 1939 and 1941 with a retorting capacity of 1,200 tons of shale per day. The new Westwood retort was to be more or less based on the Broxburn retort but offering further significant energy saving improvements.

The retort temperatures attained in the Westwood Retort were:
- Upper cast-iron section
  Top   400°F
  Bottom   1075°F
- Lower firebrick section
  Top   1150°F
  Bottom   1450°F

These temperatures were obtained by:
- external heating by burning the permanent shale gas
- internal heating by the injection of super-heated steam
- internal heating by combustion of residual carbon in the shale by air injection.

Westwood retorts produced oil of the same quality as the Pumpherston retorts and an equal yield of sulphate of ammonia.

The retorting process, whatever design was used, had two principle objectives: to ensure the complete yield of oil by uniform heating of the shale and to obtain the highest possible yield of sulphate producers, with some 93 per cent of all its electricity coming from the direct burning of the oil shale there, with an annual production of some 31,000,000 tonnes of raw shale, but China is fast catching up. Extensive shale oil fields exist worldwide, in such diverse places as Australia, Brazil, Canada, USA, Estonia, China, South Africa, Israel, Spain, Russia and, yes, here in Scotland, with immense reserves and an immense energy potential far exceeding all known conventional oil reserves.

What does the future hold for oil shale? As of 2010, Estonia, China and Brazil are all actively extracting oil from shale by the Galoter process (Estonia), the Fushun process (China) and the Petrosix process (Brazil). All these processes are *ex situ* (above ground) processes and in 2008 some 7.33 million barrels of crude oil were produced by these methods. Focus has once more turned to the vast energy resources offered by the oil shale measures existing underground and experiments with *in situ* extraction of oil and gas have been, and are being, carried out. Royal Dutch Shell have recently completed a trial demonstration project where they produced some 1,400 barrels of crude oil plus associated gas from oil shale, without the shale ever seeing light of day, by a process called the *In situ* Conversion Process (ICP) where the shale was heated underground to release the crude oil. In this manner, perhaps West Lothian will, one day, see oil being produced from shale again. Who knows? There is, now, with liquid oil reserves becoming depleted, a new worldwide interest in shale reserves and just how this potential energy source might best be exploited. Australia, Canada and the USA are all looking at *in situ* extraction and a great deal of money is being invested. Some interesting times lie ahead.

In the UK the current thrust of underground exploration of known shale reserves is concentrating not with producing oil, but rather the shale is seen as a source of natural gas (methane) instead (see pages 31 and 61). There have been a series of test bores for gas extraction using a process called 'fracking', that is, drilling into the shale seams longitudinally and fracturing them *in situ* by small controlled explosions, water is then injected at high pressure to further fracture the shale in order to release the gas. However, a recent (2011) gas extraction experiment using fracking in the northwest of England, at Banks, near Southport, was considered to be the likely cause of subsequent earth tremors of magnitudes between 2.3 and 1.5 (Richter Scale) being felt in Blackpool and currently a ban has been imposed on this form of gas extraction pending further investigation. As always, there are two schools of thought, with supporters dismissing the risks of earthquakes and water pollution, and the environmentalists who consider that exploitation of natural gas will do nothing for the carbon footprint and is indefensible. It must also be said on the negative side that, apart from the earth tremors, trials with drilling into shale seams in the USA have thrown up some extremely unpleasant and unexpected consequences such as significant water pollution, ground movement and methane gas coming out of domestic water supply taps.

The headgear at Breich shale pit.

Courtesy of Grangemouth Heritage Trust

of ammonia from the spent shale at the lower end of the retort by introducing steam to act with the residual nitrogen contained therein. The modern Scottish vertical retort produced crude oil of an incredible uniform quality and was something peculiar to the Scottish oil industry. Nothing remotely similar was ever used elsewhere.

Mr H.R.J. Connacher, later to be General Manager, Scottish Oils Ltd, said it all when he stated:

Probably the retort is the most characteristic item of equipment in the industry, and advances in its design played an essential part in the survival of the industry in the face of very unequal competition.

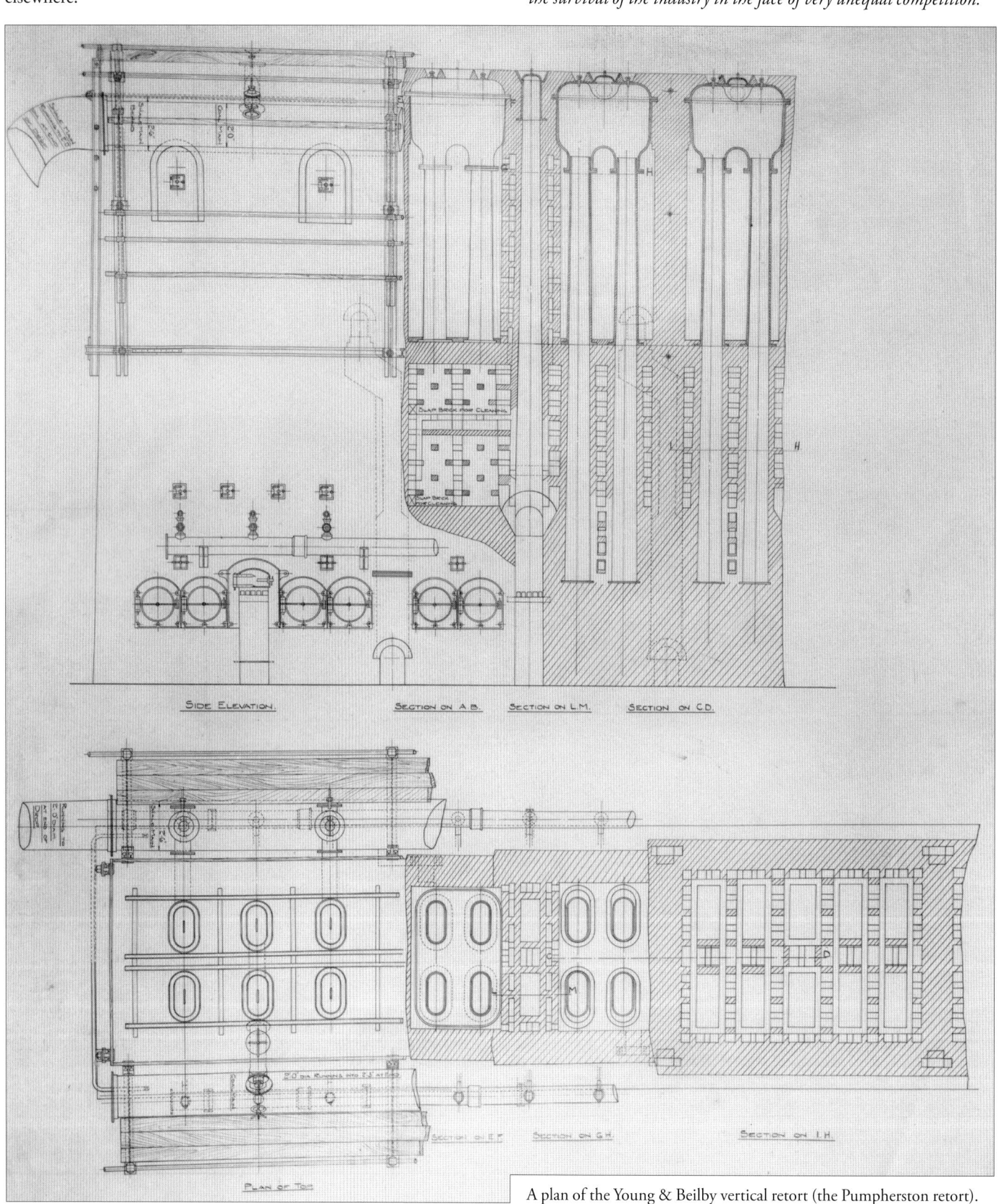

A plan of the Young & Beilby vertical retort (the Pumpherston retort).

EXTRACTION AND REFINING OF SHALE OIL

RETORTING THE SHALE

After the raw shale had been crushed, it was transported by hutch up a sloping scaffold to the top of the retort bench and loaded into a mild steel hopper. From the hopper, the shale was fed into the main upper body of the vertical retort which was of cast iron construction. This part of the retort was about 11 feet in length and 2 feet in diameter, tapering out to about 2 feet 4 inches at the lower end. The lower part of the retort was of fire-brick construction, tapering out downwards from 2 feet 4 inches to 3 feet at the bottom and was around 18 feet in length. The retorts were arranged in sets of four within a firebrick heating chamber and each pair shared a common hopper. The actual retort bench was carried on brick piers and the spent shale could be discharged directly into hutches running under the bench, from whence it was removed to the bing.

From the top hopper, which was charged every six hours or so, the shale travelled slowly down through the retort by gravity, being heated continuously at ever increasing temperatures, from around 270°F at the top of the cast-iron section with heat progressively

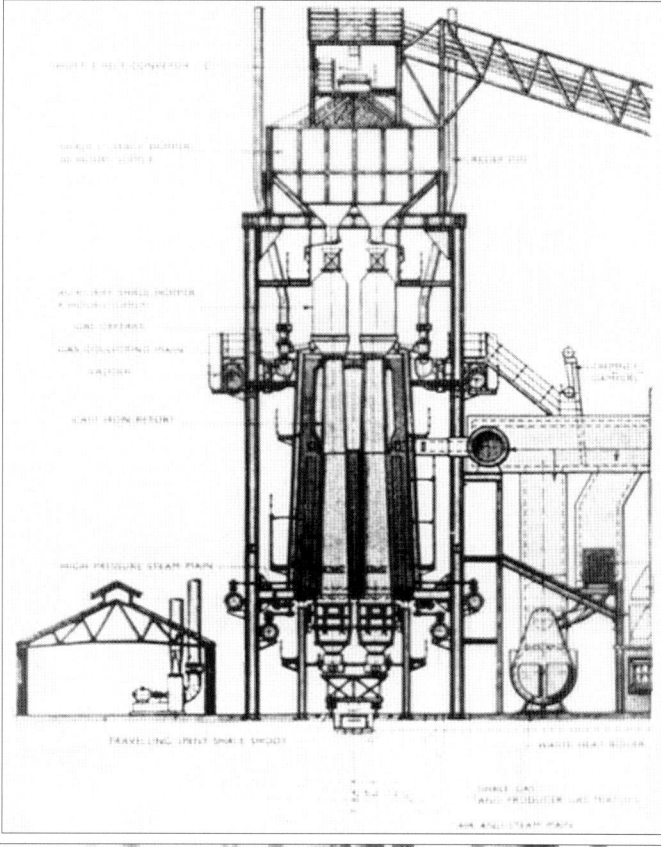

RIGHT: Diagram of the Westwood retort.

BELOW: One of the retort benches at Pumpherston Oil Works. The large-diameter pipe running the length of the bench just below the top is the pipe conveying the hydrocarbon gases given off by the heating of the raw shale to the condensers. *Courtesy of Grangemouth Heritage Trust*

EXTRACTION AND REFINING OF SHALE OIL

The condensers can be seen in front of the retorting plant at Pumpherston, the shale bing dominating the background. On the left of the photograph can be seen the incline for delivering shale to the top of the retorts
Courtesy of Grangemouth Heritage Trust

increasing to around 480°F at the foot of the cast-iron section, then from around 950°F to 1800°F through the fire-brick section. From the bottom of the fire-brick section, the shale passed through a fire-brick cone, upon which the whole column of shale in the retort was supported, into a cast-iron chamber containing a circular horizontal plate which did not fill the chamber. An annular space was left through which the spent shale was swept by a rotating metal arm into a discharge hopper which was then emptied every four hours or so into the hutches running under the benches.

The external heating of the retort was largely performed by using the incondensable gases given off by the shale itself, supplemented where necessary by gas from external coal-gas producers. The burning gases were led upwards in a spiral direction through the heating chamber with the waste gas escaping by means of a chimney at the top of the bench. A quantity of steam, amounting to around 75 gallons of water per ton of shale, was introduced into the bottom of the retort, forming, in contact with the hot shale, superheated steam (a gas) with considerable heating qualities. This provided the hydrogen necessary to form ammonia from the nitrogen in the shale. These hot gases promoted distillation of the shale in the upper portion of the retort and also served to equalize the temperature of the shale throughout the diameter of the retort; they also helped protect the ammonia and oil vapours from further 'cracking' or decomposition by sweeping them quickly up out of the lower high temperature zone through a large outlet pipe in the neck of the retort and into the atmospheric condensers. This practice increased the throughput of the retorting process quite dramatically.

Naphtha and Sulphate of Ammonia Separation

On leaving the condensers, the crude oil and ammoniac liquor flowed to a separator where the lighter crude oil rose to the surface to be drawn off via a pipe to the receiver, whilst the ammoniac liquor settled to the bottom and was drawn off from there into a separate receiver. The uncondensed gases, on leaving the condenser were cooled and scrubbed to recover any ammonia which might still be present and also naphtha which had escaped condensation.

The underside of the retorts at Pumpherston Oil Works showing the conveyor for transporting shale which has passed through the retorts, away to the bing.

EXTRACTION AND REFINING OF SHALE OIL

Courtesy of Grangemouth Heritage Trust

The condensers and stills at Pumpherston.

ABOVE: Pot coking stills at Pumpherston. *Courtesy of Grangemouth Heritage Trust*

BELOW: A bench of crude oil stills at Pumpherston with three condenser stills in the centre. *Courtesy of Grangemouth Heritage Trust*

EXTRACTION AND REFINING OF SHALE OIL

Deans Oil Works: the retorting plant. *Courtesy of Grangemouth Heritage Trust*

For this naphtha recovery, the gases passed though a scrubbing process where they were sprayed with shale gas-oil which absorbed the naphtha present; this oil/naphtha mixture was passed into a naphtha still where the naphtha was driven off by steam and condensed into unrefined naphtha. The gas-oil was cooled and recirculated, the scrubbing operation being a continuous process. In this scrubbing process, the normal recovery was around 3 gallons of naphtha per ton of shale.

The ammonia scrubbers were of a similar structure but with water being the absorbing liquid; this was mixed with the ammonia obtained from the retort condensers and sent for processing. The ammonia, expelled as a vapour, was absorbed in a solution of sulphate, formed by mixing sulphuric acid and water; when the solution was evaporated, the sulphate of ammonia was left in a crystalline form.

During the First World War, the Ministry of Munitions required vast quantities of concentrated ammoniac liquor, containing a minimum of 25% ammonia, for the manufacture of nitrate of ammonia, a key constituent in the manufacture of AMATOL, a high explosive used in munitions. One crude oil works (most likely Addiewell) started producing 20 tons of this liquor every day (7,300 tons/year) with a minimum ammonia content of 27.3%, for the war effort.

REFINING THE CRUDE OIL: FIRST DISTILLATION

The refining process was a very complicated process from the condensation of the crude oil through to the final refined products and the following is, at best, merely a thumbnail sketch in layman's terms. The general principles of refining were to:
- separate the mixture of hydrocarbons constituting crude oil into the fractions necessary to obtain quality liquid marketable products;
- rid the oil of all impurities and constituents;
- separate the crude solid wax from the liquid oils.

The crude oil, which was in a semi-solid state at room temperature (the setting point was 87°F) was heated in a still and, since the specific gravity of oil is related to its boiling point, separation was effected by condensing the oil vapours from the still and leading the different condensed liquids into different receiving vessels. Steam was introduced to assist in the distillation of the various fractions and control the temperature of the oil to avoid decomposition. The heating for the distillation process was obtained by burning the impure 'tars' removed from the crude oil. The average Scottish crude oil supplied sufficient tar in refining to meet all the heating needs in the distillation process. During this first distillation there was a certain degree of 'washing' the crude oil by using sulphuric acid (H_2SO_4). The oil obtained from this first process was known as 'green oil' by virtue of its colour.

The earliest stills were cast iron horizontal stills heated by coal but were prone to fracturing and were soon replaced by vertical stills. These again gave problems, which then led to the adoption of 'pot' stills or coking stills. These stills lasted in use at Pumpherston until the demise of the industry. Pot stills reduced the crude oil to a solid form of coke which increased the quantity and quality of paraffin. Various other types of stills were used but perhaps the most

Deans Oil Works: the charging level on top of the retort bench. The hutches in the background contain shale which has been crushed and is ready for the retorting process. The retorts are charged via the hopper openings between the rails. The rows of chimneys vent the spent gases to the atmosphere.
Courtesy of Grangemouth Heritage Trust

significant was the continuous boiler still patented by Henderson, he of retort fame. Another all-Scottish invention, this type was first used at Broxburn and reduced the distillation process to a single stage; it was also initially used at the new Grangemouth Refinery to refine the Middle East crude until replaced by a pipe still designed by Scottish Oils Ltd.

At the centre of the early refining process, and still essential in today's modern refineries, is the 'cracking' process, the first version of which was the brainchild of James Young's son who took out a patent for it in 1865. This involved heating the oil under a pressure of 20psi in an oval-bottomed pressure still; by this process the quantities of lower boiling materials were increased. The process was used to increase the volumes of paraffin and 'burning' oils as demands for these grew apace and, later, with the development of the internal combustion engine and the need for motor spirit, cracking was once more used to increase the yield of petrol. Young's early process was to be overtaken by the more modern thermal pipe still crackers, introduced at two shale oil refineries. Today, modern hydrocrackers using hydrogen at 2,000psi in the presence of a catalyst are used to crack heavy oil products to gasoline and LDF products. The 'cat cracker' at Grangemouth refinery was installed in the 1970s.

REFINING THE CRUDE OIL: SECOND DISTILLATION

The 'green oil' so obtained in the first distillation process was then run through a second distillation process in stills similar in design and arrangement to those in the crude oil plant. Again the oil was heated but no steam was employed in this second run in order to obtain a more crystalline form of solid paraffin which could be more easily extracted. In the first stage of the second distillation process, the fractions obtained and collected were:
- crude burning oil
- heavy oil and paraffin.

The crude burning oil was stirred and treated with a sulphuric acid to remove basic nitrogen compounds and to polymerise the lower olefins likely to affect stability of the product, before being washed in a caustic soda solution and, after settling, was redistilled in a series of boiler stills (known in the industry as fine old boilers). The residue was run off continuously during this part of the process. The fractions then obtained were:
- lamp oil or power oil
- lighthouse oil
- residuum.

EXTRACTION AND REFINING OF SHALE OIL

Above: Deans Oil Works: the retorts and condensers. *Courtesy of Grangemouth Heritage Trust*

Below: Deans Oil Works: the Mond gas plant. *Courtesy of Grangemouth Heritage Trust*

A view of Roman Camp Oil Works showing the Henderson retorts, atmospheric condensers and scrubber towers. In the foreground on the left is a private owner wagon belonging to Barr & Thornton, who had a coal mine at Fauldhouse, whilst on the right can be seen a number of wagons carrying oil tanks. Again the scene is dominated by the shale bing, the incline up which is marked by a line of posts. *Courtesy of Grangemouth Heritage Trust*

The paraffin wax cooling shed (probably Broxburn Oil Works) showing the racks of cooling wax.

The residuum was then blown into a separate single boiler still and redistilled, giving:
- heavy burning oil
- light gas oil
- heavy gas oil.

At this stage the lamp, power and lighthouse oils were drawn into storage tanks, ready to be marketed.

Meanwhile, the heavy oil and paraffin were cooled to very low temperatures in refrigerated vessels charged with liquid ammonia. When the required temperature had been obtained, the cooled oil was forced under pressure into a series of filter presses. These presses were then compressed to force all the remaining liquid oil out into a lower container. The filter-pressed cakes of paraffin wax still contained some oil and these were further pressed in linen or cotton cloth until, finally, only cakes of solid paraffin wax were left. These were taken for further refining and purification. The oil run obtained through this filtering process was termed 'blue oil', again because of colour.

The blue oil was again treated with a solution of acid and caustic soda. When the oil had settled it was collected and redistilled in blue oil stills, to give the following distillates:
- heavy gas oil
- cleaning oil
- lubricating oil
- residuum.

The residuum was drawn off and prepared for the market as oil residuum, being in great demand for grease-making and as a wire rope lubricant. The gas oil, cleaning and lubricating oils were further treated and, when fully settled, run off and prepared for the markets.

Crude Solid Paraffin

This crude paraffin wax was subjected to a sweating process in large, brick-built 'sweat houses', each fitted with close-fitting iron doors and steam coils along the inner walls. The houses were also fitted with a series of water-cooled horizontal pans. The melted crude wax was run into these pans from a charging tank on top of the building and when the wax had solidified once more (cooled by the water), the water was run off, the iron doors were closed and the sweating process began. When again in a liquid state, the wax was clarified and all impurities were run off. The once-sweated wax was again cooled and then sweated a second time, yielding a soft wax and leaving a hard paraffin wax of good clear colour, with a melting point of between 119°F and 122°F. This quality wax was further steamed and re-filtered to remove any residual colouring, and was then prepared for the markets. In the earlier day of the industry, most oil works had an associated candle house where high quality candles were produced from this wax. The candles burned with an extremely bright light and quite transformed domestic illumination until superseded by the ubiquitous paraffin lamp and the new 'brilliant light'.

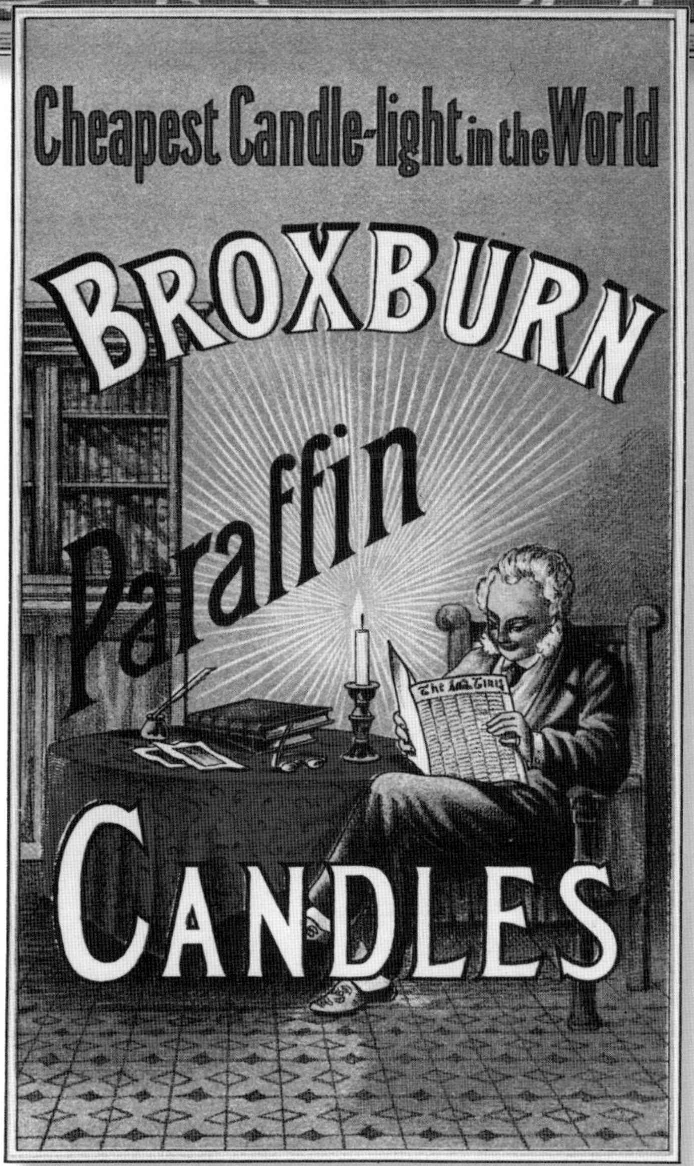

EXTRACTION AND REFINING OF SHALE OIL

Publicity material for the Broxburn Oil Company:
Top left: Petrol advert.
Above: Facsimiles of the various labels used by the company for the marketing of their candles.
Left: Publicity poster for paraffin candles.

The Refined Products

Spirits and Naphtha

Various spirits were obtained as markets changed, such as motor spirit (later to become petrol), extracting spirit, solvent naphtha (used in rubber-making and waterproofing), burning naphtha, white spirit (used in the manufacture of paint) and solvents used in the manufacture of linoleum. A much purified distillate of naphtha was used to produce oil gas, a product invented in 1851 by Julius Pintsch of Germany; when pressurised, the gas was to be much used in the lighting of railway carriages. However, the highly inflammable nature of the product was to be a contributory factor in the devastating railway accidents at Thirsk (1892), Hawes Junction (1910), Ais Gill (1913), Quintinshill (1915) and Charfield (1928) where, in each case, fire, fuelled by the escaping oil gas, broke out following collisions, with great loss of life. Oil gas was eventually replaced by electric lighting on railways but continued to be used to illuminate maritime buoys and beacons for a considerable time thereafter.

Lamp and Power Oil

Again, various grades of illuminating oil were produced and marketed, lamp oil (paraffin) being used in domestic lamps. The railway companies all took vast quantities of both continuous burning oil – for use in signal lamps which had to burn continuously for seven days at a time – and also basic lamp oil. Paraffin had become the preferred lighting oil for use in railway carriage lamps before the advent of oil gas, since it gave a much better light than common oil and was cheaper. To give one example, the North British Railway Company was contracted to the shale oil industry to take 25,000 gallons of paraffin at a time, at a cost of one halfpenny a gallon and ordered this amount every two months. The contract was shared between several oil companies. The lighthouse oil was, as the name indicates, used in lighthouses as a source of illumination. This grade of oil had a very high flash point and great illuminating power, it was thus safe for use in the remote locations where it was required.

Gas/Fuel Oil and Cleaning Oil

The fuel oils (furnace fuel oils) were widely used in steam boilers as an alternative to coal, both on land and at sea. Gas oil was just another grade of fuel oil. The industry supplied, under contract, fuel for the ships of the Royal Navy for the duration of the First World War.

The cleaning oil was highly refined oil and was in great demand by the railway companies for cleaning locomotives and other machinery. It was also used in the manufacture of axle grease.

Batching Oil

This grade of oil was specially adapted for the spinning industries. In the main it was used in Dundee for jute spinning and in Manchester in the cotton spinning industry. Mixed with water, it was sprinkled over the fibres of cotton and jute during the spinning process to keep them soft and pliable, thus preventing breakage.

Lubricating oil

As the name implies, this was widely used as lubrication for all sorts of moving machinery and was also used in the manufacture of specialised greases.

Residuum Oil

This thick oil residue was used as a constituent part of thick lubricating greases.

Motor Spirit (Petrol)

With the coming of the internal combustion engine and motor cars at the turn of the century, motor spirit (petrol) manufacture commenced, based on naphtha. The yield of the spirit was increased by employment of a 'cracking' process involving treatment of higher boiling fractions (above 200°C) at temperature and under pressure. This spirit was then subjected to a process to remove sulphur compounds using lead (known as the 'Plumbite' process (PbO+NaOH), producing 'leaded' petrol.

This petrol was selling for 1/- (5 pence) per gallon in 1910. Both James Ross at the Philpstoun Works and Young's were early suppliers of this new fuel, the latter marketing it as 'Scotch Brand'.

Diesel Oil

By 1938 there was an increasing demand for diesel oil for the internal combustion engine. Upon experimentation, it was found that some 50 per cent of shale crude oil could be converted to diesel by vaporizing the crude in a pipe still and separating the fractions by use of a bubble-cap fractioning column. By this process, naphtha, the hydrocarbon fraction with the lowest boiling point, was taken from the top of the column, mixed with scrubbed naphtha and then refined to make motor spirit.

The next fraction, with an intermediate boiling point, was removed from the upper part of the column (the wax free cut), first treated with sulphuric acid to remove the basic nitrogen compounds and reactive olefins, and then washed with 10% caustic soda to remove acid tars and neutralise it before it was redistilled to give diesel oil.

This process incurred little or no waste since all the impure materials were recycled back to the appropriate stage in the refining procedures.

Production of Sulphuric Acid

The sulphuric acid (H_2SO_4) used in the refining process described above was, in the early days of the industry, made by Young's at the original Bathgate Works and continued to be so produced until these works closed in 1956. Acid was also produced by the Broxburn Oil Company. The process used pyrites ore (iron sulphide), a material imported from Spain. This ore was burned in air in furnaces and produced sulphur dioxide gas (SO_2). This gas reacted with nitrogen obtained from sodium nitrate in a water spray contained within lead chambers, and produced a dilute sulphuric acid. Using a process to recover and re-use the nitrogen oxides, the sulphuric acid was concentrated to 80%. This 80% acid could be further concentrated, if required, by running it through a series of heated silica basins. Highly concentrated sulphuric acid was known as oleum or disulphuric acid ($ySO_3 \, H_2O$) and this concentrated form was a very convenient way to transfer sulphuric acid compounds in rail tanks between refineries and to industrial consumers.

The refining process was very complicated from the condensation of the hydrocarbon gases through to the final refined products and the foregoing is, at best, merely a thumbnail sketch in layman's terms of the whole process. Inquiring readers wishing to learn more may source a detailed description of the complete refining process in the book *Oil Shales of the Lothians* produced by the Geological Survey of Scotland and published by HMSO (1927). Whilst out of print, copies can yet be sourced.

9

THE MAJOR SHALE OIL COMPANIES AND THEIR WORKS

E.W. Binney & Co., Bathgate

In the earliest days, there was but one single company with an oil works, and indeed the first oil refinery in the world. These works (NS966672), for ever to be known as the 'Secret Works', were the result the partnership formed by Messrs James Young, Edward Meldrum and Edward W. Binney who established the Bathgate (Boghead) Works to produce oil from cannel coal (Torbanite) in 1851. The oil works lay on the south-west side of Bathgate, close to Durhamtown miners' rows. Here, the partners enjoyed a virtual monopoly of the process between 1851 and 1864, thanks to the foresight of Young who had taken out a patent for the process in 1850. This patent expired in 1864; in that same year Young bought out his partners and continued to operate the Bathgate Works under his own name. Bathgate Works was to come under the umbrella of Young's Paraffin, Light & Mineral Oil Company in 1866 and, after oil production ceased there in 1884, was to continue production as an acid works, lasting until 1956.

The floodgates opened after 1864 and it has been suggested that by 1871 there were some fifty-one retorting oil works in production in Scotland, producing some 25 million gallons of crude oil per year. Other sources indicate that between 1850 and 1890, some 120 firms, many small, had been established but these figures are unable to be accurately corroborated. Nevertheless, there was an oil boom during this period with at least half of these companies, like E.W. Binney, producing oil from cannel coal right across the Scottish coalfields but, unlike Binney's (latterly Young's), they had almost all ceased production by 1880. It is known that in 1870 there were some twenty oil works of some significance in production, but by 1875 this had dwindled to ten.

The refining of crude oil could only be carried out most economically when large quantities of raw crude were dealt with, thus it was the norm to establish the large refineries in a location convenient for the most productive shale fields. Even then, in some of the biggest fields, the mines were so widely spread around that railway lines had to be constructed to carry the raw shale from pit or mine to the oil works, and this at no little expense. In several cases, therefore, instead of incurring high costs for the transport of shale, a number of crude oil works with retorts and sulphate of ammonia plants only were established in close proximity to various groups of mines.

The Bathgate Oil Works were still in production, although with the supply of Torbanite coal drying up, Young had by now turned his attention to shale and the area around West Calder. In 1887, Young transferred much of the refining equipment from Bathgate to the works at Addiewell and Uphall, leaving the Bathgate Works for acid production.

Broxburn Oil Company

The Broxburn Oil Company Ltd was formed by Messrs Brown, Hinshaw and Faulds in 1862. An oil works was established on the north side of the Union Canal at Broxburn, consisting of thirty-six horizontal retorts. Later it became the Albyn Oil Works, to retort shale produced from Robert Bell's Stewartfield mine (Stewartfield No. 2). The oil works were closed in 1863 due to a severe fall in the price of crude oil. Re-opened by E.W. Fernie in 1864, the works were closed once more in 1866 and purchased by the Glasgow Oil Company Ltd, a company in which the same Robert Bell had the major share. This company was voluntary wound up in March 1873.

Bell had himself moved on from being a mine owner and in 1862 he had opened an oil works at Broxburn to the south of the canal, the Stewartfield Oil Works, fitted with 100 horizontal retorts and supplied with shale from his own pits. In 1865, Bell built a second oil works – the Greendykes Oil Works – with a further 100 horizontal retorts, the works again being supplied with raw shale from his own mines.

By 1863 there were a further four small companies each with oil works in the Broxburn area, namely:
- A small oil works of twelve horizontal retorts at Roman Camp, built by a Mr McLintock in 1862, passed to W. Fraser of Broxburn in 1865 who increased the number of retorts to seventy-five and added a refinery.
- Crude oil retorts erected by John Poynter in 1862, located near Greendykes Road in Broxburn.
- A small paraffin oil works to the north of the canal in Broxburn erected by Messrs Millar and Steele.

The signatures of James Young, Edward Meldrum and Edward Binney on the agreement for the lease of land upon which they were to build the world's first oil refinery, at Bathgate.

- A small refinery was erected on the west side of Greendykes Road, Broxburn by Mr T. Hutcheson in 1863/64, close by the workshops of the Broxburn Oil Company. In close proximity within this very small area, these establishments were all in full production and by 1872 were employing just over 500 men and retorting between 40,000 and 70,000 tons of shale each year.

A completely new Broxburn Oil Company was established in 1877 with the purpose of consolidating ownership of Robert Bell's various shale leases in the Broxburn area. The same Robert Bell was to be the company's second chairman until his death in 1894. As well as acquiring ownership of the shale leases, the company also absorbed Bell's oil works at Stewartfield and Greendykes, plus the other small run-down businesses, to form a cohesive oil producing company. The former Greendykes Oil Works, in much improved form, became the new Broxburn Oil Works, central to the company.

This new company was to see considerable success, not least due to the brilliant, innovative engineering skills of its general works manager, Norman McFarlane Henderson, who was to design and produce his new retort, the Henderson retort (as discussed in the previous chapter). In 1878, under his supervision, a completely new and larger Albyn Oil Works was built on the site of the former works and equipped with the new Henderson retorts. With these efficient retorts, the company was to become one of the strongest in the industry. By 1889, 784 retorts were in use here. The Albyn Oil Works produced crude oil only and lying east of the Broxburn refinery was connected to same by a quarter-mile long pipeline which led into a 50,000 gallon holding tank.

Between 1863 and 1868 a small oil works was established at Drumshoreland by McClintock & Rankine, given the name Almondell Oil Works in 1866, but also referred to as the Broxburn Roman Camp Works, operated by William Fraser. A completely new and extensive oil works was constructed in 1892 at Roman Camp near Drumshoreland by the Broxburn Oil Company, incorporating these earlier works and equipped with some 240 Henderson retorts, but of an improved design. New mines were driven here and a new sulphate of ammonia plant was provided.

A new sulphuric acid production plant was built at the Broxburn Works in 1895 to produce the acid required in the refining process and it was also used in the manufacture of ammonium sulphate.

By 1902, Broxburn Works (see Map C, p. 29) covered 250 acres and was the largest single works in the shalefields, employing over 1,800 workers, including miners, and processing over 1,600 tons of shale each day. It had an annual oil production of 13 million gallons, had its own cooperage for barrel production and had a candle works which was producing 20 tons of candles every day. The company gas works also provided gas to the town of Broxburn.

The Glasgow Oil Company (Broxburn) Ltd

Formed in 1866, the subscribers were Robert Bell, Walter MacLellan, William Scott, William Hurst (of the North British Railway) Benjamin Conner (of the Caledonian Railway), Robert

Two images from a postcard illustrating the Broxburn Oil Company's oil works:
Top of Page: A composite photograph of the crude oil-producing Albyn Oil Works, Broxburn. The right-hand view is includes the Union Canal crossing the foreground centre to right. Far right is the bridge carrying the NBR Broxburn Branch over the canal and into the oil works. The left hand image is looking north-west. On the extreme left is the tramway from Stewartfield No. 2 mine to the oil works whilst the retort benches are behind the chimneys. The structure in the foreground is thought to be the tramway carrying the spent shale for tipping on the bing.
Right: Broxburn Oil Works refinery with Greendykes Road, Broxburn running across the foreground. *John Ryan collection*

THE MAJOR SHALE OIL COMPANIES AND THEIR WORKS

Faulds, James Hamilton and James Millar. Bell, McLellan and Millar were directors of the company. The company oil works were the Albyn Oil Works, built to retort shale obtained from the pits of the same Robert Bell, a man with a finger in many pies! In March 1873, after two closures of the Albyn Works had been forced upon the company by the falling price of crude oil, the company was finally wound up.

THE UPHALL MINERAL OIL COMPANY

After Meldrum had parted company with Young at Bathgate, he went into a new partnership in 1866 (Meldrum, McLagan and Simpson). The company they formed was known as the Uphall Oil Company, with an oil works at Starlaw on the east side of Bathgate and another works and refinery at Stankyards close by Uphall. Other small oil works in the area, including those of Taylor & Marshall and John Raeburn, were to be absorbed into this new company. A few retorts had previously been provided by McLagan (who later was to become Liberal MP for Linlithgowshire, 1865–93) to retort

a rich seam of bituminous shale which outcropped on his estate at Pumpherston, but later an even richer seam of Broxburn shale was discovered some 25 fathoms (150 feet) below his Stankyards farm. A new oil works with a bench of horizontal retorts was constructed close by Uphall Station and adjacent to the Edinburgh/Bathgate railway.

These Uphall Works (see Map C, p. 28) had become the second largest in the industry by 1869. With a separate oil works at Starlaw, by then known as the Boghall Works, the Uphall Oil Company drew upon a combined shale field of some 5,000 acres. By 1871 the company was refining 2 million gallons of oil each year, the works had been rail connected to the Edinburgh & Bathgate Railway and crude oil from other works was brought in by rail tanker for refining. In the 1920s this oil works was, along with Pumpherston, equipped with a new thermal-pipe still cracker to increase yields of various oil products.

In 1871 the company was incorporated as a limited company and renamed the Uphall Mineral Oil Company. This step was taken so that the company could acquire the rich Hopetoun shale fields lying close by Winchburgh. In 1872 the company built a new oil works at Niddry – later to be known as Hopetoun Oil Works, lying to the south-east of Winchburgh – in order to process the shale there. By 1878 the company was operating the following mines working the Broxburn seams: Hopetoun Nos 1 & 2, Hopetoun Nos 3 & 4, Hopetoun No. 5, Boghall, Forkneuk and Niddry.

YOUNG'S PARAFFIN, LIGHT & MINERAL OIL COMPANY

We left Young as sole proprietor of the Bathgate Oil Works in 1864, after buying out his partners for a sum of £32,000. However, as always, Young was not one to allow the grass to grow under his feet and in the early 1860s he secretly purchased six oil shale concessions around West Calder (as described in Chapter 4). He lost no time in establishing a new oil works at Addiewell, just over a mile to the west of West Calder village. Addiewell Works, believed to be the largest oil works in the world at the time (see Map A, p. 20), went into production in 1866 and by 1868 was retorting some 172,000 tons of shale per year. The works occupied some 75 acres, with retort sheds 200 yards in length, each containing a double row of retorts. Prior to the opening of the works, Young sold his interests for £400,000 to a limited company called Young's Paraffin, Light & Mineral Oil Company and joined the board of directors where he remained as general manager until 1871. By that time, Young's had taken over the original Stewart's Paraffin Oil Works at Breich Dykes, and by 1877 had abandoned these works completely.

In 1872 the company employed 1,200 workers at Bathgate and Addiewell, retorted 370,000 tons of shale supplied from only seven shale mines, and yielded 4,477,000 gallons of liquid oil products. These comprised 270,000 gallons of naphtha, 3,785,000 gallons of lighting and paraffin oil, 422,000 gallons of lubricating oil and 2,300 tons of paraffin wax; the whole process involved the use of 2,000 tons of sulphuric acid and 113,000 tons of coal. This refining capacity was to increase to an output in excess of 8,000,000 gallons of crude and, as the output increased, so the candle house was expanded to cope with the increasing demands. The Addiewell Works was employing over 2,000 workers at its peak, but the slow decline was to begin around 1885 with the closure of two of the mines.

In 1879, Young's Paraffin, Light & Mineral Oil Company bought

Broxburn Oil Works and refinery looking south. *John Ryan collection*

over the interests of the West Calder Oil Company in the South Cobbinshaw Oil Company, including the rows (houses) which had been built there.

In 1884, Young's absorbed the Uphall Mineral Oil Company including the mines working the rich shale deposits in the area, Hopetoun Nos 1 & 2 and Forkneuk. Young's then moved a significant part of its Addiewell operation to both the crude works and refinery at Uphall and the Hopetoun Oil Works (previously Niddry Oil Works) in a bid to exploit the new shale field at Newliston. By 1885 the three works' combined output was a staggering 120,000 gallons of crude per week, equating to some 6 million gallons of lighting oil and 5,500 tons of paraffin wax per year. Between the three works there were by now some 580 retorts of Henderson's design processing around 800 tons of shale every day. New mines were being sunk around this period, with a new mine coming on stream at Glendevon (Hopetoun No. 6), to the west of Winchburgh. Another new and very productive pit (No. 35) was sunk at Threemiletown (Redhouse) in 1897 and in 1906 the internal railway serving No. 35 was extended northwards to Hopetoun No. 41 (Fawnspark) mine to access the shale there.

In the same year, the Newliston shalefield was opened with three pits. One was Newliston No. 29 near Newbridge, the others were Ingliston No. 33 and Nos 36 & 37, near to where the western end of the main runway of Edinburgh Airport now stands.

Linlithgow Oil Company

Formed in the main by Fife coal masters in 1884, the company leased the minerals on the Linlithgowshire estates of Ochiltree (Lord Rosebery) and Champfleurie (Robert H.J. Stewart). From the outset, with a potential production of around 1,000 tons of shale per day, the scheme was well subscribed. Oil works and a refinery with an associated candle house were built at Bridgend, some 1½ miles east of Linlithgow. The oil works were equipped with four benches of Henderson retorts, with fifty-two retorts to each bench. Mines were sunk both at or near Ochiltree (Ochiltree No. 1, Ochiltree Nos 2 & 3, Ochiltree Nos 4 & 6 and Ochiltree No. 7) and on the estate of Champfleurie (Champfleurie No. 1 and Champfleurie No. 4). A railway line was constructed to connect the works with the NBR at Linlithgow. Houses for workers were built at Kingscavil and in Bridgend. By 1885 the mines were producing 500 tons of shale per day from the Broxburn seams, although latterly one mine was working the Dunnet seam.

The oil works, known both as Bridgend and Champfleurie, also refined crude oil from the Holmes Oil Works and the Hermand (old) Oil Works. The need for re-equipping with the new improved Beilby & Young retorts was identified in 1890 – the refinery was capable of a throughput of an annual 6 million gallons of oil, but the old retorts were capable of producing just over half this quantity of crude. However, with the increasing effect of the USA oil industry on the paraffin markets, by 1897 wages had to be reduced to save costs. A disastrous fire in 1899 destroyed the paraffin shed which contained German filter presses and twelve hydraulic presses, although the candle house was saved. Five men were injured, one fatally so, and this loss, coupled with a wage increase needed for parity with the coal and iron industries, meant that the plan for new retorts was shelved. Low prices continued to take their toll and the company was wound up in 1902.

THE MAJOR SHALE OIL COMPANIES AND THEIR WORKS

J. Ross & Sons (Philpstoun Oil Company)

In 1883 James Ross, a Falkirk coal master, established an oil works at Philpstoun, east of Linlithgow, to retort the rich shale seams lying in the surrounding area – the very productive Champfleurie/Philpstoun shale field, containing shale of the Broxburn, Champfleurie and Dunnet seams in quantity. The oil works were built adjacent to the then NBR Edinburgh–Glasgow railway line and astride the Union Canal, providing ready transport links. Philpstoun was a crude oil works only, with all the crude produced being transported to other works for refining. Philpstoun amalgamated with the other remaining oil works in 1919 to form Scottish Oils Ltd, but was closed in the 1931.

In earlier days, Ross responded quickly to the new market for motor spirit (petrol), the shale oil was found to produce a spirit which was almost free of detonation and was quickly in demand. Ross established the first 'service station' at Newton Village, near South Queensferry. Philpstoun No. 1 and No. 6 mines worked the rich Dunnet seam and the mined shale was transported underground right into the oil works. These mines closed in 1932, but Philpstoun No. 6 mine was re-opened and, along with the new Whitequarries mine, supplied shale to Niddry Castle Oil Works until closure of the industry in 1962. When the Linlithgow Oil Company closed in 1902, the Philpstoun Oil Company extended the private railway serving Philpstoun No. 4 mine immediately behind Bridgend village to transport the still available shale from the Ochiltree mines to Philpstoun Oil Works. A retort, based on the Pentland retort, was designed for use at Philpstoun by Mr A.H. Chrichton and called the 'Philpstoun' retort. It had a similar throughput as the Beilby & Young retort but, whilst it remained in use until closure at Philpstoun Oil Works, the design was never used by any other oil company.

Hermand Oil Company

Two small oil works and mines were established in the mid 1860s by Thomas and James Thornton on the Maitland estates at Hermand, immediately to the south-east of West Calder. The Thorntons were coal masters and farmers at Crofthead near Fauldhouse. In 1871, the operation was taken over by Dunnet and Brown, Dunnet being a chemist who was the first to mine the seam of shale named after him. These early oil works were demolished later in the 1870s. Thomas Thornton resumed shale mining at Easter Breich, north-east of West Calder, employing just over 100 men. James, however, purchased Hermand House and the estate with associated farms, and in 1885 issued a prospectus for the Hermand Oil Company, outlining the rich reserves of oil shale under his estate. The company was floated in that same year and an oil works was built at Mid Breich and fitted with 120 of the latest Beilby & Young retorts. At this point in time, James Thornton died and the company chairmanship went to one John Armour, with his brother becoming Works Manager and Secretary.

In 1889, another new oil works was constructed at Hermand, just south of West Calder (Shuttlehall), to retort the shale being mined

Uphall Oil Works at Uphall Station. The railway curving away to upper left is the NBR branch which served Middleton Hall workshops and Uphall goods station. Thereafter, as a purely internal mineral line, it crossed Uphall West Main Street on the level and, serving several shale mines en route, ran to Hopetoun Oil Works at Faucheldean, via Ecclesmachan.   *John Ryan collection*

at Hermand No. 6 and Birniehill No. 5 mines. Eighty-five houses were constructed for the workforce. This oil works was fitted with a modified version of the Beilby & Young retort, which resulted in the designers taking the Hermand Oil Company to court for breach of patent rights. The claim failed, as did, it must be said, the modified retorts, with the quality of the oil being produced deteriorating due to continuing weaknesses in the retorts. The company was in dire straits in 1890 and negotiated an amalgamation with the Walkinshaw Oil Company of Johnstone, Renfrewshire, in order to obtain refining capacity. The company name was retained, but the Johnstone refinery had been non-operational since 1886 and was in need of renewal. Trading conditions rendered this impracticable and both the company's works were closed in 1892, with Hermand Oil Works and houses being sold off. All the products from this works were taken to the Linlithgow Oil Company's works at Bridgend for refining.

THE MAJOR SHALE OIL COMPANIES AND THEIR WORKS

A view of Philpstoun Oil Works looking north-west, with the Union Canal running through the centre of the operation.

The Mid Breich (New Hermand) Oil Works was re-opened by the company under new chairmanship in 1899, but the works required modernisation and new retorts. Only a few new retorts had been installed when, in 1902, the company had to reduce the throughput at the works after the Linlithgow Oil Company, who refined the crude from the works, failed. In 1903 the company was put into liquidation, never having recovered from the setbacks.

In that same year, the Mid Breich Oil Works and houses were purchased by the Pumpherston Oil Company who sunk a new mine at Breich (Mid Breich) to work the Dunnet seam of shale, this being transported by a tramway to the company's Seafield Oil Works.

Over the years the following mines worked the Fells seams in this area, Easter Breich No. 1, Wester Breich No. 3 and Hermand Nos 5 & 6. Breich Nos 1, 2 & 3 worked the Broxburn seams.

A vintage horse-drawn oil tanker for deliveries of paraffin to households. The tanker was owned by Scottish Oils Limited who had it restored and kept at Middleton Hall, Uphall. Here it is seen in an equally historic setting in the restored National Trust village of Culross, in Fife.

West Calder Oil Company

Established by a partnership of Alexander Fell (after whom the Fells shale seam is named) and Robert Russell (who leased mineral rights at Gavieside, north-east of West Calder) in 1862, the Gavieside Oil Works were built there and were in operation by 1863. The company also leased mineral rights at South Cobbinshaw which were then sub-let to the South Cobbinshaw Oil Company who mined the shale there and set up a small oil works on site. Fell left the company in 1870 and the remaining partners (the Findlays) floated the company in 1872.

The company was the largest and most successful of the smaller companies in the area and by 1869 Gavieside Paraffin Oil Works (see Map B, pp. 24–25), consisting of 100 retorts, employed around 300 men and was producing just under one million gallons of oil per year. Further investment was made in 1872, raising oil production to just over the one million gallons mark – all being produced from shale from two mines, Gavieside No. 1 working the Fells seam and Gavieside No. 3 working the Raeburn seam, with coal coming from the Woolfords coal pit.

From 1874, the company was in trouble for pollution of the River Almond (via the West Calder burn) and in 1877 the adjacent landowners obtained an injunction (reluctantly, it has to be said, granted by the judge). In 1878 the company ceased trading and its assets (oil works and pits) were sold off by public roup (auction) in September 1880 for an upset price of only £17,500.

Mid Calder Oil Company

This company was established by a partnership formed by Sir James Young Simpson (of chloroform fame) and others. An oil works was opened, with an associated mine working the Pumpherston seam, on the Calderhall Estate of the Hare family at Oakbank, to the south of Mid Calder village (see Map F, p. 166), in the mid-1860s. In 1869 the operation became the Oakbank Oil Company Ltd. Initially, there were but ten shareholders, however the company was extremely fortunate to obtain the services as manager of Mr N.M. Henderson, formerly of the Broxburn Oil Company. By 1872 Oakbank was employing around 350 workers and was retorting about 42,000 tons of shale annually. By 1877 only Oakbank and Addiewell works were left in the area of the Calders.

In 1870 the company purchased the small 200-retort oil works at Dumcross, near Bathgate. This had been built by Lester & Wylie, owners of the Forth & Clyde Oil Paraffin Works & Refinery at

Kirkintilloch but, after its purchase, the Dumcross Oil Works was closed and dismantled.

BATHGATE OIL COMPANY

Formed in 1883 to take over the Seafield Patent Oil Works and the mineral leases thereof by Messrs Wood and Hamilton-Beattie, both of Edinburgh, and with around forty shareholders, the company was soon unable to meet its liabilities. It was wound up in 1885, being taken over by the Pumpherston Oil Company.

WEST LOTHIAN OIL COMPANY

Formed by the Provost of Bathgate, Donald Sutherland, and one Archibald Simpson to take over the assets of George Simpson (Archibald's father), which included the mineral leases at Deans, Boghall, Caputhill and Drumcrosshill (near Bathgate), and his paraffin works and refinery at Braehead, near Fauldhouse. The new company built an oil works at Deans, about 1½ miles east of Bathgate. The company was to be short-lived, going into liquidation in 1891 and being bought by the Pumpherston Oil Company in 1894.

However, prior to 1883 a small paraffin oil works had been established at Drumcross and operated by Lester & Wyllie until 1868. The works passed into the hands of Robert Fraser in 1868 and on to Robert Calderwood in 1869 when it was purchased by the Oakbank Oil Company, but the works were closed and site was to be subsequently cleared.

TARBRAX OIL COMPANY LTD

Tarbrax Oil Works was to have a convoluted ownership from the outset. A crude oil works was established just inside the Lanarkshire border with Midlothian, at Tarbrax, near Cobbinshaw, in 1865 – close by the West Calder Oil Company's Cobbinshaw South Oil Works – by the trustees of E.W. Fernie of Mid Calder. Fernie had leased the minerals at Tarbrax from David Robertson of Lawhead and then died shortly after the transaction. The trustees kept the operation going until 1873. Ownership then passed to one William Black, a farmer at nearby Easterhouse Farm, in 1873. The ownership of the works was again transferred when the British Oil and Candle Company Ltd bought the operation in 1880. This was a Scottish company which had been incorporated in 1880 from the remnants of the North British Oil and Candle Company (1865) and the assets included the candle factory and refinery at Whitelees near Lanark.

In 1883, by a special resolution, the company property was sold off and a new company was set up to purchase and take over both the operation and the mineral leases at Tarbrax, from one George Simpson, already mentioned with reference to the West Lothian Oil Company, plus the candle factory and refinery at Lanark. This new company was named the Lanark Oil Company Ltd and formed by a partnership of James Thornton of Hermand near West Calder, George Lorimer and John Thornton. Tarbrax Oil Works was expanded but the whole operation was losing money and was wound up in December 1886, the Lanark works and refinery and Tarbrax Oil Works being sold separately. The assets lay unused for two years until, in 1889, the Caledonian Oil Mineral Company, an English firm formed in 1889, took over the former operations at Tarbrax and the works were reconstructed. The company was liquidated in 1898 and reformed yet again, under the same name but with reduced capital. In 1903 the company defaulted on payment of wages and was again wound up.

Mr Fraser, the manager of the Pumpherston Oil Company, recognised the potential of this shale field. He purchased the mineral leases of Lawhead, Tarbrax, Wester Crosswoodhill, Greenfield and South Cobbinshaw, and formed the Tarbrax Oil Company Ltd in 1904. This company was to be the only company to work the shale profitably in this very remote area. By 1912, however, the company was again in difficulty and was bought over entirely by the Pumpherston Oil Company. Tarbrax Oil Works was finally closed in 1925.

Tarbrax was to be the unlikely site for an riot in 1912. Woolfords No. 5 pit supplied the coal for the oil works and the coal miners there came out on strike in support of the national coal strike. Infuriated oil workers, around 700 in number, locked out as a result of a lack of coal supplies, broke into Woolfords No. 5 pithead and stoned the buildings. They then flung weights down the shaft to damage the cage and set fire to the winding head frame and winding house, causing extensive and expensive damage.

OAKBANK OIL COMPANY

The former Mid Calder Oil Company works at Oakbank was one of the smallest of the oil works to survive. It was fortunate to have the services of a young man, George Beilby, a graduate in chemistry from Edinburgh University, who joined the company in 1869 aged just nineteen. By 1883, in association with William Young, a gas and oil expert working with the Clippens Oil Company, Beilby had jointly developed and patented the Young & Beilby retort, better known in the industry as the Pentland retort. Beilby was to go on to greater things, eventually being appointed as President of the Royal Institute of Chemistry from 1900 to 1912 and was knighted in 1916.

In 1885 the oil works, by now the Oakbank Oil Company's works, were reconstructed and re-equipped with the new Pumpherston retort.

In 1902 the company expanded its horizons by developing and opening the first all-electric oil works, near Niddry Castle close by Winchburgh village (see Map D, p. 148). Niddry Castle Oil Works were to prove revolutionary in many ways, not least by the use of electricity and the construction of an all-electric railway to transport shale from outlying mines to the oil works, an aspect which is described further in Part 2. Niddry Castle retorted shale from the Duddingston pits, Totleywells and the Philpstoun/Whitequarries complex.

At Mid Calder, Oakbank mine (Oakbank No. 1) closed in 1908 and a new mine was sunk at New Farm (Newyearfield) – close by an area which was to become known as Dedridge, about 1 mile west of Mid Calder village – to work the Lower Dunnet seam. The shale from New Farm was then to be transported to Oakbank Oil Works by a quite unique (for the Lothians shale fields) overhead aerial ropeway of German design, in suspended buckets (see Chapter 17).

Twenty-eight new houses (rows) were built at Dedridge for the miners. The mine, New Farm 3 & 4, failed and was closed in 1919 although the houses remain until this present day, now sitting in the midst of the new housing of Livingston New Town.

Oakbank Oil Works closed in 1932, but Niddry Castle Oil Works at Winchburgh survived until 1960.

THE SCOTTISH SHALE OIL INDUSTRY & MINERAL RAILWAY LINES

ABOVE: A view of Oakbank Oil Works. The railways and environs of the works were always a source of interest for small boys.

BELOW: The Oakbank aerial ropeway was also proving an attraction for a number of children when the photographer paid a visit.

DALMENY OIL COMPANY

In 1868 the Dundas Shale Oil Company set up an experimental retort near Kirkliston (Almondhill Paraffin Works) to exploit the shale reserves which had been found in the grounds of Dundas Estate and which were worked by opencast mining. The company fell by the wayside in 1870, but in 1871 the Dalmeny Oil Company was formed by Robert & Alexander Pattison to retort the shale reserves which had been discovered on Lord Rosebery's Dalmeny Estate near South Queensferry. The crude oil produced at Dalmeny was taken by rail tanks to Oakbank Oil Works for refining. Never to be amongst the largest oil works, Dalmeny Works employed around 170 workers and was served by Dalmeny Nos 2 & 3 mine which worked the Broxburn and Camps seams. Some 30,000 tons of shale were mined each year, producing around 950,000 gallons of crude oil for refining.

The company was proving profitable by 1877 and was reconstructed around 1896. In 1905 there was serious explosion in Dalmeny mine and the mine itself was closed in 1914. A new mine was opened, Rosshill Nos 1 & 2 mine, working the Dunnet seam but the financial position of the company became unstable and the whole operation was taken over by the Oakbank Oil Company in 1915. Rosshill mine was closed in 1921 and Dalmeny Oil Works closed in 1925.

PUMPHERSTON OIL COMPANY

The development of the Beilby & Young Pentland retort had provided the industry with a retort which greatly improved the production of the lucrative sulphate of ammonia, a product which held its price as the price of crude oil fluctuated, and led to new players entering the shale oil industry. One such was the

Pumpherston Oil Company, formed in 1883, with works built to exploit the potential of the Lower Pumpherston No. 3 seam of shale which, although not a high yielder of crude oil, gave higher yields of ammonium sulphate. The driving force behind this new company was William Fraser who had been the manager at the Uphall Oil Works. Pumpherston Oil Works were built on land leased from local landowner Peter McLagan (who was later to become the MP for Linlithgowshire). The works started production in 1884 with 440 Beilby & Young retorts processing about 700 tons of shale per day. The company also built some 165 houses (rows) for its employees, numbering 700 at that time, in what was to become Pumpherston village.

In 1892 the company took over the operations of the Bathgate Oil Company at Seafield, near Blackburn, and two years later took over the works and mines of the West Lothian Oil Company at Deans, later to be known as Livingston Station. The oil works at Deans were refurbished with 140 of the new and efficient Pumpherston retorts in 1896 at a cost in excess of £20,000 and redevelopment of Seafield Works followed in the next year, with a further 140 new retorts being provided. The company was to see great financial benefits from this investment and profits increased year by year up until the First World War.

The company opened a new mine at Starlaw (Deans No. 3) in 1901 to serve Deans Works in addition to the existing mines, Deans Nos 1 & 2 (Caputhall) and Deans No. 4 (Barracks). All worked the Dunnet seam, although an opening had been made into the Raeburn seam at Starlaw. A further two mines, Starlaw/Deans No. 5 and Deans Nos 6 & 7 worked the Fells and Jubilee seams. Seafield was supplied by its own mines, Seafield No. 1 and Seafield No. 3, and also took the shale produced in Breich Nos 1 & 2 pit, which worked the Broxburn seam, by means of a linking surface haulageway.

By 1902 the company, with its three oil works, employed 1,200 men, 800 of whom were miners. Pumpherston Oil Works was re-equipped with an additional 180 new Pumpherston retorts, using shale from Pumpherston Nos 1 & 2 and Pumpherston Nos 3 & 4 mines. The refinery at Pumpherston was compact and was an industry leader, employing ammonia compression refrigeration (Bryson coolers) for the production of paraffin wax – this being sold on to other oil companies – and in the 1920s was equipped with a thermal-pipe still cracker. Pumpherston also produced a high quality railway signal oil, 'Pearline', and another product for bleaching soiled and dirty clothing called 'Laundrine'.

To source the insatiable demand for coal, in 1907 the company, through association with the Tarbrax Oil Company (described above), secured the Woolfords coal field near Cobbinshaw. In 1908 it invested further in Deans Works, doubling the size with a further three benches of retorts, 156 in all, bringing the total to 328 retorts in production. The Tarbrax Oil Company was in difficulties and failing by 1912 and in that year was taken over in entirety by the Pumpherston Oil Company.

CLIPPENS OIL COMPANY

In 1866 at Straiton, near Loanhead in Midlothian, one Peter Brash, a partner in the company of William Taylor of Leith, set up a small oil works of only twelve retorts with William Young, he of retort fame, as manager. The following year Brash and Young

Dalmeny Oil Works, South Queensferry with, on the extreme left, the retort benches with (centre rear) the inclined tramway carrying hutches loaded with crushed shale to the top of the retorts.
John Ryan collection

A bird's eye view across Niddry Castle Oil Works, looking north-west over the retort benches. The tramway on the right is for spent shale and other waste going to the tip on the bing.

patented a new design of retort and re-equipped the oil works at Straiton. Brash died in 1876 and the running of the company passed to the Straiton Estate Company Ltd which, in 1880, was renamed the Straiton Oil Company. The company suffered financial problems from the outset and, in 1882, was taken over by the new Midlothian Oil Company Ltd. The oil works were refitted with 288 of the original Beilby retorts, but these failed to live up to expectations and the company was soon incurring big losses. In 1885 the company sold out to the Clippens Oil Company Ltd.

The original Clippens Company was formed in 1870 by a Glasgow Merchant. A large oil works containing 270 retorts and a refinery was constructed at Clippens near Linwood in Paisley to work the Lillies shale (cannelloid coal) field. By 1878 the oil quality being produced was poor in comparison with the Lothian shale products and in 1880 Clippens leased oil shale reserves at Straiton. Here, a new mine, Pentland No. 1, was sunk to work the Dunnet, Broxburn and Fells seams and a new oil works was erected equipped with the new Pentland (Beilby & Young) retorts with a throughput of 300,000 tons of shale per year. Further mines (Straiton Nos 3 & 4, Straiton Nos 7 & 8 and Mortonhall Nos 9 & 10) worked the Dunnet seams and supplied the oil works with the raw shale.

The company employed around 1,000 workers by 1897 but problems were looming. The company became involved in litigation in that year owing to damage the mining was causing to the Edinburgh water-supply pipes which passed through the shale fields. The Edinburgh Water Trustees obtained an injunction

A general view of the oil works and refinery at Pumpherston.

against the company, causing it to cease operations and close. The litigation process was to last another ten years, the company finally being awarded a poor £27,000 compensation against its original claim for in excess of £135,000.

HOLMES OIL COMPANY

Robert Bell, whom we have met in connection with the other developments in and around Broxburn, leased the shale reserves on his Goshen Estate to another Wishaw coal master, who then formed the Holmes Oil Company in 1884. Bell owned a considerable interest in this new company. An oil works was constructed between Broxburn and Uphall; equipped with Beilby & Young Pentland retorts, the works was soon retorting some 200 tons of shale per day from the Pumpherston shale seam worked by the Holmes Nos 1& 2 mine. By the 1890s the company was suffering from the poor trading conditions, by 1900 the works closed, the mine was allowed to flood and in 1901 all the equipment was sold off.

BURNTISLAND OIL COMPANY

The oil bearing shale of the Dunnet seam, after passing deep under the River Forth, outcropped once more at Binnend, just above Burntisland, in Fife. In 1878 the Binnend Oil Company Ltd was established to exploit this rich seam, with a lease of sixty years. Three inclined mines were driven down the seam, retorts were set up in a fully integrated oil works and production commenced soon afterwards. The retorts had a throughput capacity of 200 tons of raw shale per day. The initiative failed to prosper and went into voluntary liquidation in 1880.

The land and works was then purchased by one John Waddell of Edinburgh, who sold it on to the Burntisland Oil Company Ltd, the latter being formed in 1881. There was a major reorganisation on site by the new owners, with new retorts being installed and a new village of rows built at Binnend for the much expanded workforce. When the new retorts came on stream, some 500 tons of shale was being processed daily with a yield of about 15,000 gallons of crude oil. Transportation to and from the oil works was initially conducted by road, utilising a large steam traction engine and six large road wagons, but the constant stream of heavy road traffic soon annoyed the local inhabitants of the town. A possible rail connection to Kinghorn was surveyed in June 1886 and the proposed railway connection was costed at £15,000 for all construction works. Work on the rail connection commenced in September of the same year, and was completed in June 1887.

In 1892 there was a roof collapse and a disastrous underground fire in Binnend No. 1 mine, which cost in excess of £10,000 to rectify. This, coupled to the perilous trading position of the company, forced the directors into liquidation. The liquidator obtained authority to carry on production in the hope that the overall situation would improve, but finally, in January 1893, the works was put up for sale.

There was a further abortive attempt to recover the situation with a new mine being opened at Newbiggin, but this proved to be unproductive. Thus the works and candle works were again put up for sale by public roup (auction) in 1894, but with no offers forthcoming. The mines were immediately closed and all machinery sold off. The Oakbank Oil Company purchased the candle works and fourteen years of oil production in Fife ceased in 1894. The ground upon which the oil works stood was sold to the British Aluminium Company in 1905.

An early (1912) road oil tanker belonging to the Pumpherston Oil Company. This was most likely a 2-cylinder, 16hp Albion motor lorry and is fitted with solid tyres.

Other Oil Companies in the Lothians

Over the early part of the industry's history there were many small concerns which came and went and were never to be taken over by larger concerns. The list is incomplete and many records have been lost, but the 1897 publication *A Practical Treatise on Mineral Oils & Their By-Products* by Iltyd I. Redwood, a contemporary history of the early days of the Scottish oil industry, is possibly the most accurate record now in existence. Included in this latter category were the following works.

Burngrange Oil Works: West Calder

Also known as Blackbraes, this small oil works was erected immediately to the west of West Calder village by the Glasgow Scottish Oil Company in 1866 and worked for only one year. In that short time it produced 10,000 gallons of crude oil but nevertheless was closed in 1868 and put on the market. It was taken over by Young's Addiewell Works but never went back into production.

Handaxwood Oil Works

(See Map E, p. 161.) Located to the south east of Fauldhouse, lying immediately adjacent to and on the north of the Caledonian Railway's Mid Calder Junction to Cleland railway line, this was a small oil works which was worked intermittently from around 1860, using shale recovered from the limestone mining activity at Levenseat. The seams of shale here were extremely thin and overlaid the Levenseat limestone which was the principal commodity being worked. The mineral lease, however, included the shale at a royalty of 1/- (5 pence) per ton and, being oil shale, it was not covered by Young's patent. Partners Hamilton and Ross were the first owners but sold the concern on to one George Gray of Levenseat. By 1872 the works were retorting some 7,000 tons of shale per year with an oil yield of 240,000 gallons of crude oil. The Levenseat Oil Company took over the works from 1895, however the works folded soon afterwards, although the company existed until 1903.

The site of Levenseat (Handaxwood) Oil Works is still in some dispute, but the author believes that the works lay in the 'V' on the north side of the Caledonian Railway main line and east of the North British Railway branch line from Croftshead, and that it was not located at the Levenseat complex to the south of this railway, which was purely a limestone operation.

Grange Farm Oil Works

(See Map B, p. 25.) The Charlesfield Estate. Started in 1864 by John Raeburn (grandson of the famous painter, Henry Raeburn) who also operated one of the Stankyards Oil Works near Uphall. The Grange Farm Works retorted the shale from the Dunnet seam, mined at Charlesfield No. 1, which was to be abandoned in 1878. The works closed at or about the same time. The site of the oil works lies beside the present B7015 road close by Guns Green Toll (NT019650).

North Cobbinshaw Oil Works

This was a small oil works lying to the north side of the Caledonian Railway main line between Edinburgh and Carstairs close by Cobbinshaw station. Opened in 1867, it was operated by Mungle & Thornton, shale being produced in three pits nearby, North Cobbinshaw Nos 1, 2 and 3. In 1871 ownership was vested in J. & A. Mungle; in 1872 the Caledonian Mineral Oil Company acquired the leasehold interests in the minerals and land, but J. & A. Mungle was dissolved in 1881. No records have been found regarding the three pits, the seams they worked or the closure date, but certainly all were gone by 1881.

Westwood Paraffin Oil Company

Also known as Breich Dykes and owned by R. Stewart, later absorbed by Young's Paraffin, Light & Mineral Oil Company, these works were closed by 1877 having had a working life of just seven to eight years. The location of the works is still somewhat uncertain but it is now generally accepted that it lay at the end of a short branch of railway (NT002633) near Breich Dykes Farm. Shale from the Mungle seam which was mined here was most likely supplied from local pits, since old shafts are now marked on OS maps.

A second Westwood Oil Works was later constructed slightly further east and to the north of Breich Water, built and operated by Scottish Oils Ltd. This stood on a site close to the location of Westwood Nos 12 & 13 pit, adjacent to the West Calder Loop Line of railway (NT013644). Opened in 1941, it retorted shale from a new Westwood pit and from Hermand No. 4 mine. The most modern of all the shale oil works, it became a casualty of the withdrawal of the tax preferential arrangements and was closed in 1962. It has left its own legacy in the shape of the five conical bings, known as the Five Sisters, which are now preserved for posterity.

Hartwood (Paraffin) Oil Works

Located on Hartwood Estate to the south of West Calder and operated by Andrew and Mrs Walker, the local landowner between 1871 and 1873. Little else is known of this activity.

In October 1887, Calder District Committee granted permission for the construction of a tramway connecting Hartwood mine with Baads No. 22 mine, crossing the West Calder to Baads road on the level, but it believed that this tramway was never in fact constructed, the mine closing soon thereafter.

The Final Years

By the start of the twentieth century, only ten oil companies and works were still in production. These had, in the main, been formed by takeovers, outright purchasing or amalgamations with earlier undertakings. The ten companies are shown in Table 9.1.

By 1912 the number of companies was down to seven, these being Young's, Broxburn, Pumpherston, Oakbank, Dalmeny, Philpstoun and Tarbrax. Both employment within the industry and mined shale peaked in 1913, with a total of 10,000 persons employed and a 3,280,000 tons of shale being supplied to the retorts. By 1935 this tonnage was down to 1,408,000 tons, by 1961 it was a mere 468,000 tons. Whilst the total number of pits and mines was, over the life of the industry, around 100, the average life-span of many was to be five years. Between 1910 and 1920, forty-five were recorded as being in production.

Post the First World War there was an interesting experiment conducted at Oakbank oil works, when a large consignment of shale from Japan, estimated to be between 500 and 1,000 tons, was shipped across and brought by rail into Oakbank Oil Works for processing. This raw shale was accompanied by a large contingent of Japanese workers, eager to learn best-practice retorting and refining processes, with a view to establishing a similar operation in Japan. The Japanese shale proved to be of the highest quality with a heavy yield of oil per ton, and the exercise lasted some considerable time,

THE MAJOR SHALE OIL COMPANIES AND THEIR WORKS

TABLE 9.1: THE OIL COMPANIES AT THE START OF THE TWENTIETH CENTURY

COMPANY	WORKS AND REFINERIES	FACILITIES
Broxburn Oil Company	Broxburn Refinery Albyn Oil Works Roman Camp Oil Works	Crude oil, refinery, candle house, sulphuric acid house
Caledonian Oil Company	Tarbrax & Cobbinshaw Oil Works Lanark Refinery	Crude oil, refinery, candle house
Clippens Oil Company	Pentland Oil Works, Straiton	Crude oil, refinery
Dalmeny Oil Company	Dalmeny Oil Works	Crude oil
Linlithgow Oil Company	Bridgend (or Champfleurie) Oil Works	Crude oil, refinery, candle house
New Hermand Oil Company	Hermand Oil Works	Crude oil
Oakbank Oil Company	Oakbank Oil Works Niddry Castle Oil Works	Crude oil, refinery
Pumpherston Oil Company	Pumpherston Oil Works & Refinery Seafield Oil Works Deans Oil Works	Crude oil, refinery
James Ross & Sons	Philpstoun Oil Works	Crude oil
Young's Paraffin, Light & Mineral Oil Company	Bathgate Chemical Works Addiewell Oil Works Uphall Oil Works Hopetoun Oil Works	Crude oil, refinery, candle house, sulphuric acid house

Middleton Hall, Uphall. This was the headquarters of Scottish Oils Ltd.

Courtesy of Grangemouth Heritage Trust

The scientific side of the shale oil industry. A view of the busy laboratory at Pumpherston Oil Works.

with the Japanese workers becoming part of the daily landscape around Mid and East Calder.

Because of increasing foreign competition, with the importation of crude oil from the Middle East in particular, every economy and labour-saving device had to be employed. In 1919, Young's, Broxburn, Pumpherston, Oakbank, Dalmeny and Philpstoun merged to form Scottish Oils Ltd with the headquarters being established at Middleton Hall in Uphall. The headquarters provided a central location for the laboratories, stores, workshops and administration services for the new Scottish Oils, with superior housing provided for many of the workers.

During the Second World War Pumpherston had a significant role to play in oil refining, and one which was highly secret at the time. By August 1942, Britain was some 2 million barrels of oil below the considered safe level of the resource. Indeed, it has been recorded elsewhere that the country was actually just months away from a negotiated deal with Nazi Germany, so dire was the situation. At an emergency meeting held in that month, Mr Philip Southwell of the D'Arcy Oil Company, an American company which had been active between the wars, drilling at eleven sites within the UK for liquid oil (petroleum) and natural gas, was in attendance. It was generally understood that, whilst natural oil did exist, it had been found in such small quantities as to be non-viable; but Southwell dropped what was a bombshell at this meeting when he revealed that vast quantities of natural oil had been discovered below Sherwood Forest in Nottinghamshire. This source was quickly developed for greater production under the utmost secrecy, output for the duration of the war being:

Table 9.2: Location of the Shale Bings

Bing Name	Oil Company	Grid Reference (to centre)	Height ASL
Addiewell North	Young's	NS 990630	180m
Addiewell South	Young's	NT 005627	210m
Albyn	Broxburn	NT 084729	135m
Binnend	Burntisland	NT 244873	n/k
Bridgend	Linlithgow/Ross	NT 037758	125m
Contentibus	Oakbank	NT 074658	175m
Clapperton	Pumpherston	NT 079697	160m
Dalmeny	Dalmeny	NT 014766	n/k
Deans	Young's	NT 015683	175m
Drumshoreland	Broxburn	NT 075700	180m
Fauchledean	Young's	NT 085740	120m
Gavieside	West Calder	NT 029653	n/k
Greendykes	Broxburn	NT 082729	185m
Green Bing	Uphall	NT 070710	160m
Hermand	New Hermand	NT 009650	n/k
Mid Breich	New Hermand	NT 009646	145m
Niddry Castle (Winchburgh)	Oakbank	NT 097747	150m
Oakbank	Oakbank	NT 077664	175m
Pentland	Clippens	NT 267664	n/k
Philpstoun North	J. Ross	NT 058768	100m
Philpstoun South	J. Ross	NT 058766	125m
Pumpherston	Pumpherston	NT 077692	n/k
Seafield	Pumpherston	NT 005667	200m
Shuttlehall	Thornton's	NT 029625	n/k
Stewartfield	Broxburn	NT 087727	n/k
Tarbrax	Tarbrax	NT 025556	250m
Uphall East	Young's	NT 065711	160m
Uphall West	Young's	NT 061708	160m
Westwood (Five Sisters)	Scottish Oils	NT 009641	120m

1940   16,689 tons (125,167 barrels)
1941   29,992 tons (224,940 barrels)
1942   81,298 tons (609,735 barrels)
1943   112,760 tons (845,700 barrels)
1944   94,570 tons (709,275 barrels)
1945   71,542 tons (536,565 barrels).

Total production to the end of 1949 was 601,553 tons (4,511,647 barrels), the main production sites being Eakring and Dukes Wood. This oil was loaded to rail tankers at Bilsthorpe and carried by rail to Pumpherston for refining. The refined spirit was found to be exceptionally well suited to the Rolls Royce Merlin aero engine which was fitted to both the Spitfire and the Lancaster bomber, amongst others. Thus Pumpherston played a leading role in the course of the war, a fact which was one of the best-kept secrets at the time. This role adds credence to a tale which was related to the author many times in his younger days, which told of Howden House, near Livingston village, remaining lit and without blackout precautions at night as a decoy for Pumpherston Oil Works during these difficult times. Howden House was indeed bombed and lives were lost!

Interestingly, recently-released Luftwaffe aerial photographs taken in 1937 show that each and every oil works and refinery was secretly filmed, a clear indication that the Scottish shale industry was seen as an important target in the hostilities which were to follow. Perhaps they show that Pumpherston Oil Works were not so secret as people thought at the time!

When the Government of the day increased the level of duty preference, there was a degree of renewed confidence in the industry and Scottish Oils built a completely new, modern oil works at Westwood, just to the north-east of West Calder. This was both a crude oil producer and refinery and, being equipped with modern retorts based on the Broxburn retort, had a daily throughput of around 1,200 tons of shale per day. It also was to later take over the refining role previously carried out at Addiewell but, with the ending of the tax preference in the 1950s, Westwood and the remaining works were closed by 1963. The shale oil industry was dead.

The Legacy of the Shale Bings

A bing is very much a Scottish term describing the spoil heaps of spent shale (or colliery waste) following extraction by mining and retorting, the word having its origins in Old Norse, *bingr* meaning 'heap'. The shale bings were (and still are) a distinctive part of the Lothian shale fields. The bings which were to blight the shale fields, and West Lothian, numbered twenty-nine in total and were to be found as detailed in Table 9.2.

The spent shale (blaes) has been in great demand for infill material in big road and other civil engineering undertakings over the years. Spent shale has a good weight to volume ratio, is an all-weather material, compacts well, is cheap and there is an awful lot of it

Inside the engineering workshops at Middleton Hall. *Courtesy of Grangemouth Heritage Trust*

around. Some bings have been more or less removed completely, or are in the process of being worked for this same purpose. Others have been landscaped and are now valuable nature reserves, while the famous Five Sisters, the five Westwood bings, and Greendykes bing at Broxburn have been deemed to be Historic Industrial Monuments.

One bing at Dalmeny has even been cunningly hollowed out to form a bund within which the giant oil holding tanks containing North Sea oil are hidden. One of the more imaginative uses! Interestingly, at Shuttlehall, south of West Calder and site of the Old Hermand Oil Works, the small bing became the site of an Air Reconnaissance Unit and later a Civil Defence Post, the concrete construction of which can still be seen where the spent shale has been taken away over the years. All in all, the bings, large and distinctly red, mark, even yet, the scope and scale of the shale oil industry in West Lothian.

Coincidentally, in the middle years of his railway career, the author served as Area Manager at Bathgate at a time when the BMC (later British Leyland) was the new major employer. He was also there during 1975 when British Rail, in partnership with William Griffiths (Whitburn Ltd), the biggest handler of shale blaes in the country, and Fairclough's, the Civil Engineering Contractor for the project, set up an operation to have spent shale recovered from Deans bing and transported by rail over the intervening 40 miles to Shieldhall in Glasgow. This blaes was required for the massive infill and base construction works in connection with the M8 motorway western extension. A new railway siding was driven into the heart of Deans bing, where Griffiths was responsible for loading the wagons. Bathgate-based drivers and guards then worked a total of nine loaded trains per day via Airdrie to the former docklands at Shieldhall where a distribution depot had been specially set up, and worked nine empty trains back for reloading. Each train consisted of thirty-nine 21-ton hopper rail wagons with bottom door discharge and carried a payload of around 800 tons. Some 7,350 tons of shale blaes was moved out of Deans each day during this operation, well over 1¾ million tons of spent shale overall. BR also started up a similar operation around this time involving the Scottish road haulage contractor, W.H. Malcolm (Brookfield) Ltd, moving shale by rail from Contentibus bing at Oakbank to Malcolm's distribution centre at General Terminus in Glasgow. This was a much smaller operation involving only two trains worked by Glasgow train crews, with 1,500 tons being delivered each day, but some 2 million tons were moved out by rail during the seven-year contract.

Winchburgh bing (Niddrie Castle), a major source of blaes extraction, has provided a considerable tonnage of spent blaes for major road schemes over the past years, and now, in 2012, finds itself sitting almost on top of the motorway interchange works associated with the new Forth Crossing road bridge. A handy source of the huge amounts of the infill which will no doubt be required by this civil engineering project, right on the doorstep as it were!

In 2003–5, West Lothian Council commissioned a study under the umbrella of the *West Lothian Local Bio-diversity Action Plan: Oil Shale Bings*, seeking professional guidance on what future policy should be and just what role the bings might have. The study was undertaken by Dr Barbra Harvie of the School of Geo-sciences at Edinburgh University. She published her report in 2005.

In her study, Dr Harvie identified the bings as being a unique habitat, quite peculiar to West Lothian and unknown anywhere else

THE MAJOR SHALE OIL COMPANIES AND THEIR WORKS

4217 Deans Oil Works, Livingston Station.

ABOVE: Deans Oil Works, Livingston Station, the largest of the West Lothian oil works. In front of the bing are the retort benches and condensers, with a modern office block in the foreground. This block also housed the laboratories. The bing in the background was removed in the early 1970s by being taken, in trainloads, to Shieldhall in Glasgow and used in the construction of the M8 motorway western extension. Around 2 million tons were removed and transported in this manner in two years. *John Ryan collection*

RIGHT: The chimney at Deans Oil Works falls in a controlled explosion as the site is cleared.

in Europe, and home to rare and protected species of plants and animals. More than 350 plant species were identified as growing on the various bings and Addiewell North bing has already been declared a Scottish Wildlife Trust Nature Reserve. She considered (correctly) that the bings were a focus of community identity, recommending that they be recognised more widely as the unique structures they are and that they should be afforded adequate protection status to remove any future threat. Accordingly, the bings have been designated as to future status, as shown in Table 9.3.

Strangely, the bings have evoked some interesting public reactions over the years. Just after the industry closed, the general opinion was that West Lothian had been left ravaged, with giant eyesores which would blight the district for evermore. As the years rolled by, and with bings being landscaped or, as described herein, removed for civil engineering projects, there was almost a collective sigh of relief as the skyline was altered for the better. Then there was great public support when the Five Sisters (Westwood bing) and Greendykes bing (Broxburn) were deemed to be Historic Industrial Monuments. Now, as this work was being written in 2010, and with not two but four bings destined for preservation, the author has become aware of a community campaign in Philpstoun to save the South bing, currently the subject of a Planning Application for development. '*Save our Bing*' read the banners! Here's to their success.

Table 9.3: Future Status of the Shale Bings

Status	Bings	Comments
Bings where blaes extraction has current planning permission.	Drumshoreland North & South, Clapperton and Niddry	
Bings where extraction may be approved, but subject to planning permission.	Philpstoun North & South	There is clear evidence that some previous extraction has taken place at Philpstoun South.
Intact bings where extraction has been resisted and now scheduled as Historic Monuments.	Five Sisters, Greendykes, Fauchledean and Oakbank	
Restored bings with further extraction prohibited.	Addiewell North, Deans, Green bing and Stankyards	Strangely, Deans bing is included in this category although the bing was largely removed in the 1970s.
Restored bings, further extraction resisted but potential for reworking not excluded.	Addiewell South and Seafield	Addiewell South has largely gone and what remains has been landscaped. Seafield bing has now been rehabilitated as open space for recreational purposes and is now officially known as Seafield Law.
Abandoned bings where extraction has been exhausted.	Bridgend and Mid Breich	What's left of Bridgend bing now forms part of a rather splendid golf course.

LEFT: The decline of the industry. The retort bench at Deans immediately before the demolition charges are set off.

BELOW: The same retorts at Deans as the charges are fired and the structure begins to fall.

10

THE SOCIAL ISSUES

In considering social issues, researchers have to cast a wide net to obtain an accurate picture of life in the shale communities. It is, essentially, a pointless exercise to carry out research of an industry, any industry, confined to that industry and its communities, since the results will be inevitably skewed. In many of the social histories of shale, the story unfolds of poverty, poor housing, ill-health, overcrowding, hard and dangerous work and low earnings as being the norm of everyday life within that industry. In reality what they are describing was life across many of the new and heavy industries spawned by the Industrial Revolution. Thus the author has, for comparisons and as a control measure, looked at the coal industry (also involving mining), the iron and steel industry and the railway industry to assess just what life was actually like in the shale industry. In this way, comparable pictures emerge of life in other areas created by the Industrial Revolution as being similar to or, indeed in several cases, worse than the shale oil industry in terms of earnings, housing, working conditions, safety and health.

With the commencement and rapid development of the Scottish shale oil industry there was a considerable and immediate demand for labour. There was a small pool of experienced men employed in adjoining coal-mining and quarrying activities who were quickly attracted to the new industry, and there were also the navvies fresh from being employed on canal and railway construction to swell the numbers. Scots from other areas, particularly the Highlands, migrated towards West Lothian for work but, in the main, there was

A group of oil workers at Oakbank Oil Works circa 1929.

to be a massive and ongoing influx of Irish immigrants fleeing the problems of poverty and famine, and seeking to better their lives. This influx was to change the face of both West and Mid Lothian for all time.

Prior to the start of the shale industry, Bathgate (Baedd Coed) was probably one of the larger settlements in West Lothian – a village, but also a royal burgh dating from the twelfth century, where weaving was the main trade. Smaller villages around, such as West Calder and Mid Calder, were rural, agricultural and self-contained communities, whilst other places such as Broxburn and Uphall were little more than hamlets. West Calder was a 'kirktown', that is, it was a community developed around an established kirk or church, the Auld Kirk (1643) being the original and the ruins of which can still be seen. Uphall, a hamlet lying on the grounds of Strathbock Estate, was also founded around a church, St. Nicholas, which pre-dated the Auld Kirk of West Calder, going back to the twelfth century. Again, the small population, numbered in hundreds, worked the land. Broxburn originated around the mid-1300s and the hamlet of that time was known as Eastertoun. Renamed Broxburn in 1600, it was to continue as a small agricultural community until the coming of the shale industry. However, Broxburn did see some activity when the Union Canal, which passes through the village, was being constructed between 1818 and 1822.

Then came shale. As an indication of the scale of the influx of labour, consider Bathgate in 1831 when the population was but 3,593; by 1861 this had grown to 10,134, by 1891 to 11,359 and by 1921 had swelled to 18,862. The effect on a smaller community such as Broxburn, finding itself smack bang in the centre of the industry, was even more dramatic. In 1861 the recorded population was but 660 persons; by 1871 this had grown to 1,457, by 1881 to 3,066 and by 1891 there were 5,898 persons recorded as being resident in the town. In the landward part of West Calder parish, the population had swelled from a mere 1,927 in 1861 to 5,390 by 1881. This story was repeated throughout the shale area, where whole new shale villages quickly grew up, such as Addiewell, Seafield, Livingston Station, Deans, Niddry, Winchburgh, Roman Camp, Threemiletown and Newton.

During the Second World War, a great number of Poles and Lithuanians came to the area, escaping from Nazi Germany after their countries had been overrun. Many of the Poles were extremely well-educated men from professional backgrounds but few could speak English with any degree of fluency. As an example, the author had a neighbour who had been a university lecturer and was then a drawer underground in Burngrange pit. Jobs for such men were difficult to come by, but they were welcomed with open arms by Scottish Oils Ltd, since manpower was at a premium. These men proved to be popular, they were extremely hard workers and were much sought-after by facemen, as drawers. They assimilated well into the local community and many eventually settled down with Scottish wives.

Many of the social aspects of the shale industry have been well illustrated in that excellent book by Alistair Findlay, *Shale Voices* (Luath Press), and it is certainly not the intention to try and rewrite that particular story. Enough to say that life was never to be easy in the industry, particularly for the wives. It was they who had to make the money spin out for food, clothes, heating and rent, and, sad to say, many of the miners and oil workers of the day took their share to buy drink before handing over their meagre earnings. It was the women who had to wash and mend, clean boots and generally run the home from setting fires and making breakfast and pieces (sandwiches etc.) in the early mornings, to having meals and baths ready for late-shift workers coming home in the late evening. The

Retort workers at Tarbrax Oil Works in the 1920s.

women, in reality, had a much harder time than the men. Even where leisure facilities were provided by the company, these were a 'men only' facility. Yet, amongst the women, there was almost a general acceptance that this was what life was all about.

Considerable professional research has been carried out into the social aspects of life and relationships in the shale communities, but, unsurprisingly, there has been much contradiction. Much of the evidence given, all of it retrospective, has presented, on one hand, a picture of good neighbourliness, shared hardships and a strong community spirit. On the other hand, evidence of domestic and public violence, religious bigotry, hostility between neighbours and child neglect emerged. Hence a considered interpretation of each of the emerging stories is essential in an attempt to establish the true picture of community solidarity and co-operation. In some of the aforesaid research, the shale industry has been painted as an industry typified by poor housing and poverty, that shale communities were bottom of the scale in terms of morbidity and general ill-health. This is a seriously distorted picture; what was actually being described, as stated, was life across the whole industrial spectrum, but from a shale perspective. As someone who was brought up in a shale community the author can only add his own personal memories.

He grew up during the war years and, with his father in the army, his mother had to make ends meet on a corporal's pay. The family never went hungry although the food was basic and, of course, rationed. The one foodstuff there was in some abundance was cheese. Miners got an extra ration of cheese during the war and in a mining community cheese was thus plentiful and was shared around. Indeed, the author personally considered it a privilege and good fortune to have grown up in a close-knit community where everyone knew each other and where people, by and large, cared for each other – and here the word 'cared' is used in its truest sense. It was a warm, safe upbringing and whilst there was little in the way of material possessions or luxuries, all had the basics and were, by and large, happy. Children made their own entertainment, with games in summer lasting for weeks at a time. They were never bored, there simply was never the time; perhaps most of all, children never felt in any way threatened. There never was any feeling that one was either underprivileged or poor. Yes, at school, the author had classmates who (in retrospect and with a greater understanding of reality which comes with age) were indeed ill-cared for, ill-dressed and ill-shod, and were, to all intents and purposes, sadly neglected, but they were the unfortunate minority.

In all the research recorded in Findlay's *Shale Voices* and the research reports, as well as the oral histories now recorded for posterity and researched for this work, the main memories dwell on hardship and poverty and the dangers of the industry – but the reality is that the shale industry was no worse, and was in many ways eminently better, than other comparable industries.

Coal mining was by far and away an industry where conditions were much worse. In the early days, the coal miners were tied to the pit owners and could not resign or leave. It was, in essence, a form of slavery, with children of miners being obliged to follow 'down the pit'. There was, quite simply, no way out. The work in coal was worse by an order of several magnitudes and one has just to study the Annual Accident Reports of HM Inspector of Mines throughout the life of both the coal and the shale industries to realise just how hazardous the act of winning coal actually was. These records detail accidents in both shale and coal, and whilst there were considerably more coal mines and pits, the accident rates and fatalities in coal mining far and away outstripped the shale accidents, both in terms of sheer number and multiple fatalities (see Chapter 6).

Whilst it is true to say that there had been many improvements instigated within the coal industry, certainly relative to the employment of young children and women underground, the actual work, working conditions or associated risks, had changed little. The West Lothian coal masters were to continue employing children under the age of ten years and women post-1840 – in which year a Parliamentary Commission was set up to examine this situation – and continued to do so even after legislation was passed in 1842 outlawing this practice. The old feudal order was to prove strong in West Lothian and it was to be the 1850s before the coal companies finally stepped into line.

One shale miner, who had also at some time worked both in coal and stone mining, is recorded as stating that shale miners were 'toffs' whilst coal miners were the 'tramps'. By this he meant that, as a shale miner, he could come home from work reasonably clean and not caked with coal dust in every pore, that he never had to crawl into 2ft and even 18in. seams and lie on his side or back to work, and the risks of serious chest complaints were much reduced.

The author's own chosen industry was the railway where, prior to the First World War, engine drivers, firemen, signalmen and guards had long, long working days – 18 to 20 hours per day was not unknown, as were 100 hour weeks. The risk of injury and death whilst at work was considerably higher and the same problems of sub-standard housing was a recurring theme.

Such poor conditions as did exist did, however, throw up many characters. Mention must be made of one such character, a lady who was a well-known and highly respected name in the history of shale. She was Mrs Sarah Moore, well-known locally as 'Ma Moore', daughter of an Irish immigrant family, who was to feature large in pre-Second World War days. Married and living in Addiewell, she rebelled against the social injustices of the time and the economic evils which were the legacy of industrialisation generally, but, in her case, she focussed within the shale community. In short, she cared. It was she, with her two sisters and two helpers, who, during the 1926 strike, obtained and prepared food and fed every school child, every day of the strike. She was an active member of the Labour Party and was to become County Councillor for the Addiewell Ward of Midlothian County Council, a position she was to hold for twenty-one years. She became a Justice of the Peace and was latterly Convenor of the Public Health Committee. She also, in this busy life, found time to bring up her own family of six daughters and three sons. As stated above, she cared, and was able to translate this caring into practical actions. She died, the peoples' champion, on 19th September 1947 and even today, although then just a young boy, your author can remember her funeral in West Calder.

One other great worthy of the time was Doctor Young (no relative of James Young), who was the GP in West Calder. It was Young's Paraffin, Light & Mineral Oil Company which brought Doctor Young to West Calder, and each employee paid a weekly contribution (2d for married men and 1d for single men) for his services. In this way, Young enjoyed a secure practice, supplemented by his private patients, and shale workers had the peace of mind of medical care when required. He did not suffer fools gladly and was noted for his outspoken ways. He also was the champion of the miner and his family, and it was he who cared for the children as best he could through all the epidemics of scarlet fever, diphtheria and the like which regularly swept the area.

Pay and Conditions

In the early days of the industry, many were glad to obtain some permanent work and indeed work which offered the possibility of advancement. The work was hard, both in the mines and the oil works – but with little competition, the oil and associated products sold well and the companies were making reasonable profits. It was not to last. After liquid oil (petroleum) had been discovered in the USA, a price war commenced with imports of large quantities of American kerosene (paraffin or burning oil), a commodity which was already in big demand and the mainstay of Young's company. Some 4,200,000 gallons per year were being imported by 1863, and this had risen to a staggering 29,460,000 gallons in 1878. This was serious competition with prices falling from the 3/6d per gallon (17½ pence) enjoyed by Young during the duration of his patent, to a mere 10d (around 4.2 pence) per gallon, causing large losses and consequent closures across the Scottish industry. The problems facing the remaining companies were compounded by the fact that technological advance, essential to the future of the industry, was capital intensive and thus the companies which could survive were to be those who could obtain the economies in scale to remain competitive. The ten or so companies surviving by 1875 were able to offset the low burning oil prices against the more lucrative by-products such as sulphate of ammonia, naphtha, paraffin wax and lubricating oils.

It is against this background that the pay and working conditions of the oil workers have to be viewed. Falling prices in the late 1800s had led to wages being reduced. Over-capacity in the industry, in no small part due to the establishment of new oil companies at this difficult time, led to the imposition of wage reductions, and of course wage reductions triggered strikes. Attempts had been made by the workers to establish trade unions after a series of strikes in 1874, and in 1880 a number of small unions were established. The wage reductions of 1887 resulted in a three month long strike – which was to be a twenty-one week long strike in the Broxburn area where there were lock-outs in both the mines and works. Some forty-three families were evicted during this time and the strike was ended by the offer of a wage increase of 2d per ton. The Broxburn Oil Company also recognised the union for the first time, and the union quickly spread its influence across much of the industry. The first leader was one John Wilson, who was able to negotiate the appointment of union checkweighmen, who could independently check the tonnage of shale being recorded for each miner.

By 1898 the industry was again in chaos, all production ceasing because of pay demands. In 1921 a strike in the coal industry caused lay-offs in the shale industry with communal soup kitchens being set up.

Up to the First World War, oil workers had a working day of 12 hours and a 6½ day working week, although the shale miners by that time had an 8 hour day and a 5½ day week (the Miners Eight-Hour Day Act, 1908). Today, these may be considered to be long hours, but they were not when compared with the railway industry of the time – and here the author can compare with confidence – where drivers, firemen, guards and signalmen were working even longer hours and 100 hour working weeks were not unusual. Days of 15 to 18 hours were a regular feature and it was not until 1910 that drivers won a 10 hour day. It was to be 1919 before the railways won a 6 day 48 hour week. Thus the lot of a shale worker was in reality, no worse, and often somewhat better, than elsewhere at the time. The oil workers, as opposed to the miners, won a 49 hour week in 1926.

Tarbrax village showing the old miners rows in their bleak moorland setting.

THE SOCIAL ISSUES

A view of the main street in a typical mining village. This is Oakbank in 1922.

Like other industries, the employment of children had been a matter of concern. By the First World War, the minimum age limit for underground work in the shale industry was set at fourteen years. Paid leave was unknown but unpaid leave, although a rare occurrence, was granted on one or two days each year. This situation lasted until just prior to the Second World War when employees were granted one week's paid holiday a year.

The joint Shale Miners & Oil Workers Trade Union was created in 1924 through an amalgamation of the earlier unions. The first General Secretary was one Walter Nellies who hailed from Kelty in Fife. A Labour Movement activist, he was to bring both dignity and personality to a post which was anything but a sinecure. In 1926 there was a general strike; Scottish Oils Ltd and subsidiary companies announced closures of some mines and oil works, with the impact that would have on several thousand employees. The remaining employees were asked to take a 10 per cent reduction in wages. A full strike in the shale fields followed and Nellies had to argue for his members at a later Court of Inquiry. His stature as a seasoned negotiator was recognised by all, but to no avail, and a period of depression followed. More closures followed in 1931, but the company introduced a spread-over scheme whereby employees worked three weeks out of four, thus spreading the available work over the greatest number of men.

In terms of wages, in the late 1880s a shale miner earned 25/- (£1.25) per week, by the early 1900s this had risen to around 36/- (£1.80) with a peak being reached in 1910, when 8/- per day (48/- per week) was the norm. By 1921 the average earnings were 10/- per shift equating to around £3 per week. This was because, up to 1914, there was parity in the payment per ton of shale with the tonnage payment given to coal miners, meaning that in many instances the shale miners could earn more because of higher production. These earnings compared favourably with wages paid in other industries. As an example, a top grade engine driver was earning 7/6d per day in 1910 compared with the shale miner's 8/-.

Everything is relative, and it has to be recognised that the landowners, under whose land much of the shale lay, received rich rewards from mineral rights, and that for doing absolutely nothing. The miner, on a contract basis, was paid per ton of shale he produced, out of which he had to purchase explosives, all the tools of the trade, and also pay his drawer(s) – so yes, life was unfair from that point of view, but many had life much harder.

The severe competition and cheap oil imports were to have a continuing effect on the shale industry and, in an effort to reduce direct costs, wages were often targeted with reductions, normally around 10 per cent being imposed. By 1925, average wages paid were £3 19/- for a placeman, £2 15/- for drawers and £2 9/- for on-cost surface workers. In the oil works, the average wage was also about £2 15/- per week. Once again, this was not exactly below the average for workers in other industries and, whilst amongst the shale community it was considered subsistence living, poverty was relative and equally as poor, or indeed poorer, living and working conditions existed outwith the shale villages.

Before the disastrous strike of 1925, some 6,800 were employed in the industry (4,250 in the mines and 2,550 in the oil works) and the strike was to cause many redundancies. Some oil workers emigrated, many going to Australia where there was a developing shale oil industry. Those who retained jobs had to accept, yet again, a reduction in earnings and by 1936 the average wage in the mines was about 8 per cent below the national average.

However, average earnings are not an effective way of measuring poverty since this gives no indication of the actual amount of money coming into individual households, nor does it offer an indication

TABLE 10.1: THE DISTRIBUTION OF MINERS AND OIL WORKERS ROWS

SHALE FIELD	LOCATION	NUMBER OF HOUSES	PROVIDER
Cobbinshaw	South Cobbinshaw	52	owned by South Cobbinshaw Oil Company
	Tarbrax Old Rows and New Rows	250	owned by Tarbrax Oil Company
	Woolfords Old Row	22	(old Woolfords) Woolfords Coal Company
	Woolfords New Row	46	(new Woolfords)
	North Cobbinshaw (Kipsyke)	13	owned by Mungle & Thornton
West Calder	Happy Land	170	built by Young's
	Mossend	140	built by Young's
	Addiewell Village	362	double row and single row houses built by Young's
Polbeth/Gavieside	Gavieside	116	acquired by Young's
	Raeburn Row	6	built by Charlesfield Oil Company
	Westwood Row	25	leased by Young's (abandoned 1900)
	Hermand Old Rows	41	owned by Dunnet & Brown
		13	owned by Mungle & Thornton
	Hermand New Rows	85	built by Hermand Oil Company
Breich	Mid Breich	43	acquired by Pumpherston Oil Company
Cousland/Seafield	Seafield	120	provided by Pumpherston Oil Company
Livingston/Deans	Livingston Station	160	acquired/built by Pumpherston Oil Company
	Deans	36	leased by Pumpherston Oil Company
	Starlaw	30	leased by Pumpherston Oil Company
Mid Calder	Mid Calder	335	provided by Pumpheston Oil Company
	Oakbank	165	built by Oakbank Oil Company
	Dedridge (New Farm)	28	built by Oakbank Oil Company
Pumpherston	Pumpherston Old Rows	84	built by Pumpherston Oil Company
	Pumpherston Works Rows	32	built by Pumpherston Oil Company
	Pumpherston North Rows	101	built by Pumpherston Oil Company
	Pumpherston Erskine	5	built by Pumpherston Oil Company
	Pumpherston Letham	23	built by Pumpherston Oil Company
Broxburn	Greendykes	16	built by Steel and acquired by Young's
		311	built by Broxburn Oil Compnay
	Stewartfield	92	built by Broxburn Oil Company
		8	owned by William Fraser
		10	owned by John Poynter
	Albyn & Greendykes	26	acquired by Broxburn Oil Company
	Holygate	66	built by Broxburn Oil Company
	New Holygate	64	built by Broxburn Oil Company
	Westerton	41	built by Young's
	Stankards	52	built by Uphall Oil Company
	Uphall	118	provided by Young's
	Roman Camp	34	
	Uphall Station Row	20	acquired by Young's
	Beechwood Cottages	34	acquired by Young's
	White Row	21	acquired by Young's
	Stankards (Randy Raws)	36	acquired by Young's
	Holmes Rows	32	built by Holmes Oil Company
Winchburgh	Winchburgh	230	built by Oakbank Oil Company
	Niddry	96	built by Young's
Ecclesmachan	Redhouse	30	built by Young's
	Burnside	6	built by Young's
Champfleurie/Philpstoun	Kingscavil	108	jointly owned by Lord Rosebery and Linlithgow Oil Co.
	Bridgend (Auldhill Entry)	86	jointly owned by Lord Rosebery and Linlithgow Oil Co.
	Philpstoun Garden City	115	built by J. Ross & Co.
Dalmeny	Dalmeny	74	provided by Dalmeny Oil Company
Duddingston	Newton Village	13	built by Oakbank Oil Company
Burntisland	High Binnend	56	built by Burntisland Oil Company
	Low Binnend	30	built by Burntisland Oil Company
Loanhead	Pentland Cottages	64	built by Clippens Oil Company
	West Straiton	10	built by Clippens Oil Company
	Meadowbank	73	built by Clippens Oil Company

of the variations involved. Many houses had several wages coming in, or rent from lodgers, and the reality of the situation pertaining was that oil shale workers were, in the most part, well above the generally accepted poverty line. In fact the real poverty of those times was to be seen amongst those who really suffered deprivations such as the sick, the widowed, the unemployed and those fit for only casual or part-time labour.

Housing

So! What was life like growing up in a shale mining village in the heart of the Scottish shale industry? Again, everything is relative and compared with life in the twenty-first century, life in the days of shale may appear to have been particularly harsh and difficult. However, as stated, there is no real evidence today to indicate that shale communities were any worse off than other contemporary working class communities, nor were they individually worse off than other non-mining families in the same areas.

Housing was the great common problem created by the Industrial Revolution with the ensuing mass migration of people attracted by regular employment in a labour-intensive era. In examining shale housing, one must consider Glasgow and the infamous Glasgow tenements. There, high density housing was provided using the smallest ground footprint to ensure that the greatest number of homes could be crammed in. These then were Glasgow's answer to the burgeoning demand for housing. Yet the tenements were virtually the equivalent of miners' rows, but arranged vertically instead of horizontally and with living conditions in many cases in these tenements greatly sub-standard to the 'poor' conditions highlighted in the shale houses.

The houses provided by the larger oil companies were, in the main, built on greenfield sites, better constructed and eminently better in comparison with many of those constructed by smaller concerns. Quality did vary widely, however, and for every row considered to be reasonable housing, there were rows which, from the outset, were less than acceptable in terms of basic amenities and living conditions. The larger companies then inherited the serious and significant problem of inferior housing through mergers, takeovers etc., and this proved to be an insuperable problem in many cases. The demand was forever to greatly exceeded supply.

In many areas, not only housing was provided but reading rooms, games rooms, small libraries and bowling greens were also established, for use of which the oil workers paid a small weekly fee to the company.

The distribution of miners and oil workers' rows on a shale field by shale field basis is shown in Table 10.1. This list is by no means complete, but gives a fairly accurate representation of the sheer scale of house building entered into by the various oil companies to accommodate a rapidly increasing workforce. The majority of oil shale workers, miners, oil workers, managers, foremen, craftsmen and labourers lived in houses built by the various companies between 1860 and 1870 and known as 'rows'. The typical house in the rows consisted of two rooms only, although single-roomed houses were also provided (single ends) in the earlier days of the industry. The common design was a 'through house' with the main living room at the front and entered from the street, with a smaller 'back' room to the rear, which served as the kitchen, having a door out the back. Bed recesses were provided in both rooms, the latter serving as both living and sleeping quarters. A cast iron 'range' was provided in the smaller rear room for cooking and heating purposes.

At Addiewell and elsewhere, two storey houses were also built with the upper house being accessed by an outside stair, but the norm across the industry was for single-storey terraced houses, some of which were built back-to-back as found at Pumpherston. Some houses were perhaps somewhat smaller than normal, but in the main they were well-built and, indeed, Young himself had decreed that the roofs of his houses be clad with slate rather than the tarred felt used elsewhere.

Young's provided a large number of houses around the shale fields and also what was, in effect, a completely new village for the Addiewell workforce, adjacent to the works with streets of both double rows and single rows, totalling some 360 houses. The main street, Livingstone Street – so named after Young's great friend, missionary and explorer, David Livingstone – with its distinctive double rows, lasted into the 1960s. Other streets of double-row housing included Bank Street, Cross Street, Watt Street, Davies Street and Stevenson Street, whilst Graham Street, Simpson Street, Campbell Street, Baker Street and Faraday Place were low rows (single-storey houses). One single-storey row is still extant today, Faraday Place; the houses here were of somewhat superior quality, built to house foremen and other more senior staff, and each had two bedrooms, one public room, with sculleries and (eventually) inside water closets. They also had extensive gardens to the front of the house. Good houses they proved to be over the years, and they all remain inhabited today. Young's provided 861 houses with gardens out of a grand total of 946 houses.

Most of the earlier houses, until the 1920s, had shared outside (dry) toilets or privies, no running water, shared common wash-houses and middens (ashpits). Water was supplied via a standpipe in the street and had to be carried home. Compared with present-day standards, this does appear to be unacceptable – but compared with the private housing then available for rent, this was largely the norm. Many shale houses were overcrowded but, for living accommodation at that time, they were still somewhat better (and cheaper) than much of the available private rental housing in the older industrial areas around.

The quality and standard of comparable rented accommodation of the time was little better, and was in many cases significantly worse, since these properties were already very old buildings. Irrespective of condition, houses for rent were a scarce commodity.

Other workers, generally the single men, did as non-shale workers had to do. They lived in rented accommodation, often as lodgers, which generally provided a similar quality of accommodation as the rows but, on balance, often had slightly better space, with some rented houses containing as many as three to four rooms and, quite often, a garden. The oil companies generally did not provide electricity; lighting was thus by oil lamps and candles until the 1930s, and although not subsidised, supplies were always readily available. Every house had a coal fire and some companies supplied cheap coal, this being mined alongside the shale for this very purpose.

The houses built by the Pumpherston Oil Company were possibly of the higher quality amongst the company houses although the Oakbank Oil Company also provided good quality housing for its work force at Winchburgh. The other housing ranged from adequate to poor.

In the latter part of the nineteenth century and the early twentieth century, some better quality houses were built, particularly in the period between 1900 and 1910, with three rooms, scullery and WC and an internal water supply. Examples of this better type of housing

Above: Workers' houses at Middleton Hall. *Courtesy of Grangemouth Heritage Trust*

Below: Houses built for the officials at Middleton Hall. *Courtesy of Grangemouth Heritage Trust*

THE SOCIAL ISSUES

Miners' rows in Greendykes Road, Broxburn.

were to be found at Tarbrax, Livingston Station and Deans. Indeed, the Livingston Station Rows were considered to be amongst the best planned and laid-out villages in the whole industry. The best of the older houses were, from around this time, also being modernised with running water and toilet facilities, whilst the oldest and most insanitary houses were demolished.

At Tarbrax, the company had provided flush toilets (one between two) from the outset and from around 1910 other companies started a series of improvements which included provision of running water, flush toilets inside and sculleries. The rows in Greendykes Road in Broxburn, other than the 'wee rows', were improved in 1910. Oakbank Rows were modernised in 1914. Most rows had been brought up to a better standard by the 1920s. At Winchburgh, where Niddry Castle Oil Works was opened by the Oakbank Oil Company in 1902, the rows provided were newer, dating from 1900 to 1910 and were thus much superior to elsewhere. Limited street lighting by electricity was also a feature of the Winchburgh Rows from the earliest days. The rows in Winchburgh are also extant and lived in today, as, it must be said, are other surviving rows all around the county. Some of the very best housing in the lifetime of the industry was to be provided at Middleton Hall in Uphall for workers in the Headquarters for Scottish Oils which was located there. These semi-detached houses, superior to begin with, are still highly desirable properties in Uphall.

Other houses, known as 'wee rows' – such as those at West Calder, in the contrarily named Happy Land, this location being anything but a happy land for the unfortunate inhabitants – were sadly acknowledged as an example of the poorest quality of shale housing. Similar rows existed at Gavieside, Greendykes and Stankards, and these wee rows also represented this poorer quality of housing. The wee rows consisted of a two-room 'but and bens' and many had earthen floors. Some were 'single ends', some had sculleries but with no running water, all had outside dry toilets. Water was supplied by a standpipe but it has to be said that the quality of the water was excellent. The rent (in 1926) was 3/6d (17½ pence) per week. In all, a total of 172 houses were crammed into a very small area. The one important fact which must be emphasised, however, is that the oil companies did not set out to supply the workforce with inferior housing but built what were for the time houses of a similar or better quality when compared with the rental properties available. With amalgamations and take-overs, many of the larger oil companies did inherit some basic housing stock of a poorer quality, constructed by the smaller companies, with the problems these brought. These lacked what were later to be considered most basic amenities, but were, when built, somewhat better than most of the workforce had ever previously known. Despite all the new company housing, overcrowding was to become, and remain, a major problem as the industry attracted many immigrant workers, particularly from Ireland. These immigrant workers, when allocated a weather-proof house, perhaps for the first time in their lives, immediately sent home for the rest of the family. Big families, extended families, and the fact that many occupants also took in lodgers, meant that there never were to be sufficient houses available, particularly in the heyday of the industry. Overcrowding was thus to be a perennial problem which was never really resolved until the latter days of the industry in the post-war years.

The general opinion was that much of the shale oil housing was basically 'good housing' and, as stated, was often better than much of the rented accommodation of the time. In 1925, for a rent of 7/6d (37½ pence) in Broxburn, a company employee of the Broxburn Oil

William and Janet Morrison stand outside the family home at 23, Main Street, Livingston Station, a typical house frontage in the rows.

THE SOCIAL ISSUES

Company had two bedrooms, one public room, kitchen and toilet, front and back door and a garden. Other company houses could be rented for round 5/- (25 pence) per week The companies generally maintained the houses well and attended to repairs as required. In 1930, a Ministry of Health report into shale houses talked about damp houses, bucket lavatories, bad drains, insanitary conditions and defective roofs, plus, for many, the proximity of bings causing a dust nuisance. However, the residents responded with a counter report stating that the houses were in good condition and basically those houses without toilets only needed flush toilets provided to be wholly satisfactory accommodation.

For comparison, consider that in 1940 the author's family home (rented) in West Calder consisted of two rooms in an upper flat, a living room with a sink and cold water tap at the window, and a small bedroom to the rear. The house did have electricity (installed by his grandfather as a wedding present for his parents), but the family had to share an outside WC with three other families. He lived in that same house until 1949, when the family moved away. That house remained *in situ*, and without any significant modernisation, until 1960 at least. The oil workers houses were, in many cases, superior, even at that time!

Much of the non-company rented accommodation of the time, with no legal standards to comply with, was deteriorating quickly, with some becoming no longer fit for habitation. This drove many single men, and young married couples, back to seek lodgings in the rows. Uphall and Broxburn both had a particularly big Irish immigrant population and conditions such as four to a bed, and four beds to a room were common. Indeed, it is reputed by reliable sources that 'hot bedding' became almost the norm in some houses where, as one man rose to go to work, another took the bed.

To put the issue of oil company housing, whether it was considered good or bad, into proper perspective, we should consider Armadale, a coal mining town lying about two miles west of Bathgate and a non-shale town. In 1911, Armadale housing, using normal comparison indices such as overcrowding, people per room, number of families in a single room, etc., was amongst the worst of all towns or villages not only in West Lothian, but also in Scotland, with 37.1 per cent of all houses consisting of one single room. In Armadale at that time, over 50 per cent of the population lived three to one room and only 17.2 per cent lived in houses where there were two rooms or more.

In 1930, the Housing (Scotland) Act became law. This Act, perhaps better remembered now as the 'Slum Clearance Act', set minimum standards for new housing, and indeed, all housing, giving local county councils the power to impose closure and demolition orders on sub-standard housing and requiring each authority to provide new homes to replace the old. Midlothian County Council and West Lothian County Council, now legally faced with housing problems of immense proportions, were somewhat unwilling to issue Closure Orders on such properties,

Uphall Station looking north. The bridge in the background carries the NBR Edinburgh & Bathgate Railway over the roadway and Uphall railway station lies on the right of the bridge (just out of sight).

A view of Pumpherston village with the oil works in the background.

whether oil company housing or private housing, since alternative suitable accommodation just did not exist. Nor were they, it must be said, exactly in any hurry to meet their legal obligations by providing new housing. There were good reasons for this, since both councils had responsibility for so many one-industry towns thrust upon them very quickly, stifling any prospect of effective forward planning and development for a future diversified economy. It must be recognised that the same county councils were facing exactly the same problems in the coal communities where similar housing problems existed. They were, in short, confronted with housing and social problems of a considerable magnitude.

Using evidence given to the Royal Commission on Housing by the Shale Workers Union and experts hired by them, the Report concentrated on the houses provided by six of the oil companies and, for comparison purposes, also considered the private rental housing available in Broxburn. It has to be said that, by the time of the Commission hearing, the shale houses were, in many cases, deteriorating quickly and were providing less than satisfactory accommodation for the tenants. The Royal Commission was to hear the same story over and over again. The houses were very poor, not fit for purpose and should be condemned, sanitation was appalling and sewage disposal arrangements archaic, a positive hazard to health and thus totally unacceptable. But it was also acknowledged at the hearing that houses were a scarce commodity and there really was no alternative housing available. In Winchburgh, there were 313 houses for 1,752 people. In Broxburn there were 1,610 houses for 8,026 people. The conditions described in the private rental housing sector in Broxburn proved to be no better. The report concluded with the words:

> But even with all these conditions, there is still a great demand for houses in Broxburn, West Calder, Mid Calder, East Calder and Seafield and houses are very difficult to get. There are hundreds of houses which should be condemned as unfit for human habitation. They cannot be made sanitary or modern. They should be pulled down and rebuilt. But, because other and better houses are not to be had, the county councils are reluctant to issue closure orders, nor will the council fulfil their duty by building houses to meet requirements.

It was not until the mid to late 1930s that new house-building programmes were widely established over both counties and good quality council housing became available quickly thereafter. Midlothian County Council built some 1,193 new houses between 1930 and the onset of hostilities in 1939, with West Lothian County Council similarly keeping pace. Places like the Happy Land in West Calder were by the 1930s becoming renowned (or rather, notorious) for the degrees of poverty and squalor as the houses aged, but rows like that were even then the exception rather than the rule. The poorest sections (in terms of housing) of the oil working communities, such as the residents of the Happy Land and the Gavieside Rows were moved *en masse* to new houses at Parkhead in West Calder and Polbeth, but these moves did not, in fact, please all. Many felt that the old community spirit which life in the rows had engendered was being broken up and destroyed for all time. New council housing estates sprung up at Polbeth, Pumpherston, Uphall, Broxburn, West, Mid and East Calders and Loganlea to name but a few areas, and many, but not all, of the old company rows were eventually demolished.

Further local authority housing schemes were embarked upon post-war, with the building of the likes of Addiebrownhill, near Addiewell – which then accommodated almost all the residual families left in the Addiewell Rows – and also the significant expansion of Polbeth, with a new type of Swedish timber house being widely employed. The ubiquitous 'pre-fabs' also sprang up all around, bringing luxuries such as refrigerators to the lucky occupants.

Nevertheless, despite the rehousing exercise, it is testament to how well-built and good many of the shale rows actually were when one looks at just how many have survived today, most in a modernised form. Former shale rows can still be seen, such as those in Broxburn, Roman Camp, Winchburgh (Hopetoun Rows), Faraday Place in Addiewell and, indeed, the Dedridge Rows which once sat in glorious rural isolation between Mid Calder and Bellsquarry, and can be found extant, still occupied with some having been extended, now sitting amongst the modern housing of Livingston New Town. When the USAAF occupied the aerodrome at Kirknewton in the 1950s and '60s, the former shale houses at Tarbrax were in high demand as accommodation for both officers and other ranks, travel between the two locations being straightforward via the 'Lang Whang' (the A70 Edinburgh to Lanark trunk road).

A good representation of the inside of an average rows house can be seen and experienced at the Scottish Shale Museum in the Almond Valley Heritage Centre at Livingston.

Health

Sir James Young was, for the time, a relatively enlightened employer, caring for the well-being of his employees and their families by providing good (for the time) housing and schools, the latter under Government Inspection. Potential risks to health were addressed and as early as 1872 the company employed a full-time doctor. A voluntary accident benefit scheme was set in place and expenses incurred in the treatment of injuries were paid to Edinburgh Royal Infirmary by the company. In the early days, workers set up their own friendly societies which, for a weekly contribution, would provide some financial assistance during periods of sickness or injury, sick-pay being non-existent in these times. As has already been mentioned, the company provided the services of a doctor. More recent critics have argued that, since the workers were obliged to pay a small weekly amount for this service, suggesting that the ensuing health care was 'free' was a distortion of the facts. Well, the National Health Service of today is anything but a 'free' service, but the important fact remains that the oil workers then, like the general public today, had peace of mind when illness or injury struck.

All of the oil companies adopted a fairly paternalistic approach to their workforce from the outset. In 1919, the newly-formed Scottish Oil Company acquired the ordinary shares in the six remaining independent oil companies – Dalmeny Oil Co., Pumpherston Oil Co., Broxburn Oil Co., Oakbank Oil Co., Young's and J. Ross & Co. – which had amalgamated to form the new company. In 1920 it established a new provident fund for immediate miners and oil workers (but not management or white-collar staff) which was a non-contributory fund and open to all who completed a minimum of two years service in the industry.

Pollution was another perceived problem, in terms of both the atmosphere and local streams. Whilst, as mentioned, the quality of the water supply was good, the fact is that in several places it had to be carried from standpipes and stored in buckets inside the home, thus increasing the risk of water pollution.

There is little doubt that emissions from the oil works could cause a small degree of air pollution but, strangely, people were divided in their opinion as to what constituted pollution and the scale of same. In Broxburn, some thought air pollution was bad, whilst others considered that there was no problem. The rows at Roman Camp were considered to suffer from dust, and in this respect Oakbank village was also very badly affected. Indeed, where houses lay to the east of the bings there was an inherent dust problem, since the prevailing winds blew from the west. However, the same prevailing winds ensured that any air pollution in shale areas was nowhere near as bad as in the coal mining areas. Within a short distance of the shale towns lay Bathgate, where the bing at Easton Colliery burned for years and houses lying down-wind were subjected to the all-pervading, acrid, sulphurous smell of burning colliery waste, day and night. Whitburn was another such location, where the bings at Polkemmet Colliery also burned for years and the obnoxious smell was never absent. Shale waste (burnt shale), unlike colliery waste, was totally inert and slightly alkali, and there was nothing left to burn. On the other hand, colliery waste contained a lot of combustible material and spontaneous combustion was an ongoing problem. So, air pollution was, as in many other aspects of life in the shale villages, relative.

We looked at health issues in respect of miners in Chapter 6, but oil workers generally let their personal views be known about the working conditions in an environment directly exposed to dust and oil fumes, and the information gathered and collated indicates the following opinions:
- the shale crushers generated a considerable amount of dust;
- the retort tops exposed workers to a moderate degree of dust and oil fumes;
- working underneath the retort meant exposure to considerably lesser amounts of dust than working at the tops;
- working around the retorts involved little exposure to dust and fumes;
- working at the tip on top of the bings produced a very dusty environment.

Other pollution was self-evident. West Calder Water – a minor stream (burn), formerly known as Killin Water – rose in the area around West Mains to the south-west of West Calder, ran immediately to the north of the village, passing right through the centre of the public park and running into the River Almond further downstream near Livingston village. The water therein was a bright ochre yellow and the burn was always known, in the author's young days, as the 'yellow burn'. The pollution came from the waste water being pumped out of the workings in Baads pit. Although the author and his friends never actually played in the burn, they certainly played alongside it and fell in on occasions, but without any adverse effects it has to be said. It was the pollution of this same burn further to the east which led to the demise of Gavieside Oil Works.

The cramped housing conditions also resulted in diseases such as diphtheria and scarlet fever being fairly common, as was tuberculosis (TB). However, the incidence of such diseases within the shale communities was no worse than in other industrial communities and was somewhat better than most. It has again been inferred that TB was rife and responsible for many deaths in the shale community, but this merely reflects how statistics can be skewed. The author does not believe that the shale villages would stand out when measured, statistically, against the industrial heartlands of the infamous Black Country of the Midlands, or the East End of London, on a *pro rata* basis. Again, everything is relative!

Whilst coal fires did add to overall air pollution, since none of the major towns or villages lay in hollows the prevailing winds kept the air quality good. Around the retorting works there was smell and some fumes but, as the aforementioned Dr Young of West Calder stated, *'it was the fumes from the oil works which kept the Addiewell bairns free from scarlet fever and diphtheria'* and there certainly was some truth in this since, during one of the many diphtheria epidemics, Addiewell was by-passed completely. Indeed, it was the practice at that time that where people had suffered from infectious diseases, houses and clothing were systematically chemically fumigated as a matter of course. It is a short step indeed, as Dr Young inferred, from cleansing chemical fumes to the beneficial attributes of the chemical fumes of the so-called industrial pollution. The author certainly cannot recall any significant air pollution. Yes, there was always a smell, not overpowering in any way, but still the faint smell of crude oil – however West Calder sat high up (around 600 feet above sea level) on the edge of the Midlothian/Lanarkshire moorland plateau and was never exactly a warm place. 'Snell', that good old Scottish word meaning bitingly cold, describes it wonderfully, but the air was always fresh.

The truth is that overall, despite the presence of cramped housing conditions in places, the shale area was never densely populated neither was it particularly low lying, thus air pollution simply did not accumulate. Although it might now be considered by many that shale oil production equated to pollution, it was in fact no worse in this respect, and indeed was somewhat better when compared with those adjacent to coal and iron workings, as discussed above. However, in today's political climate there is little doubt that the whole process was environmentally unfriendly and would not now be acceptable.

There was one particular study which revealed some surprising statistics about the shale villages. A study in fertility transition in the UK revealed that in 1861, West Lothian had the overall highest fertility rates and the second highest rate of marital fertility. This trend continued up until the 1930s when West Lothian still topped the overall fertility rates and was the eighth highest county in marital fertility with the highest index of married women in Scotland. With the closure of the industry, these figures were not sustained and by 1981 the Lothian region was to have the lowest fertility rate in Scotland. The results of this period of high fertility in West Lothian put greater pressures on family resources since, up to the mid-1930s, average family sizes were significantly larger and thus more money was required to keep such families above the minimum poverty levels. Families of up to sixteen were not unknown and big families, with many mouths to feed, sadly often inhibited any prospect of betterment for the older children, many of whom had to leave school early and seek work to help support the younger members of the family. As a result, many able children were often denied the prospect of higher education. This was, as discussed, also a significant reason for the overcrowding issue.

Education

Since the Reformation and through the influence of the great Scottish Reformer, John Knox, an enlightened attitude was adopted regarding the rights of a free and basic education for all

Young miners at rest (and play) at Tarbrax. Note just how young they are!

children, and each parish was thereafter to have a School Master or 'Dominie' to oversee this education. With education came knowledge and knowledge was a powerful tool. Basic literacy and numeracy skills amongst Scottish children was thus to be almost universal. The results of this enlightened policy were to be clearly demonstrated in the outpourings of intellectual and scientific achievements in Scotland's Age of Enlightenment in the eighteenth century, with an estimated 75 per cent of the population being completely literate at that time. The Scots were to be the most literate citizens in Europe.

As stated in the Introduction, within the shale community, and amongst the immigrant workers in particular, great store was placed on this right to education, since most had no desire to see sons follow them underground. James Young and his company, in keeping with the policy of education for all, had a school built at Addiewell for the children of the village, for which the company deducted 2d (less than 1 pence) per week from each employee towards the ongoing upkeep and staffing. In 1872, when the Education Act of that year assigned responsibility for the elementary education of children to the statutory School Boards, the oil workers intimated that they wished the deduction to carry on and this be paid into a fund to support the education in the new schools. Most of the other large oil companies also built similar schools around the county.

From that time, very good public schools (and in Scotland, a public school is a Local Authority-run school) were established, especially those providing a higher level of education as the leaving age was raised. In West Calder for instance, a new high school was opened in Hartwood Road in the early 1890s, where, by 1899, students were able to sit Scottish Higher Leaving Certificate examinations. West Calder High School, being in the county of Midlothian at that time, was to become one of the four senior secondary schools in that county, alongside Musselburgh, Lasswade and Dalkeith. In the former county of West Lothian, across the shale fields, the same pattern was emerging with excellent high schools being established in Broxburn, Bathgate, South Queensferry and Linlithgow. At the latter location, whilst the new academy only dated from 1894, in terms of modern education there had been strong educational links in Linlithgow from as early as 1187.

These schools produced a great number of highly talented young people over the years and one particular son of West Calder was later to make his mark on the world's stage. He was one Lawrence Ennis. Born in the Gavieside (Raeburn Rows) in 1871, he went to Gavieside primary school, where the author's grandmother was later to be a teacher. After leaving school, he graduated as a civil engineer and went to work for Dorman Long, a steel construction and engineering firm on Teesside. He thereafter went, as the resident engineer-in-charge, to Sydney, NSW, Australia, when that company won the contract to construct the Sydney Harbour Bridge, the fourth longest spanning-arch bridge in the world, and he supervised the building of this famous bridge from the start of construction in July 1923 right through to opening in 1932. He was a product of the high standards of education existing within the shale areas and was but one of several eminent people who started life in such humble beginnings. Ennis died in 1938.

Another distinguished product of West Calder High School in later years was Dr John McKay. He hailed from Pumpherston where his father was a foreman plumber at the oil works there. On leaving West Calder High School aged eighteen years, McKay followed in his father's footsteps before joining the Excise service. He entered local government in 1975 as a Labour councillor in Edinburgh and became Lord Provost of that city in 1984. He obtained an Open University Degree in 1975, followed by a PhD from Edinburgh University in 1985 for his work on the Social History of the Scottish Shale Oil Industry. McKay died in November 2011, aged 82.

So, the youngsters of the shale area had teaching establishments of the highest order available for their education and the area was to produce many fine academics and professional people. It was in the Education Act of 1918 that the establishment of Roman Catholic schools was first permitted and this had, in an area with a significant Irish population, another big impact in educational trends, with separate Catholic senior secondary schools being provided for the first time.

School days in a mining community also provided much merriment and not all children were angels even then, when school discipline was much more severe than it is today. Calcium carbide has been mentioned (page 55) and this commodity was a common purchase by schoolchildren. The sum of 3d (1 pence) bought a small paper bag full of carbide. And carbide, when slipped into school inkwells provided a dramatic spectacle, and much entertainment, as the ink foamed out and covered an unsuspecting pupil's desk (and often the exercise book), accompanied by the distinct and rather unpleasant smell. Retribution was severe for anyone caught in the act.

Leisure

The cramped housing conditions meant that much of the social life was conducted outside the home. It was not at all uncommon to see groups of men standing or, more commonly for miners, squatting on their hunkers (heels) just talking, and of course the women gathered round the wash houses or drying greens for a good blether (gossip) most days. Many of the older oil workers and miners remember and speak with great affection of the 'meeting corners', quite simply the many street corners where men regularly met to put the world to rights.

Whilst paid holidays had yet to happen, one or two days unpaid leave each year was the norm and up until 1918 unemployment did not exist in the area. The workers made full use of any leisure time to follow a variety of activities. 'Kites' (quoits) were a popular pastime, being widely played on specially laid out areas of ground. Every group of rows had their own quoiting pitch; in West Calder the author can remember the quoiting pitch lying near the top of the Cleuch Brae. Quoits involved the tossing of metal rings weighing some 5 to 15lbs towards a metal pin some 25 yards distant. Points were given relative to just how close to the pin the quoit landed and of course, ringing the pin accrued maximum points. The quoit pitch was generally of clay and was kept soft to increase the skill required. The West Calder pitch was one of the last remaining in the shale area but in earlier times inter-village competitions were regularly arranged. The earlier Gavieside Quoiting Club were Scottish champions in two successive years, 1918 and 1919, and even today there is a Scottish Quoiting Association. Once more, a quoiting pitch has been recreated at the Almond Valley Heritage Centre.

Bowling was another popular summer pastime in shale communities, with many of the oil companies providing bowling greens for the use of their workers. Another popular but seasonal sport was curling, with several curling ponds lying close to every community. There were some four or five curling ponds surrounding West Calder, with one pond located within the village, lying on the left-hand side of Harburn Road, close by the bowling green; by the author's days the site of this had been transformed into a very active tennis club. In the shale communities, quoiting (just like the winter sport of curling) was a 'drouthy' pastime – that is, an activity which worked up a thirst amongst the players and requiring the frequent consumption of suitable 'refreshments'.

Whippets and greyhounds were another interest (the 'dugs'), as were the racing pigeons (the 'doos'), and well the author remembers women being asked to take in the washing as racing pigeons circled to land, so that they would settle quickly, and people did so. These interests involved a fair degree of betting. Gambling was thus always present in varying degrees and the author and his friends, as youngsters, all knew where the card schools and 'tossing schools' (pitch and toss) were and sensibly gave them a wide berth. The breeding and showing of cage birds was also a common pastime.

Gardening was another activity widely followed by oil workers. Many of the better rows had garden ground provided and great pride was taken in the floral displays on view. Growing vegetables and flowers for exhibiting was common and there were many annual flower shows. Indeed, the author remembers the lengths to which such enthusiasts went to create a suitable liquid feed for show leeks and the like, and the murky (and odorous) concoctions contained in pails countersunk into the gardens are best not discussed.

Music was another major interest and a great mining tradition. Each village had a brass band, or, as in West Calder, Broxburn and Winchburgh, very good championship silver bands. Pipe bands flourished and there were many choirs. The music tradition of both the shale, and coal villages in the county, despite the demise of both industries, is still proudly carried on today by the West Lothian Schools Brass Band, this band being well-known worldwide as a very good band. The author, whilst working in Sydney, Australia, was walking in Darling Harbour one Sunday morning, when he heard brass band music being played; on investigation, he found this band of very talented young musicians from West Lothian giving a free concert, having filled the Sydney Opera House the previous evening. Needless to say, he felt very homesick indeed.

Football, cricket and tennis were well supported and often-played sports, and the bowling greens were busy. Most shale villages had football clubs and one, Mossend Swifts of West Calder, actually played against Glasgow Celtic in a cup match in the late 1890s. Then there were the very popular annual Gala Days for the children, each village hosting its own. Holidays were virtually unknown and travel outwith the local community was a rare occasion indeed. On the Gala Day the children were dressed in their finery and the day commenced with a procession of all the children, generally with a tin cup secured round their necks and led by at least one band, through the village to a suitable field where a picnic was held, generally involving 'Co-operative purvey' of a scotch pie, a cake, an apple or orange and lemonade, after a sports afternoon where there were races for children and parents. This celebration was doubly important for the children of the more remote communities like Tarbrax, sitting at 880 feet, high on bleak moorland.

Pumpherston village could boast an Institute with a library and reading room, a cricket club, a cycle track, a 9-hole golf course, an

Miners' rows in Livingston Station showing the well-tended garden belonging to Mr Joseph Morrison at 23, Main Street, Livingston Station.

angling club, a football field and a bowling green. Indeed, bowls became a very popular pastime within the oil companies with the companies not only providing the bowling greens but the bowls as well.

And then, in West Calder, there was the Cauther Fair. This fair had been a long-established feature of old village life (since 1685) which had largely passed into the annals of history, but in 1860, partly as a result of Young's and other oil companies' social services policies, in conjunction with the re-establishment of the old Farmers Union, this event was successfully revived in a more modern format. It became more of a sports day, the famous West Calder Annual Sports day, when professional athletes took part. This fair had traditionally been held on a Thursday and formed part of a long weekend holiday, but in latter times it was held on a Saturday. The highlight of the day, apart from the athletics events, was the professional road race from Edinburgh Saughton Park to the West End Park in West Calder, a distance of 17 miles.

On Sundays, another leisure activity, very common across central Scotland, was getting dressed up in Sunday best and – after Sunday School and church in the morning, and before evening service – going out, or being taken out for a walk … often around the cemetery. A very typically Scottish touch this, an afternoon spent quietly meditating amongst gravestones, a popular practice which persisted well into your author's lifetime.

In 1875 a group of miners formed a local co-operative society which was to become the West Calder Co-operative Society. Despite initial opposition from local shopkeepers, it soon grew and flourished becoming one of the largest and most successful co-operative societies in the central belt. Eventually it had branches at Cobbinshaw, Tarbrax, Haywood, Forth, Addiewell, Philpstoun, Pumpherston, Blackburn, Livingston Station, Oakbank, Uphall, East Calder and Mid Calder. A job in the Co-op was much prized and a sinecure for some. Appointment to the Co-op Board, any Co-op Board, was a stepping stone for those with political aspirations, either at local level or nationally.

The Broxburn Co-operative Society was formed in 1879, being probably the biggest at the time, with a branch opening later at Winchburgh. Bathgate Co-operative Society, formed in 1881, was also another large and successful undertaking. The Co-op played an important part in village life since it could, by the sheer scale of purchasing power, offer a greater variety of products at reasonable cost, plus the fact that a dividend was paid to members (they being shareholders) each year, based on the year's trading performance; as much as 5/- (25 pence) dividend was recorded at one time as being paid for every one pound's worth of purchases made over the year. This created a powerful incentive for shopping at the Co-op. Not only food-stuffs, but clothing, shoes, prams, furniture, ironmongery and paraffin oil – indeed almost everything was stocked, or could be obtained. The various Co-ops were the normal venue and caterers for wedding receptions; and even in death the Co-operative Undertakers promised a dignified funeral. The Bathgate Co-operative Halls in Jarvey Street were for many years the venue for a whole season of formal balls, all 'black tie' affairs, such as the Golf Club Ball, the Tartan Ball, the Farmer's and Young

The men of Tarbrax Brass Band are not wearing a uniform, as some other brass bands did, but nevertheless pose proudly with their instruments in everyday wear, including the miners' favourite headgear, the cloth cap or, as more commonly known in the area, the 'bunnet'.

Above: Gala day in Pumpherston in 1911. The children carry their 'tinnies', tin cups which will be filled after the sports with milk or juice. The West Calder Co-operative supplied all victuals – biscuits, milk and juice – and presented prizes at the sports.

Below: The children's Gala Day at Tarbrax in 1923. The children are dressed in their finery and are parading through the village *en route* to an afternoon of races and a picnic. In the background is the bing, with the Tarbrax branch of the West Calder Co-operative Society on the left.

THE SOCIAL ISSUES

Tarbrax old and new Co-operative building standing, apparently in glorious isolation, on the moors above West Calder.

Farmer's formal dances, the Round Table and Rotary Club balls, to name but a few.

In the nineteenth century, many of the new shale villages were 'dry', in that they had no public houses, mainly as a result of the terms of lease dictated by the landowners. However, some of the villages which were established pre-shale could boast licensed premises, with West Calder boasting six public houses and one licensed grocer in 1844, but with a total of sixteen such establishments existing by 1880. Gradually, licensed establishments were to be opened across the industry. The evils of alcohol were no better and no worse in the shale society, but it did cause the inevitable social problems where money was often a scarce commodity. Drinking was considered, of course, a working man's prerogative and drinking to excess, although not general, was, sadly, fairly prevalent. As a youngster of around nine or ten years, the author and his friends would make a point of being around the Main Street in West Calder of a Saturday evening, to watch the fun as the pubs emptied after 9.00pm.

The temperance movement was also well represented, with such groups as the Band of Hope, the Hallelujah Society, the Rechabites and the Temperance Society being well supported. Freemasonry was another activity well represented around the shale villages with most having a lodge within a Masonic Hall.

By 1877 the spiritual needs of the communities were also well represented, with the establishment in West Calder of a Catholic church to accommodate the large Roman Catholic influx from Ireland, the village also boasting a parish church, a Free church and a United Presbyterian church. The same story was true of the other shale villages such as Addiewell, Winchburgh, Uphall and Broxburn. Even the smaller shale communities were well served. The various churches throughout the area also provided many social and recreational activities, particularly for the young, and amateur dramatics was one particular activity promoted by the church. The author's own church, Hartwood UP, in West Calder, had a well-known and long-lived amateur dramatic society, the Hartwood Players, of which he was, for a time, part. The society took part in festivals and competitions, and put on regular productions in the village. Despite the influx of a big Irish community, religious bigotry was certainly not an issue of any significance in West Calder, although it was believed to be somewhat more prevalent elsewhere in shale communities and the Orange Order was large in Broxburn.

West Calder, at the end of the Second World War, could also boast two cinemas or, as locally known, picture houses. One, on the right-hand side as one went down the Cleugh Brae, was the Peoples Palace – commonly known as 'Nannie Mullen's', so-named after the proprietress – part of a grand conglomeration of buildings which included a billiards hall, the Polytechnic Hall (the Poly) where Saturday evening dancing was held over many years, and shops on street level. The whole lot was owned by another West Calder worthy, 'Pie Jock' John Thomson, who was a baker of note and both occupied a shop and had his home in the buildings. The other cinema, the Regal, was a far grander affair designed by the Leven-based firm of A.D. Haxon. It was located on the Main Street near the old Happy Land. It was opened in 1939 and was built in the then standard 1930s symmetrical 'Art Deco' style of the Regal cinema chain. Operated by Messrs Millar & Walker, with Regal cinemas also in Armadale, Broxburn and Bathgate, they lasted well into the 1960s, becoming part of the ABC chain before closure and demolition. The local lads did much of their courting here, author included. Happily, the Regal was preserved in Bathgate and now, although no longer a cinema, still provides a venue for many forms of entertainment.

The shale industry had a dynamic effect on West Lothian, most of the villages rising to become centres of economic importance in the late nineteenth and early twentieth centuries. Shale, however, was to also leave a not so proud legacy, the legacy of one-industry towns. Scottish Oils objected as a matter of course to any ingress of new industry, especially in the post-war years, by claiming that they could offer employment to all, including school leavers, for the foreseeable future. This was, in fact, the case. Up to the late 1950s, school leavers had quite a choice, dependent upon academic prowess, or otherwise. For the high flyers, Scottish Oils offered 'sandwich' courses in chemistry and chemical engineering where students were employed by the company and split their time between work and university classes, with a degree as the prize. The author was offered, but turned down, such a course. Then there were clerical positions, apprenticeships in various trades including brick-laying, joinery, plumbing, wagon-building/repairing and, for the less fortunate, the general labouring jobs in the oil works and, of course, the mines and pits.

The industry was to leave a legacy of unemployment when it folded, but it also left the legacy of red bings, polluted ground and streams, and a veritable maze of underground workings which caused serious subsidence problems over many long years. Sadly, the shale oil industry, with its demise, ripped the heart out of all the oil communities and left a large workforce with no alternative employment.

Livingston New Town, established in the early 1960s and the last of the five Scottish New Towns constructed mainly to take Glasgow overspill, was built smack bang in the centre of the shale fields and on some of the finest farming land in the county. By 2010, Livingston has grown apace and has now taken over the role of the other, once major, towns in the county. Bathgate, once a town noted for fine shops has been badly hit and West Calder, Mid Calder, Broxburn, Uphall and Pumpherston have also suffered – some of these actually now more or less absorbed into the boundaries of the new town. Even the old county seat of Linlithgow, one-time home of West Lothian County Council, has seen nearly all the council functions move, firstly to Bathgate and now to Livingston. In August 2009 the historic Linlithgow Sheriff Court was also finally moved to Livingston. Many of these towns mentioned have also seen considerable house building, becoming dormitory suburbs for Edinburgh and Glasgow and now form part of a large and widespread commuter belt. Such is progress!

The Pumpherston branch of West Calder Co-operative Society Ltd.

ABOVE: Oakbank village. The shale industry has now gone and the miners' rows lie deserted and awaiting demolition.

BELOW: A modern day view of West Calder Main Street looking east. The red sandstone building behind the traffic lights is the former headquarters of the Co-operative Society. On the immediate right-hand side can be seen the clock from the former Co-operative building which is now incorporated into the memorial to the fifteen miners who lost their lives in the 1947 Burngrange pit disaster.

The Winchburgh tramway. A train of shale being propelled back towards the foot of the retort bank where the train will be split and individual hutches will be connected to an overhead ropeway by the shale 'runner' (see photo, p. 142) to be hauled up the incline to the top of the retort bench. The wagons on the higher level on the far left of the picture are the empties waiting return to the mines.

PART TWO
THE MINERAL RAILWAYS OF THE SCOTTISH OIL INDUSTRY

The new Westwood Oil Works at West Calder with a number of Oakbank Oil Company rail tanks sitting in the foreground. Behind are two bings created by the spent shale from the retorts. By closure, these bings had grown to become five in number, they are now a local landmark known as the Five Sisters and preserved as Historic Monuments.

11
AN INTRODUCTION TO THE RAILWAYS

From the 1850s onwards, wholly private industrial railways, worked by steam traction, were becoming extremely common in Scotland. The various shale oil companies were amongst several Scottish industrial undertakings developing such railways and the reasons for doing so were twofold:
 i) as a means of transporting large tonnages of oil shale from mines/pits to the oil works, and
 ii) to establish a link between the oil works and the nearest national rail network, essential for getting the finished products distributed to the market place in the quickest possible time.

However, unlike many other Scottish industrial undertakings of the time, the Scottish oil companies from the outset adopted the standard 4ft 8½ins gauge for their internal lines worked by steam locomotives, although wide use was also made of narrow gauge haulageways for other purposes. Being highly dependent upon good transport arrangements, the industry was from the earliest days served by an extensive network of internal railway lines, many of which were constructed to interface with the national railway system. At oil works lying close to the national railway system, direct connections were provided either by long (private) sidings or proper branch lines, and, in some places the privately-owned steam engines (pugs) were authorised to run on these branch lines of the national system. These branch lines and long sidings were in some cases provided at the railway company's expense and in others the cost was passed on to or shared with the oil company concerned.

In the following chapters we shall look at both the narrow gauge (non-standard gauge) and standard gauge lines in some detail. Abbreviations used in the context of national railway companies hereafter are:

CR Caledonian Railway
E&BR Edinburgh & Bathgate Railway (worked by the NBR until 1923 and the L&NER thereafter)
E&GR Edinburgh & Glasgow Railway Company (absorbed by the NBR in 1865)
LM&SR London, Midland & Scottish Railway
L&NER London & North Eastern Railway
NBR North British Railway
WM&CR Wilsontown, Morningside & Coltness Railway (absorbed by the E&GR in 1850)

As stated, all the main mineral lines serving the industry are known to have been constructed to the standard rail gauge of 4ft 8½ins, making interface with the national network a simple process. However, there were a number of differing requirements within the shale oil extraction and refining process which called for several different methods of transportation, and these can be classified thus:
 a) Conveyance of raw mined shale from mines and pits to the oil works. In the early days, many of the small oil works were set up adjacent to the shale workings thus avoiding transportation problems but as the industry consolidated, the shale was required to be conveyed over longer distances. This was achieved by five methods:
  i) surface haulageways, generally rope-worked and consisting hutches running on 2ft 6ins gauge tracks;
  ii) wholly internal standard gauge mineral lines worked by company locomotives ('pugs');
  iii) underground haulageways. Either rope operated (one known instance between Philpstoun No. 1 mine and Philpstoun Oil Works), or pit pony/locomotive worked with hutches running on 2ft tracks;
  iv) the national railway system on goods branch lines constructed and worked by the railway companies;
  v) surface haulage by electric locomotives (one known instance between Duddingston mine and Niddry Castle Oil Works).
 b) Conveyance of crushed shale to the retort benches. Generally carried out by hutches running on continuous rope-worked 2ft 6ins tracks.
 c) Conveyance of burned shale to the bings. By side-tipping and end-tipping hutches on rope-worked 2ft tracks.

A pit pony with pony boys at Tarbrax.

Inwards traffic consisted mainly of coal (by both internal and national railway lines), timber pit props (mainly imported timber from the Baltic countries generating a significant amount of traffic between docks and pits/mines) with other general materials. Some oil works, like Broxburn and Addiewell, employed their own coopers for barrel making and this required a regular supply of raw materials. In the early years of the industry, most of the processed oil was dispatched in wooden barrels and rail tank wagons were not at all common. This was due in part to one of the major wagon finance companies, The North Central Wagon Finance Company, initially being loath to agree to the funding of construction and purchase of specialised oil tank wagons. This problem was to resolve itself later and tank wagons became a regular part of oil works scenery.

Outward traffic from oil works consisted of either the transfer of crude oil and sulphuric acid to oil works with refineries, or the refined oil products (both in rail tanks and barrels), sulphate of ammonia (bagged) and, in earlier days, significant quantities of candles (boxed).

Part Two of this book has, therefore, been split into several chapters looking in some detail at the narrow gauge railways, the mineral railways connected to the national rail system and those standard gauge internal mineral lines which served only the industry.

Many oil companies eventually had their own fleet of rail wagons and, in particular, rail tank wagons, used for the transfer of crude oil from retorts and also the carriage of the refined oil products. These wagons had to be registered to run upon the national railway system, as will be discussed in the following chapters. Each oil company also had their own large fleets of open wooden wagons for the transportation of shale. These were, in the latter days, the equivalent of the standard railway five-plank or six-plank open mineral wagon and most were fitted with grease axleboxes and 3-link couplings. The 'pugs' hurried these wagons up and down between pits/mines and oil works, working big trains over lines with fairly heavy gradients and the noise of the thunderous exhaust was a joy to listen to. Many of these wagons were registered for main line work.

The early James Russell mining operation at Boghead, Torbanehill and Hopetoun mines supplied the original Bathgate (Boghead) Oil Works with supplies of the cannel coal (Torbanite) and the works were rail connected from the outset, with all traffic to and from the chemical works passing by rail. This was a measure adopted in no small part to maintain the secrecy which surrounded Young's Bathgate operation and his Secret Works. With the retorting and refining partly transferred to Addiewell, the works at Bathgate continued as an oil works for a few more years before finally becoming purely an acid producing facility. It was rail-connected to the E&GR extension of the ex-WM&CR branch line from Longridge to Bathgate in 1851, the access and egress being controlled by Bathgate Chemical Works signal box.

Many of the branches and sidings provided by the railway companies serving the various oil works were worked by railway company engines, some by both railway company and oil company locomotives. In the former case, most were worked under One Engine in Steam arrangements with or without a Train Staff, and the method of working of each branch published in the respective Operating Appendices. In the locations where both railway and oil company provided locomotives and drivers, occupation of the running line(s) was shared on a fixed time basis, each operator having set hours between which they could operate; again, the method of working was detailed in the Operating Appendices. In one or two cases, the oil company's drivers were passed as competent to enter and work over railway company metals and in these situations the drivers involved were actually named in the working instructions, having been passed as competent by railway company locomotive inspectors. At exchange sidings where both operators interfaced, yard working was generally applied and this required that all movements were conducted at speeds of no more than 5 mph.

On the internal mineral lines between oil works or pits/mines and an oil works, all movements were conducted by oil company-owned locomotives, the 'pugs'.

A rather crumpled photograph taken at the foot of the retort bank at Niddry Castle Oil Works. It shows the loaded hutches being connected to the overhead rope by the 'runner' so that they can be hauled up to the retort benches for discharge.

12

NON-STANDARD GAUGE RAILWAYS

Underground Haulageways

The two-axle tubs were known locally within the industry as 'hutches'. These hutches were of steel construction and ran on a track gauge of, generally, 2 feet, on rails weighing about 14 to 16lb per yard. Each hutch carried from 15 to 28cwt of shale.

The underground lines were subject to much lifting and extending as the shale faces progressed, thus the necessity for a light structure. Their purpose was to transport the vast quantities of shale recovered in the mining operation, from the working face to the pit or mine bottom, and from there to the surface, either directly to the shale crushers, or to be reloaded into larger standard gauge railway wagons for onward transportation to the oil works. These lines have been described in some further detail at Chapter 3.

From the working face, loaded hutches were normally pushed by hand (by the drawer) out to the main haulage road for the level in which his workplace was located, before being pulled by pit pony to the pit or mine bottom. The ponies were gradually replaced by diesel mechanical locomotives or electric (battery) powered locomotives for this purpose and these could haul considerably higher tonnages in a single load from the various levels to the pit bottom. Continuous, or endless, rope working was also employed for this purpose in some mines.

Some fairly unique techniques had to be developed to address the many problems associated with the overall transportation of mined shale from the face to the pit or mine bottom given the many severe inclines, and underground haulage operations were implemented to suit the conditions in each mine or pit and shaft. The various methods employed in the shale mines and pits have already been described at Part 1, page 52 in some detail.

Transferring the mined shale to the surface also posed its own problems and methods of working differed from place to place. In mines, the inclined shaft winding operation could consist of:
- *Single hutches*. Generally worked up the incline by endless rope haulage. This system was economical in operation and was particularly useful where an inclined mine served a series of

A Ruston Hornsby diesel locomotive, one of four allocated here, at work underground at Philpstoun No. 6 mine.

levels. An attendant was located at a special station (or bench) and put each loaded hutch on the rope with a distance of 40 feet being allowed between hutches. 500 tons per shift and even more could be raised by this method.
- *Trains (rakes) of hutches*. Drawn up to the mine head by direct rope haulage operated by a winding engine.
- *Carriage operation*. An inclined carriage could carry either two or four loaded hutches sitting transversely across the carriage on two or four tiers. This was pulled to the surface, again by direct rope haulage operated by a winding engine. This was the preferred method where the gradient of the incline exceeded 20° and used a double line of a 3ft gauge. This wider gauge permitted higher speeds with lower risk of derailment. One line was confined to loaded carriages whilst the other was used by the carriage for empty hutches going back underground. This system could easily raise 300 tons of shale per 7 hour shift from a depth of 2,000 feet.

In pits, vertical shafts replaced the inclined shafts and there was but one method of raising the mine's shale to the surface and that was by cage, either single-decked or double-decked, where the loaded hutches were raised to the surface and empty hutches lowered in the same single operation. The various winding systems have been discussed above in Chapter 5.

Surface Haulageways

When the loaded hutches were delivered at the pithead, a shale inspector (known as the craw-picker) examined them and would reject unsuitable material. The hutches, if approved, were then weighed and the weight credited to the placeman according to the 'tally' used. Mistakes in crediting were rare in the extreme. As the refining of the crude oil could be most economically carried out where large quantities of crude were being produced it became customary to establish the large refineries convenient to the most productive shale areas. However, especially in the early days of the industry, the shales mines were so widely spread that, to avoid the high costs of transport, a considerable number of crude oil retorts were established in close proximity to groups of mines. The large central refineries dealt with the oil from their own retorts but with the full throughput being augmented by the supplies of crude oil from the nearest crude works.

The common purpose of the internal surface narrow gauge railways was to transport the mined shale up to the retort benches after crushing – and the spent shale from the underside of the retort benches to the top of the spoil heaps or bings after burning, where this was tipped as waste. As these bings spread, so the rails had to be lifted and relaid. It must be said that one of the very worst

A view of the well-kept and well-lit pit bottom at the Fraser pit near West Calder. The hutches on the right are being run into the cage to be taken to the surface whilst the hutches immediately to the left are awaiting the next cage.

Hopetoun (Glendevon) No. 6 mine west of Winchburgh. Empty hutches are going underground on the left whilst loaded hutches come up on the right-hand side. Note the neat pile of pit props. *Courtesy West Lothian District History Library*

jobs undertaken in the industry was that carried out by the men on the tips, working at considerable height above normal land levels, exposed to all the elements and the constant dust as the spent and hot, very hot, shale was tipped out. Endless rope haulage was used to convey the hutches up the bing to the tip.

The hutches used on the bing for tipping the spent shale were again designed to operate on a 2ft gauge just like the mine hutches, but were somewhat larger than the latter and sat on extended underframes. The hopper was fitted with both side doors and end doors to facilitate the discharge of hot shale. Again, these hutches were fitted with grippers which held the haulage rope securely but which released automatically both at the top of the bing and on return to the bottom. Each hutch weighed around 1½ tons when loaded and as they came off the rope at the tip, the tip men had to physically push them to the tipping point and, once there, manually decant the contents. The hutches were very hot and most of the men employed in this unpleasant task wore old caps, suitably stuffed with waste or paper, and used their heads as well as hands in pushing the hutches to the tip. On the way down after discharge, the wheels of the hutches passed through a reservoir containing thick grease which lubricated them. This was a precautionary measure given the high temperature conditions under which these hutches operated.

There were, therefore, many narrow gauge haulageways. Some were very short, others of some length and involving civil engineering work of commendable standards. The extent of the usage of these narrow gauge lines can be judged by the fact that in excess of a thousand tons of shale could be passed through the retorts every day at the larger oil works.

Some of the longer and more significant narrow gauge lines are described below. There was one quite unique exception to this general usage, however, which was the narrow gauge railway constructed to transport shale from the Duddingston and Whitequarries mines to Niddry Castle Oil Works at Winchburgh. This has, quite properly, been given a chapter to itself (Chapter 13).

The longer haulageways were:

- *Breich Nos 1 & 2 Pits to Seafield Oil Works.* (See Map H, p. 181.) The shale produced by Breich Nos 1 & 2 pits was taken by a narrow gauge railway on the continuous rope system to Seafield Oil Works, a distance just short of a mile. This line crossed the River Almond by a bridge specially built for the purpose.

- *Seafield No. 3 Pit to Seafield Oil Works.* (See Map H, p. 181.) The shale produced in Seafield No. 3 pit, lying to the east of the works, was carried by a narrow gauge tramway some 600 yards in length to Seafield Oil Works. The pit closed in 1917.

- *Breich No. 4 Pit to New Hermand Oil Works.* Also known as Wester Breich, this pit working the Broxburn seam had a tramway some 570 yards long, running more or less due east, conveying mined shale right into the Hermand Oil Works complex. The pit closed in 1887 and the oil works followed in 1903.

- *Rosshill Nos 1 & 2 Mine to Dalmeny Oil Works.* The shale mined at Rosshill mine near Dalmeny village was taken by a narrow

Possibly one of the worst jobs in the shale industry. A workman tips a hutch of red hot, spent shale at the top of a bing. *BP Archives*

Hermand No. 4 mine, West Calder, showing the haulageway from the mine bottom, with four loaded hutches carried transversely on the special carriage used for conveyance of both men and hutches up and down the mine.

The Whitequarries mining complex in a sylvan setting. This is the Philpstoun No. 6 mine end of the continuous rope haulageway linking into the Winchburgh electric tramway at Duddingston Nos 3 & 4 mine. Loaded hutches of shale are worked away under the overbridge in the rear, to Duddingston mine where they will be reloaded into the larger hutches on the electric tramway, for onward conveyance to Niddry Castle Oil Works.
Courtesy West Lothian District History Library

gauge railway to Dalmeny Oil Works, the distance involved was just under ½ mile.
- *Roman Camp No. 4 Mine to Roman Camp Oil Works*. A narrow gauge railway of some 1,320 yards in length connected Roman Camp No. 4 mine to the Roman Camp Oil Works. This line crossed over the Beugh Burn (south of Broxburn) by means of a bridge and passed below the NBR Bathgate to Edinburgh railway line some 760 yards east of the present Uphall Station.
- *Pumpherston No. 5 Mine to Pumpherston Oil Works*. A narrow gauge railway connected Pumpherston No. 5 mine (NT079692) to Pumpherston Oil Works. The distance was just over ½ mile.
- *Cobbinshaw No. 5 Mine and Tarbrax Oil Works*. (See diagram on p. 156.) A narrow gauge, rope-operated railway of just over 1 mile in length across the moorland connected Cobbinshaw No. 5 mine (NT036572) with Tarbrax Oil Works.

There were, however, many more smaller haulageways. In addition to the above, in the earliest days, the mining of Torbanite in the Bathville/Armadale area saw around a dozen pits and mines working this material at various times; whilst a standard gauge branch line served a number of these mines, most were inter-connected by a network of narrow gauge tramways, all long gone by the late 1800s.

THE SCOTTISH SHALE OIL INDUSTRY & MINERAL RAILWAY LINES

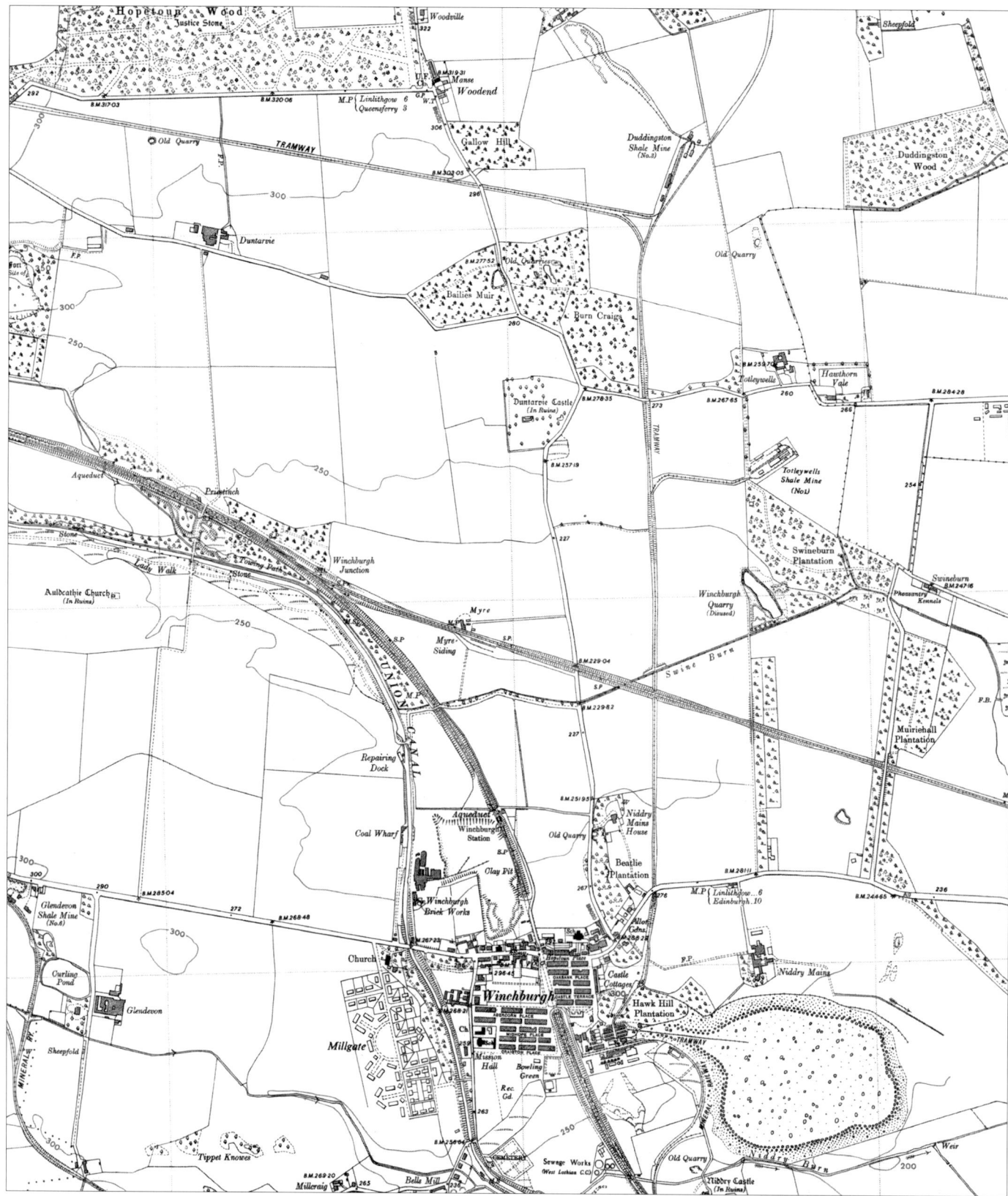

Map D: Winchburgh (1914 with additions in 1948). Bottom centre lies Winchburgh village with Niddry Castle Oil Works and the rows clearly shown. The Edinburgh & Glasgow Railway passes under the village in tunnel, whilst the Union Canal runs to the west; both run north before swinging away westwards. At Winchburgh Junction the 1890 chord line from Winchburgh to Dalmeny runs off to the left. The Winchburgh Electric Tramway is clearly marked, running due north from Niddrie Castle Oil Works and passing under the Dalmeny Chord Line to reach Duddington Nos 3 & 4 mine. The branch to Totleywells mine runs off to the east. Striking off almost due west from Duddington mines is the rope-operated tramway connecting the electric tramway and the Whitequarries shale mining complex which is off the map. (Scale: six inches to one mile, scaled to 70% to fit page.)

Courtesy of the Trustees of the National Library of Scotland

13

THE WINCHBURGH TRAMWAY

Undoubtedly, this narrow gauge haulageway was the most unique of the many mineral railway systems employed by Scottish shale oil companies, and thus pride of place must go to this relatively short (2½ mile) line linking Niddry Castle Oil Works at Winchburgh with the considerable shale oil reserves lying to the north of the Ochiltree fault line in and adjacent to the Hopetoun Estates. This line was operational for more than sixty years.

This railway was of 2ft 6ins gauge and consisted of a single line running north from Niddry Castle Oil Works at Winchburgh to access the feeder mines: Duddingston Nos 1 & 2 (NT099775) and Duddingston Nos 3 & 4 mines (NT096772) and, somewhat later, the Whitequarries and Philpstoun No. 6 mines (NT074774) (see Map D, p. 148). Both the new Niddry Castle Oil Works and this railway line were constructed by the Oakbank Oil Company and opened in 1902, electricity being the power source employed at the oil works from the outset. Naturally, the use of electricity was also extended to the operation of this railway which was, from the start, electrified by an overhead line 500V DC system.

A total of six locomotives were constructed through the life of the system:
- *Nos 1 and 2*. Built 1902. 4-wheeled locomotives. Rated 50hp. Built by the Baldwin Locomotive Company (USA) and fitted with Westinghouse electrical equipment. Open-decked as constructed, No. 2 was fitted with a makeshift cab in order to provide some protection from the elements. No. 2 locomotive

ABOVE: The Winchburgh tramway with electric locomotives Nos 5 and 6, each heading loaded shale trains from Duddingston mine, at Niddry Castle Oil Works.

RIGHT: A view taken from partway up the incline to the retort benches at Niddry Castle Oil Works showing arrival and marshalling sidings and the overhead ropeway which was the means of working the retort incline. The sidings to the right of the picture are the empty hutch roads where trains of empties are made up and worked back to Duddingston/Totleywells/Whitequarries for reloading; on the outside (furthest right) siding, an empty train is ready to depart.

R.C. Nelson/Hamish Stevenson collection

is now preserved in its original cab-less condition at the Scottish Shale Museum at the Almond Valley Heritage Centre at Livingston Mill Farm, Livingston.

- *No. 3*. Built 1907. 4-wheeled locomotive. Rated 100hp. Built by British Westinghouse. Fitted with a central cab.
- *No. 4*. Built 1929. 4-wheeled locomotive. Rated 100hp. Built by the English Electric Company. Fitted with an end cab.
- *Nos 5 and 6*. Built 1943. 12 ton, 4-wheeled locomotives. Rated 100hp. Built as a joint venture by Metro Vickers and Andrew Barclay of Kilmarnock. Both had end cabs fitted from new.

In the latter days of operation, three locomotives, Nos 4, 5 and 6, remained in general use, with a fourth, No. 3, available as and when required. The last two locomotives were powered by a 60hp, 500V DC motor driving a central shaft via intermediate gearing running in an oil bath. Large cranks were fitted to the ends of the shaft and the final drive was taken from these cranks to massive coupling rods on the wheels. The driving wheels were 34 inches in diameter, with the tyres pinned as well as shrunk on to the wheel boss. Overhead current was collected via simple trolley poles, each pole being fitted with a revolving pulley collector shoe, and a nine stud controller was provided in each cab together with a speedometer calibrated up to 20mph. Normal operating speed was around 10mph and, even when running light, the locomotives were generally never driven faster than about 15mph. The locomotives were painted in unlined olive green.

There were two different kinds of hutches used in the complete operation. Those used in the underground operation were smaller and ran on a 2ft gauge, light T-section steel rail track. Constructed

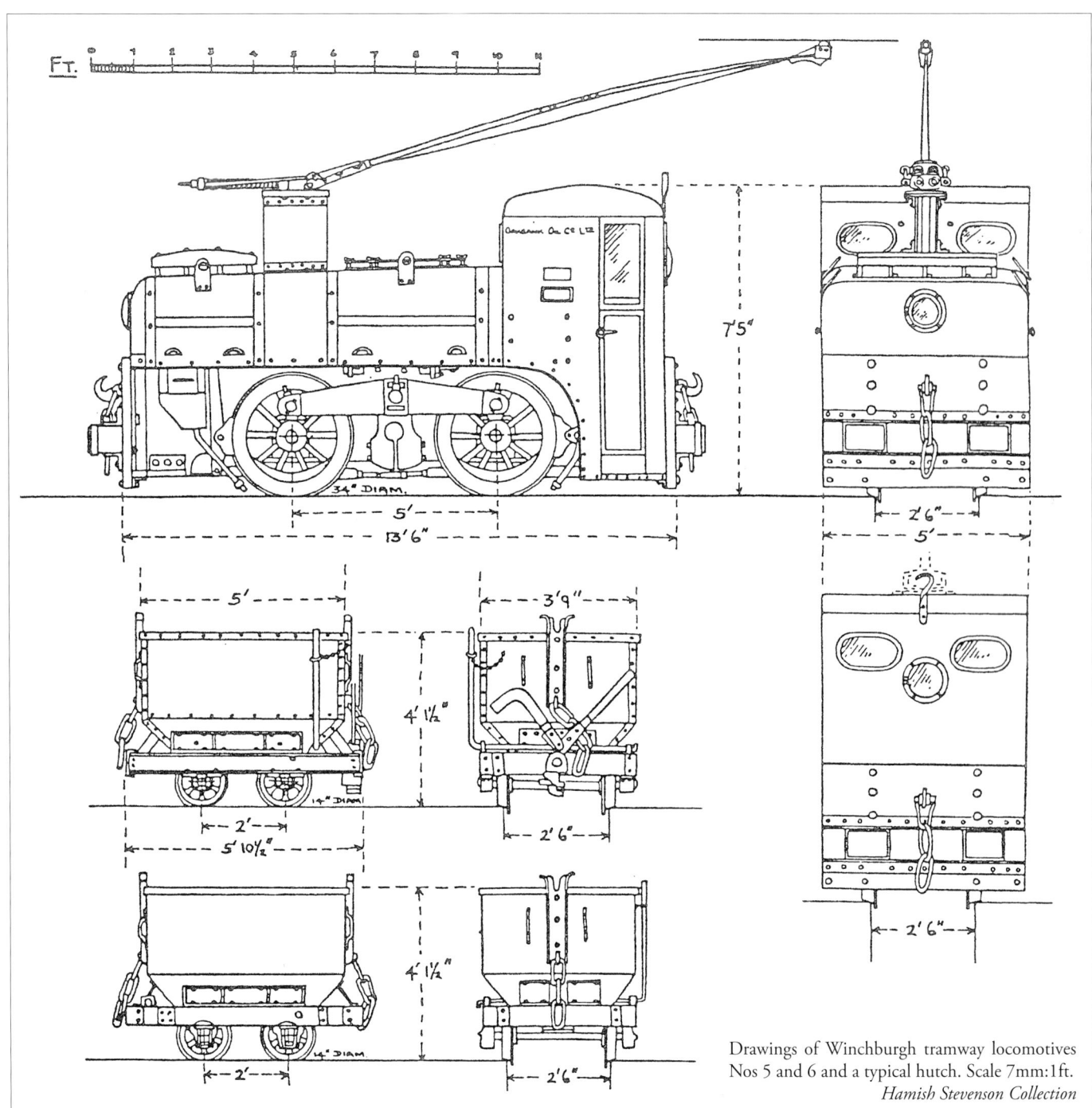

Drawings of Winchburgh tramway locomotives Nos 5 and 6 and a typical hutch. Scale 7mm:1ft.
Hamish Stevenson Collection

with plain steel bodies, they ran on 12ins diameter wheels and had a single-link coupling at either end. They were run together in rakes of four bringing shale to the surface and were coupled together by a 'ram hook' chain. The wagons were fitted with grippers which were merely spring-loaded plates which gripped the haulage rope and worked in conjunction with lineside 'kickers' or trip-ramps which caused the grippers to release at the destination point. At the trailing end of each rake of loaded hutches, a pointed iron stake was hung on the last coupling, trailing along the track. This was the emergency 'brake' in the event of a run-back and was a simple but vital safety feature. If the rake, by any chance, started to run back down the mine, the pointed stake would dig into the track and derail the hutches. The ropeway at Duddingston Nos 3 & 4 mine was somewhat different in that whilst the hutches were similar, the means of haulage to the surface was effected by a long chain which was wound round the haulage rope and secured by hooks on the hutch.

On the surface, each hutch was tipped into crushers to crush the shale before onward transport to the oil works. The hutches which formed this part of the electrified operation were somewhat larger and were constructed as small hoppers of ¼in. thick mild steel, drilled and riveted to end plates of similar thickness and designed to operate on the 2ft 6ins gauge railway. The underframes were made of oak beams although steel channelling was used on later builds. The floor consisted of a sliding steel hopper door which could be slid open or shut, and when shut with the hutch loaded there was little or no chance of this door sliding open when running. The wheels were of chilled cast iron with six spokes and were 14ins in

ABOVE: No. 5 at the head of two of the four passenger-carrying coaches, at Winchburgh on 31st August 1959.   *Hamish Stevenson collection*

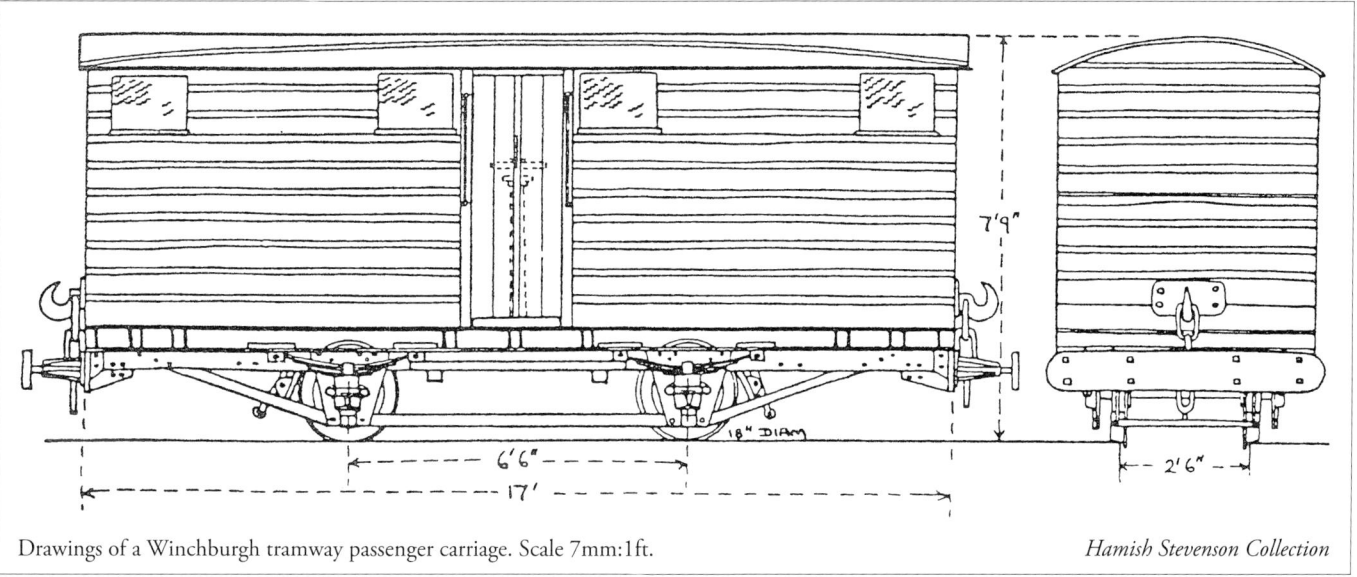

Drawings of a Winchburgh tramway passenger carriage. Scale 7mm:1ft.   *Hamish Stevenson Collection*

Former Oakbank Oil Company electric locomotives Nos 3 and 4 outside the shed at Niddry Castle Oil Works. Note the notch cut into the brickwork above the doorways to accommodate the overhead power wires. The gentleman in the left background with suit and soft hat is the Works Manager, Jimmy Peutherer. *Hamish Stevenson collection*

diameter. The works had a fleet of some 360 hutches, with 240 of these being fitted with Timkin roller bearings, the remainder having the normal oil axlebox bearings. The first hutches were built by the North British Steel Foundry in Bathgate but in the later years the works built their own wagons. Around 220 were in use daily. Hutches were rebodied after a working life of ten years or so and scrapped after twenty years. Each railway hutch carried around 28cwt of shale and trains were made of up to thirty loaded hutches. The capacity of the railway was somewhat in excess of 1,000 tons of shale per day, roughly the daily output of the mines served.

These hutches were also fitted for an overhead wire rope haulage system after running off the electrified lines. The grippers took the form of plain steel channel girders bolted to the ends of the wagons and extending someway upwards above the cape. They were splayed out and slotted in order to catch the overhead chain which pulled the hutches up the incline to the retort benches at Niddry Castle. At each end of the incline working, the haulage chain was raised, allowing it to slip out of the slots and letting the hutch run free.

The other feature which made this railway somewhat unique was the fact that it also operated trains with passenger-carrying coaches. Originally, there were five coaches in total, but one was written off after an accident, leaving four in service for the rest of the life of the railway. Three of these four were built in the workshops at Niddry Castle. They were all of similar design consisting of 17ft long wooden bodies with doors provided on one side only. Lighting consisted of eight small windows located high up in the sides. Each coach had brake gear fitted, controlled by a hand-wheel set centrally in each coach. The coaches were fitted with 18ins diameter wheels running in ordinary wagon axleguards. Buffers had been fitted when new, but were soon removed and replaced by a long wooden bar extending over the length at the end of each coach. This was done to eliminate the significant risk of buffer-locking on the tight curves abounding on the line. The coaches were used to transport the mine workers to and from the mines which were some distance away from the oil works and also oil workers from Newton village to the works.

The electrified railway, always a single line, originally ran right to Duddingston Nos 1 & 2 mine close by Newton Village. This mine was closed in 1941 and the line was then cut back to Duddingston Nos 3 & 4 mine. Later, a new, non-electrified, mile-long tramway was laid from Duddingston No. 3 to the mines at Whitequarries. The hutches on this latter tramway were conveyed to and from Whitequarries and the electrified line via an endless-rope operated haulage system, the electric locomotives being too heavy for the trackwork. Another, later, spur line, which was electrified, was constructed to serve a new mine at Totleywells, lying to the south-east of Duddingston No. 3. The Whitequarries complex was by far the most important in terms of production, producing about 1,000 tons of shale per day.

The author has been told of a practice whereby the locomotive, with train, would be set in motion by the driver, who then dismounted. The train then proceeded on its journey unmanned

THE WINCHBURGH TRAMWAY

RIGHT: A view of the Winchburgh tramway which is interesting in the extreme. Here, a train conveying shale and passenger coaches, hauled by locomotive No. 3, approaches Niddry Castle from Duddingston, but the miners coming off shift appear to have eschewed the dubious comforts of the coaches, preferring to ride 'alfresco' on the shale hutches. What would today's Health and Safety Inspectors make of that?

BELOW: No. 5 arrives in the sidings at Niddry Castle with a train of shale from Duddingston. Once again, a few miners have travelled on the hutches, going home after their shift down the mine.
Hamish Stevenson collection

until a point close by the destination, where another driver would join the locomotive and take control. This story has not yet been validated, but it would not have been impossible!

The actual railway was constructed to robust standards on a well-drained and sound base, and was to be well-maintained throughout its lifetime. The track consisted of light-section, flat-bottomed steel rail throughout, spiked directly on to wooden sleepers (approximately 53 inches by 12 inches) spaced approximately 30 inches (centre to centre) apart. The rails were in both 30 feet and 24 feet lengths and were joined by 18 inch long fishplates. All joints were electrically bonded. All points were the standard switch variety controlled by hand levers located alongside each set. Wooden poles of around 9 inches in diameter set along the line of route carried the overhead electrified wire on wooden brackets and along the top of these poles ran the telephone wires. The lie of the overhead wire 'frogs' could also be altered manually by means of chains operated from the trolley wire posts. At Niddrie Castle, the overhead wires in the works yard were suspended from headspan wires which spanned the five tracks.

Immediately on the north side of Niddry Castle Oil Works, the locomotives were detached at a one-way circular track system and, from the left-hand side of the curve (facing the works), the loaded tubs were conveyed individually by overhead rope haulage (as already described) up an incline and onwards to the retort benches. When discharged, the empties were run back down the incline via the other side (right hand side) of the circular track layout to the five holding sidings and made back up into trains of empties for return to the pits. From the sidings, a line went off to the wagon

Trouble on the line. The Winchburgh tramway at Niddry Castle with a hutch which appears to be 'off the road all wheels' to use a railway expression. Help appears to be at hand and the driver of the locomotive (either No. 1 or No. 2) is obviously attempting to pull the hutch back onto the rails.

repair shops and the main (old) engine shed at the rear of the works, but the operational locomotives were generally housed in a newer running shed constructed alongside the marshalling sidings.

The site of this interchange facility, including the incline, can still be found in part (NT094748) by following a grass path through a gate to the left at the end of the Hopetoun Rows (signposted for Golf Course) in Winchburgh village.

There is still much to be seen of the route of this railway; although the solum is now heavily overgrown, the line of the route can nevertheless be easily followed. It passed up the east side of the original oil company rows and crossed under the B9080 road just before the 30mph speed limit finishes (eastbound) at NT094752. The bridge is still extant although heavily flooded on occasions; considerable remedial work to provide proper stone parapets was undertaken here by the local authority in 2011. The straight line of undergrowth marking the path of the line can be clearly seen running due north, although the tunnel which carried it underneath the high embankment of the NBR Winchburgh Junction to Dalmeny Junction chord line has long been filled in. Immediately to the north of this latter railway line, the line of route has again been cut – this time by the M9 motorway – but beyond that, on the north side, the route is clearly discernible once more. There is a secondary road diverging off to the left from the B8020, some 1.5km north of Winchburgh, leading to South Queensferry via Westfield Farm. At NT094765, the line passed under this road and again the bridge is extant and can still be clearly seen. At about NT094768 the rope-operated tramway to Whitequarries diverged to the left. Here the original line was then cut back to serve Duddingston Nos 3 & 4 mine and the remains of this working can still be seen at, or close by, NT096772.

The rope-operated haulageway then ran in more or less a straight line from east to west to the Whitequarries mines complex. The solum has largely been restored to agricultural land and this, plus considerable road improvements to the A904 road, have rendered any significant traces almost impossible to find, although a double fence line traversing a field on an east/west axis south of, and parallel to, the A904 road marks the former line of route. At Whitequarries, on the old A904 road, now by-passed by a new road, the remains of the buildings (offices etc.) of the Whitquarries mines complex can still be seen and are still in use as commercial premises. The parapets of the road bridge originally carrying the road over the haulageway (NT076773) are still *in situ* just west of the New Hopetoun Garden Centre. BP has now run a pipeline along the solum of the haulageway and, by identifying the fluorescent markers of the route of the pipeline, the point where this tramway crossed under the B8020 road can be clearly identified (NT089771).

14

THE CALEDONIAN RAILWAY

The Caledonian Railway Company having, by 1869, two main lines of railway running east to west through the shale fields, was to provide a valuable but rather secondary service to the shale oil industry in overall terms, when compared with the North British Railway.

The original main line from Carstairs to Edinburgh (now the West Coast Main Line) skirted the southern edge of the main shale fields, but with a connection to the Tarbrax complex just west of Cobbinshaw station. However, high up on the lonely moorland country lying between north Lanarkshire and West Lothian, an area once associated with the Covenanters, significant deposits of ironstone, coal and limestone had been discovered, and the Wilsontown, Morningside & Coltness Railway (later to be taken over by the Edinburgh & Glasgow Railway Company) had penetrated this area in 1841 to exploit the extraction of these minerals.

In 1869 the Caledonian Railway, no doubt with an eye to the rich pickings to be had, constructed and opened the Cleland to Mid Calder Junction chord line to tap into the rich mineral deposits lying in the line of route thereof, this line cutting straight across both the coal fields of North Lanarkshire, through Shotts and Fauldhouse, and also bisecting the shale fields around Addiewell and West Calder. This Caledonian line was built soon after the construction of Addiewell Oil Works where, by their opening, the NBR had already made significant inroads. Nevertheless, between Addiewell station and West Calder station, at around the 28 milepost, a connection giving access to two sidings forming a loop from the Up main line was provided in 1890; this was controlled by Addiewell Oil Works signal box and was certainly extant up to the 1930s. These sidings were provided for inwards tank traffic, containing, in the main, crude oil retorted elsewhere and conveyed

Oakbank Oil Works with Oakbank Oil Company rail tank wagon No. 13, which was one of many registered (in 1892) to work over the Caledonian Railway metals between Oakbank Oil Works and Pumpherston Refinery.

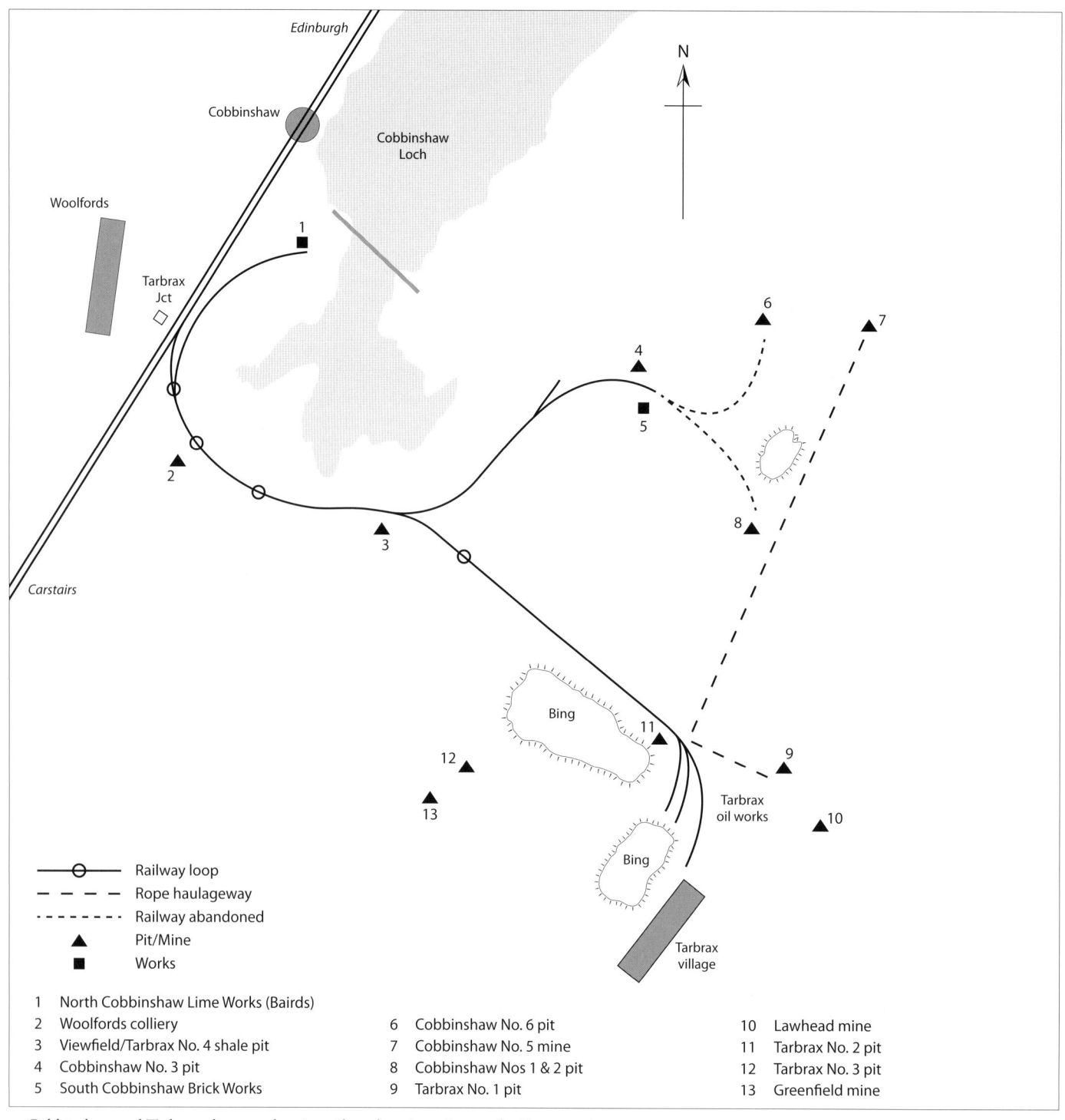

Cobbinshaw and Tarbrax: diagram showing oil works, pits, mines and rail connections.

in rail tanker for refining at the main works. Finished products also formed outwards traffic here.

The Caledonian also served the Oakbank Oil Company's oil works via a trailing connection from the Up loop at Mid Calder Junction. The final point of access to the shale fields came at Camps Junction where a trailing connection led off to run northwards to the Raw Camps Quarry, immediately east of East Calder village. Here, the Caledonian had a goods station (Camps) and an exchange facility with the NBR, this latter company accessing the quarry via a branch line from Uphall Junction. The branch was to be an important transfer link with crude oil from Tarbrax, and other products, being transferred to the NBR for onward working to Uphall Junction via the exchange facility provided at Camps.

The Caledonian Railway maintained a list of traders' wagons (private owner wagons) registered with the company between 1889 and 1920; rail tanks owned by the Oakbank Oil Company, the Pumpherston Oil Company, Hermand Oil Company and Tarbrax Oil Company were all included therein. Strangely, neither the Addiewell Oil Company nor Young's Paraffin, Light & Mineral Oil Company appeared in this list of private owners, although Addiewell was certainly, as stated, served by the Caledonian for some time via a private siding lying just east of Addiewell Station.

The Oakbank Oil Company appears to have been the largest owner of wagons so registered and in the years before 1910 some 101 tank wagons alone were registered with the Caledonian Railway. Again, Oakbank was unique in having a fleet of 40-ton bogie tank wagons in use. By 1910, only thirty-three wagons were registered but in 1912 there were fifty-seven on the register.

East Hermand Shale Company of West Calder purchased five wagons from the Caledonian Railway in 1872, upon which it mounted its oil tanks. This was a small oil works at Shuttlehall, lying just to the south-east of West Calder village and served by a branch line from Limefield Junction. The crude oil produced here was taken to Linlithgow (Bridgend Oil Works) for refining.

Hermand Oil Company had but nine wagons registered between 1892 its closure in 1902. It has been stated by one source that this company acquired additional tank wagons from the Linlithgow Oil Company in 1902, the year in which this latter company went into liquidation; this is not wholly unlikely since the crude from Hermand also went to the Bridgend Oil Works for refining, but it certainly is strange if true, since the latter oil works also closed at the same time.

Tarbrax Oil Company had twenty-one wagons registered between 1908 and 1910, in which year it was taken over by the Pumpherston Oil Company.

Tarbrax Junction (Cobbinshaw) to Tarbrax Oil Works

Since all the main seams outcropped at and around Cobbinshaw, this became a major area of development for both the mining and retorting of shale, concentrating on the productive Fells seam. In 1861 a small oil works was established at South Cobbinshaw, soon to be followed by a larger oil works at Tarbrax. Shale was also worked at North Cobbinshaw (NT009584) and a small oil works was constructed there in around 1867.

It was no surprise then, that the entrepreneurs responsible for the development of this new industry approached the Directors of the Caledonian Railway in 1866, requesting that the works be connected with the national railway system by means of new branch lines. Discussions continued apace and the Caledonian Railway, with an eye to capturing new traffic to rail, sought Parliamentary approval to construct two branch lines, the first to Tarbrax, the largest of the oil works (NT026557), and the second to the smaller South Cobbinshaw Oil Works (NT027584).

As far as can be determined, a proposed branch line to provide a connection to Cobbinshaw North (Nos 2, 3 and 4 pits and oil works) was never built – beyond the initial provision of a connection and what was to remain a short siding on the Down side of the line, close by Cobbinshaw signal box, presumably to serve the small Cobbinshaw North Oil Works. In saying that, however, there is fairly clear evidence of proper earthworks (still walkable) having been constructed all the way to the site of Cobbinshaw North pits and works; whether rails were ever laid down is not known, although it is highly unlikely. When the Cobbinshaw North operation was dissolved in 1881, the siding became a public goods siding known as Mungle's siding.

Under the Caledonian Railway (Branches and Stations) Act dated 1867, approval was given for two railways:

A Railway to be called the Tarbrax Branch, One mile Seven furlongs and Fifty Five Yards or thereabouts in length commencing by a Junction with the Main Line of the Caledonian Railway from Carlisle to Edinburgh at a point where that line crosses the Boundary between the Counties of Lanark and Midlothian and terminating near the Southern end of the Retorts of the Tarbrax Oil Works in the said Parish of Carnwath.

and

A Railway to be called the South Cobbinshaw Branch, Seven Furlongs or thereabouts in length commencing by a Junction with the said Tarbrax branch about Two and a Half Furlongs Northward from Greenfield house in the Parish of Carnwath and terminating about Five Furlongs Eastward from the Farm Steading of South Cobbinshaw in the said Parish of West Calder.

By means of a facing connection in the Up main line, the Tarbrax Branch line, 1 mile 71 chains (1,562 yards) in length left the Edinburgh–Carstairs (WCML) line immediately to the west of Cobbinshaw station at Tarbrax Junction (NT007573) and curved away to the south-east. The South Cobbinshaw Branch diverged to the east (NT009568). Both branch lines were constructed on level moorland requiring no significant earthworks in construction. The Tarbrax Branch was opened for traffic in June 1869, but apparently the Directors of the Caledonian Railway Company had second thoughts about the viability of the South Cobbinshaw Branch as a Statutory Railway and the decision was taken – a mere two years after obtaining Parliamentary approval for construction – to seek abandonment, and this was duly given under The Caledonian Railway Abandonment Act of 1869.

Around this time, oil works and mines were changing ownership on a frequent basis as the industry consolidated and in 1869 the South Cobbinshaw Oil Company started operations at their South Cobbinshaw Works. The Directors of the Caledonian Railway again had a change of heart, no doubt caused by this development, and thus completed construction of the planned South Cobbinshaw Branch, no longer as a branch line but now merely as a long siding, at a cost of £2,000. Their confidence was to be misplaced for both mining and oil production ceased at South Cobbinshaw in 1871. The brickworks associated with the works continued to operate, however, and to generate some traffic to rail, under the title of the South Cobbinshaw Fireclay and Brick Company, until it too was abandoned in 1889. The South Cobbinshaw siding itself, although long disused, lay *in situ* until finally recovered by the LM&SR in 1936. A 2ft tramway connected Cobbinshaw No. 3 mine with the original Cobbinshaw South Oil Works.

Tarbrax Oil Works was also to change ownership on several occasions and a new company, the Tarbrax Oil Company, was incorporated in 1904. Under their ownership, the old works were demolished and a larger, more modern works constructed. A new village (Tarbrax) was also built to house miners and oil workers.

Access to the Tarbrax Branch was controlled by Tarbrax Junction signal box, 1 mile 1,628 yards from Wilsontown North signal box and 1,474 yards from Cobbinshaw signal box. Originally there was a loop on the main line, but by 1911 this had been reduced to a siding. There was a loop at the commencement of the branch, and a further three loop sections were contained within its length, but all marshalling was carried out at Cobbinshaw.

On the Tarbrax Branch proper, a trailing connection and loop siding served Woolfords colliery, and the branch also served sidings at Viewfield shale pit *en route* to the oil works. There was a locked (across the railway) level crossing on the Woolfords line and a similar one on the Tarbrax line, both of which were trainman (brakesman) operated.

In 1899, the Lanarkshire steel masters, Bairds Brothers, commenced a mining operation for coal, fireclay and limestone at North Cobbinshaw (NT012576 and not to be confused with the other older North Cobbinshaw shale operation). This operation lay on the south side of the main line, but to the north of Cobbinshaw Loch. A siding some 372 yards long was constructed to serve this new operation and was provided with a run-round loop at the works end. The connection to this siding trailed into the Tarbrax Branch between the main line junction and the first loop section. The total cost of construction, some £400, was borne by Bairds. Traffic ceased in 1910 but the siding lay *in situ* for a further number of years. A gate secured access to this siding.

The main spur of the branch line ran into Tarbrax Oil Works close by the houses of Tarbrax village. A tramway ran from the oil works to Cobbinshaw No. 2 pit and Cobbinshaw No. 5 mine (the bricked up mine shaft at Cobbinshaw No. 5 can still be seen today, NT036572). The Tarbrax Branch terminated with a run round loop lying close to the housing in the north side of the village.

As a consequence of a strike across the shale industry in 1925, Tarbrax Oil Works was closed and the 600-strong workforce left out of work. All traffic to and from the branch disappeared at a stroke but the line was to remain *in situ* until 1936 when, in the April of that year, the LM&SR Scottish Local Committee moved to have all rail infrastructure removed and Tarbrax Junction signal box closed in 1937.

Tarbrax Oil Works never had an internal locomotive and the branch was worked solely by Caledonian Railway engines from Carstairs shed under One Engine in Steam with Train Staff Arrangements. Carstairs engine diagram No. 8 (the Tarbrax Run) worked the branch – taking the traffic/empties left off at Cobbinshaw into the works, shunting all sidings, and working out any loaded tanks etc., leaving them at Cobbinshaw for onward working. There was a propelling authority for inwards movements to the oil works which required a brake van leading. Outwards traffic, consisting of coal from the pits and crude oil in rail tanks, was picked up at Cobbinshaw and worked forward each morning by an Edinburgh Dalry Road engine returning with the 5.00am goods from Carstairs. Inwards traffic, generally empty rail tanks and coal empties, were again left off at Cobbinshaw by the Edinburgh to Carstairs goods workings. The crude oil from Tarbrax was taken by rail tanks to Camps Goods and exchanged with the NBR for onwards working to Pumpherston for refining.

Anecdotal evidence suggests that there was, on one occasion only, a special passenger train working over the Tarbrax Branch to convey villagers on a day's outing to Edinburgh, this taking place sometime in the first decade of the twentieth century, but no corroboration has been found.

Bairds & Scottish Steel Company No. 6, an 'Ogee' tank engine, seen here at Straiton in 1952. *Courtesy: Jeff Hurst*

THE CALEDONIAN RAILWAY

CARSTAIRS ENGINES.

Time leaving Engine Shed	Name of Run.	Particulars of Work.
		No. 1. *Special Class Tank Engine.*
5.0 a.m.	Shawfield,...	To leave Carstairs at **5.15** a.m. with Goods Traffic for Wishaw (Central) to connect at Law Junction with the 1.0 a.m. ex Buchanan Street, and be employed during the day working down empty Wagons to Mauldslie and Stravenhouse Collieries, and bringing loaded Wagons back to Law Junction; also make trip to and from Castlehill Colliery; making one or two runs to Law Depot with Wagons for the South; then take a Train of South Wagons home, or assist home some of the other Engines.
		✱ No. 5. *First Class Engine.*
5.50 a.m.	Limestone Engine, ...	To leave Carstairs at **6.10** a.m. with a Train of Mineral Wagons for Oakbank Oil Works and Camps; call at Oakbank; deliver all Traffic and bring out all empty Wagons and Goods from there; leave the empty Wagons at Midcalder Junction, and the Goods at Midcalder Station; shunt that Station; then proceed to Camps and deliver all Traffic for N. B. Line; shunt the Station, take forward the South Traffic, and lift all Limestone Traffic, leaving it at Midcalder Station; return to Oakbank and work Oil Traffic from there, leaving West-going Traffic at Midcalder Junction, and East-going Traffic at Midcalder Station; return and lift empty Wagons and South-going Traffic from Midcalder Junction and Harburn; leave empty Wagons at Wilsontown Junction; make up Train and proceed to Carstairs.
		✱ No. 6. *First Class Engine.*
7.30 a.m.	Wilsontown, No. 1,...	To leave Carstairs at **7.45** a.m. with a Train of Goods and empty Wagons for Wilson-town, deliver on Branch where necessary; work the 9.15 a.m. Passenger Train to Auchengray, and 9.35 a.m. Goods Train, Auchengray to Wilsontown; work Wilsontown Colliery, No. 9; work 11.53 a.m. Passenger Train, Auchengray to Wilsontown; work Climpy Colliery, making up Train at Haywood, No. 14; proceed to Cobbinshaw, daily except Saturdays, for Train of empty Wagons, and deliver them on Wilsontown Branch where necessary; then work following Passenger Trains:—3.15 p.m. "S O," Wilsontown to Auchengray; 4.15 p.m., Wilsontown to Carstairs; 5.14 p.m., Carstairs to Wilsontown; 6.5 p.m., Wilsontown to Carstairs.
		No. 7. *First Class Engine.*
5.40 a.m.	Castlehill,...	To leave Carstairs at **6.5** a.m. with a Train of empty Wagons for Law Junction, and load Traffic for Garriongill; lift a Train for Braidwood, calling at Law Depot, Castlehill, and Carluke; return to Carluke and shunt Station and Brick Works; call at Castlehill Iron Works and Foundry; take Train to Law Junction; then work Train home to Carstairs.
		No. 8. *First Class Engine.*
6.30 a.m.	Tarbrax, ...	To leave Carstairs at **6.40** a.m. with a Train of Goods and Mineral Traffic for Carnwath, Cobbinshaw, and Tarbrax Branch; shunt all Sidings on Branch, working outgoing Traffic to Cobbinshaw; work Train of empty Wagons, Cobbinshaw to Haywood, Nos. 4 and 14 Pits; work Train of loaded Traffic from there to Cobbinshaw; lift Traffic for and work Tarbrax, taking the Traffic to Carstairs and Lanark Oil Works, and return to Carstairs with empty Oil Tanks. NOTE.—Carnwath to wire Inspector, Carstairs, at 6.0 p.m. the previous night what Traffic there is to lift.

Extract from a Caledonian Railway working timetable showing the Carstairs engine workings including those working Tarbrax oil traffic.

BENHAR JUNCTION TO BENHAR OIL WORKS AND POLKEMMET COLLIERY

This branch was constructed by the Caledonian Railway to tap the rich mineral workings lying high up on Polkemmet Moor (also known as Fauldhouse Moor) on the Lanarkshire/West Lothian county march. The actual junction lay at the 23 milepost (from Glasgow) on the Mid Calder to Cleland Junction chord line of 1869, approximately one mile west of Fauldhouse North (Caledonian) station. The connection to, and branch proper, were situated on the Up side at Benhar Junction, where stabling sidings were provided. The single line branch, worked under the One Engine in Steam arrangements, proceeded eastwards in the first instance before swinging northwards up onto the high moorland hinterland.

The branch served the short-lived Benhar Oil Works (Simpson's Caledonian Oil Company) and the Benhar Paraffin Oil Works, both separate concerns, the former closing in 1874 and the latter in 1895. There was also an interchange facility with the NBR lines which served the pits high up on these moors. It also served several collieries such as Braehead, Starryshaw and Cultrigg, and later (post First World War) was extended eastwards to serve the new, large, Polkemmet Colliery (the Dardanelles pit) close by Whitburn.

The working of this Polkemmet Branch was latterly the domain of Motherwell-based locomotives, working trip trains to and from Mossend Yard. It was somewhat unusual in that a pooled fleet of 24½ ton steel mineral wagons made up all trains running from the colliery, with much of the coal destined for the Ravenscraig steel complex at Motherwell. Polkemmet Colliery did, however, have its own large fleet of steam 'pugs' which worked the coal up the heavy gradient between the colliery and the exchange sidings lying high up on the moor, taking empty wagons back to the colliery.

Bairds & Scottish Steel Company No. 8 seen here at Straiton in 1952. This was a strange engine, apparently having been constructed by the Shotts Iron Company at their Shotts workshops from parts held in stock.

Courtesy: Jeff Hurst

Muldron Junction to Muldron Pit

Muldron Junction was on the Caledonian Railway's Glasgow to Edinburgh via Shotts line, south of Fauldhouse (North) (see Map E, p. 161). The main line here swung to run due east after crossing over the viaduct across Breich Water (NS935600), the junction having a trailing connection from the Down main line at Mouldron (or Muldron) Junction signal box. The single line branch, some 76 chains in length, ran southwards to Mouldron West where there was an ironstone pit, ironstone having been mined here in great quantities since 1802. The line was provided with a loop for rounding purposes about half way along its length and was opened in 1871. It was out of use by the early 1900s.

Levenseat Junction to Handaxwood Oil Works

A chord line was constructed by the late 1800s to connect with the NBR branch from Crofthead to Levenseat, which passed over the Caledonian lines (NS943599) nearby (see Map E, p. 161). Controlled by Levenseat Junction signal box, it branched to the left from a facing connection on the Down line, thus giving the Caledonian Railway access to the Levenseat limestone complex on the south side of the line. This line, which jointly provided access for both companies to Levenseat, was worked under One Engine in Steam arrangements – the NBR (L&NER) worked the line between 12.00 and 13.00, with the Caledonian (LM&SR) taking over working between 13.00 and 14.00.

At around the same time, another chord line – from a trailing connection in the Up line – struck of to the north-west (Up side) at this same location, controlled from the same Levenseat Junction signal box. It again connected the Caledonain Railway with the NBR line from Crofthead, forming a triangular junction. This short branch was provided to give the Caledonian its own access, with sidings, to the short-lived Handaxwood Oil Works lying immediately to the north side of their main line.

The Caledonian Railway West Calder Loop

The Cleland/Mid Calder Junction chord line opened in 1869, shortening the overall distance by rail between Glasgow Central and Edinburgh Princes Street to just 46½ miles (see Map A, pp. 20–21 and Map B, pp. 24–25). At that same time, the Caledonian Railway Company (plans dated 1866/67) built a loop line some 4 miles 51 chains (1,122 yards) in length to serve the developing shale oil industry in the West Calder area. This 'West Calder Loop Line', as it was to be referred to thereafter, was opened on the 1st January 1869. At Woodmuir Junction (NS 977615) – a remote moorland location sitting between Addiewell Station and Breich Station at 18 miles 1,645 yards from Edinburgh Princes Street – a single line of railway diverged south-westwards from the main line (Down side) to serve United Collieries' small Woodmuir colliery (locally known as 'Blinky Pit', so small that if one blinked whilst passing, they would miss it) and the earlier Rashiehill coal pits. Woodmuir Junction had both Up and Down side sidings. The

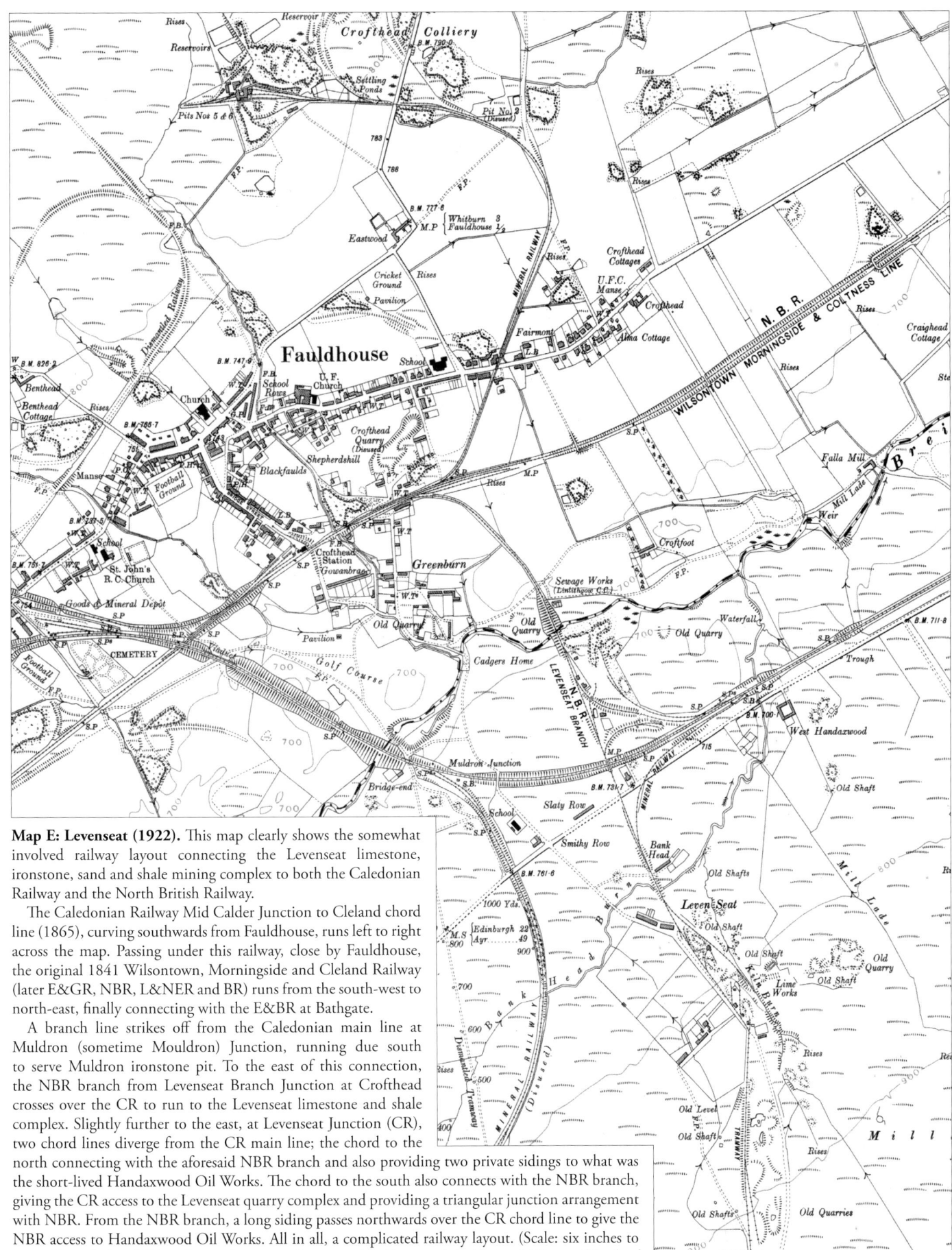

Map E: Levenseat (1922). This map clearly shows the somewhat involved railway layout connecting the Levenseat limestone, ironstone, sand and shale mining complex to both the Caledonian Railway and the North British Railway.

The Caledonian Railway Mid Calder Junction to Cleland chord line (1865), curving southwards from Fauldhouse, runs left to right across the map. Passing under this railway, close by Fauldhouse, the original 1841 Wilsontown, Morningside and Cleland Railway (later E&GR, NBR, L&NER and BR) runs from the south-west to north-east, finally connecting with the E&BR at Bathgate.

A branch line strikes off from the Caledonian main line at Muldron (sometime Mouldron) Junction, running due south to serve Muldron ironstone pit. To the east of this connection, the NBR branch from Levenseat Branch Junction at Crofthead crosses over the CR to run to the Levenseat limestone and shale complex. Slightly further to the east, at Levenseat Junction (CR), two chord lines diverge from the CR main line; the chord to the north connecting with the aforesaid NBR branch and also providing two private sidings to what was the short-lived Handaxwood Oil Works. The chord to the south also connects with the NBR branch, giving the CR access to the Levenseat quarry complex and providing a triangular junction arrangement with NBR. From the NBR branch, a long siding passes northwards over the CR chord line to give the NBR access to Handaxwood Oil Works. All in all, a complicated railway layout. (Scale: six inches to one mile, scaled to 85% to fit page.) *Courtesy of the Trustees of the National Library of Scotland*

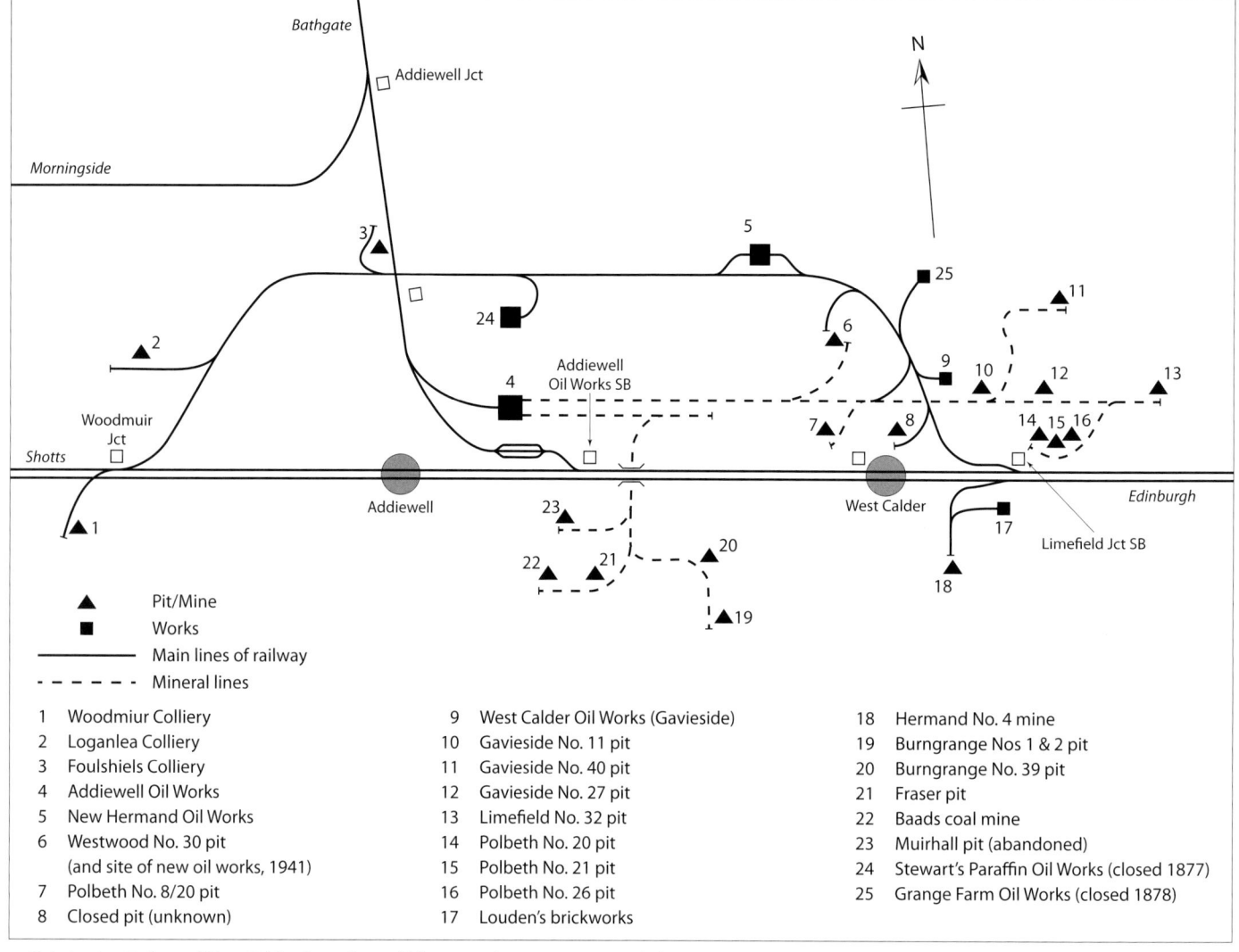

Caledonian Railway: West Calder Loop Line. (Schematic drawing: not to scale.)

signal box, lying on the north side of the main line, was probably built and opened in 1869 at the same time as the chord line.

On the Up side at Woodmuir Junction, a single line of railway diverged to the north-east (the West Calder Loop proper); crossing the intervening moorland it then swung eastwards round behind the mining village of Loganlea. At that point, some 45 chains (990 yards) from Woodmuir Junction, a trailing junction gave access to Loganlea Colliery (the Dykes pit) (NS978620), also owned by United Collieries, where much of the coal mined was taken to produce gas to heat the retorts at Westwood Oil Works.

The loop line continued north-eastwards, crossing Breich Water on a viaduct just north of the houses in Loganlea before swinging due eastwards. The NBR branch line from Addiewell Junction (on the Bathgate to Morningside Branch) to Addiewell Oil Works was crossed on the level (Cuthill Crossing) 990 yards further on, at a point just to the south-east of Stoneyburn (NS 986629), before passing under the intersection of the B7015 and B792 roads at Cuthill. There was a signal cabin provided at the flat crossing to control movements. There was also a trailing connection from the Caledonian loop swinging northwards to serve Foulshiels Colliery (NS977634), another colliery owned by United Collieries.

It is known that the West Calder Loop had been severed by 1917, and most likely considerably earlier, with the Caledonian rails in the flat crossing having been removed in their entirety. A Caledonian Railway Permanent Way Inspector walked the route on the 2nd August 1917 and recorded:

Rails and permanent way in good condition between Limefield Junction and Westwood pit (used daily by Edinburgh engines).

Rails and permanent way in fair condition between site of Hermand oil works and Limefield Junction (not used).

Rails and permanent way in defective condition between site of Hermand oil works and crossing with NBR line, rails very rusty and sleepers very much decayed. Young shrubs and trees growing between the rails, some of them 8 feet in height and indicates that the line has not been used for 10 to 15 years.

Rails and permanent way between Woodmuir Junction and level crossing with NBR line are in good condition and is worked over several times per day by both our [Caledonian] engines and by United [Collieries] engines which bring traffic from Foulshiels and places it in the sidings at Woodmuir.

The section of the loop line between Woodmuir Junction and Cuthill Crossing was worked for some time thereafter under a Time Arrangement by both the Loganlea Colliery pug and Caledonian (and LM&SR) engines – each company had set times for sole

United Collieries Loganlea Nos 1 & 2 pit locomotive. Built by Gibb & Hogg in 1899, it is seen here resting between turns at Loganlea Colliery, a colliery which supplied coal to the shale oil industry. *Hamish Stevenson collection*

occupation of the line. This was different from the early days when a through route existed, then working had been by Train Staff between Woodmuir Junction and Cuthill Crossing with a separate Train Staff from there to Limefield Junction.

It is not known when the portion of line between the connection to Loganlea Colliery and that to Foulshiels Colliery was abandoned, but it was certainly well before Foulshiels Colliery finally closed in 1957; it is probable that the line abandonment occurred after the Grouping (1923) when, under the LM&SR, it was quite likely conceded that all Foulshiels coal traffic should thereafter be worked by the L&NER via the NBR connection out to Bathgate Yard. Certainly there is no mention of the Foulshiels connection, or working thereto, in the 1937 LM&SR General Appendix.

The loop then continued eastwards on the north side of the minor road from Stoneyburn crossroads, to Wester and Mid Breich and, at 1 mile 17 chains from the flat crossing, sidings served the New Hermand Oil Works (NT012649), closed in 1903. There is evidence of an early facing (towards the east) connection and branch line leaving the loop line close by Westwood Rows (NT000637), swinging south on a raised embankment before terminating near Breich Dykes Farm (NT002633). It is believed that this was the site of the original Stewart's Westwood Paraffin Works; however, the absence of any evidence of a shale bing on or near this site – normal for an oil works, even on the earlier OS maps – makes positive identification difficult. The West Calder Loop Line then swung south to pass under the B7015 road at Easter Breich where a trailing connection (from the Woodmuir direction) gave access to the original Westwood No. 12 (NT009638) and Westwood No. 13 (NT0157644) shale pits, both gone by 1897, and later to the 1941 new Westwood Crude Oil Works complex.

A facing connection just beyond led to the former West Calder (Gavieside) Oil Works (closed 1880) (NT029650), with a further facing connection to Addiewell pit (probably Westwood No. 30/ Polbeth No. 11), again abandoned by 1900. During the short life of the Charlesfield (Grange Farm) Oil Works (1864–75), these works, lying close by Guns Green, were served by a trailing connection leading to a short branch, or more likely, just a long siding off the loop line (NT018646), all traces of which have long disappeared. A right-hand facing chord spur connected the loop to the internal mineral railway that ran from Addiewell to the Polbeth pits, a line which the loop then crossed over. Running under the next two minor roads at Breistonhill House, it crossed West Calder Burn (known locally as the Yellow Burn) on a small viaduct, passed under the A71 road between West Calder and Polbeth, before rejoining the Caledonian main line via a trailing connection on the Up side of the main line, at Limefield Junction (15 miles 396 yards from Edinburgh) (NT028637).

The line between Hermand Oil Works and Limefield Junction was still in good condition and well used in 1917, as shown in the above report, and, indeed, continued in use serving the new (1941) Westwood Oil Works of Scottish Oils Ltd until the works finally closed in 1962.

Whilst Hermand Oil Works closed in 1903, a siding known as Hermand siding, which had originally served the oil works, was left *in situ* and used for local goods traffic and farmers' traffic. The section of line between Limefield Junction and this siding was worked by a train staff arrangement, the staff being held in Limefield signal box and marked 'Limefield Junction and Hermand Siding'. This same staff was also used for the Westwood Pits Branch.

At Limefield Junction, the siding on the Up side was capable of holding forty-eight wagons. Also at Limefield Junction, a trailing connection on the Down side crossed over Harwood Water (also referred to in Caledonian documentation as Hermand Water) and swung south and west to serve the original oil works of Old Hermand (Shuttlehall) (NT029625) and, latterly, Hermand No. 4 mine (NT031628) on the south-east side of West Calder. Just after leaving the main line, there was a one-time private siding serving Messrs Louden's brick works, accessed by a bridge over Harwood Water. This bridge was, even in Caledonian days, in frail condition and locomotives of the Caledonian Railway were prohibited from crossing it. Nevertheless, the private siding, despite its inherent bridge problems, survived for some considerable time and was certainly still in operation in 1915. This branch was also worked by train staff key held by the signalman in Limefield Junction. In latter days, when the branch was extended to serve Hermand No. 4 mine, it was worked under One Engine in Steam arrangements.

It is known from evidence gathered that the West Calder Loop Line had a very short life as a through line. Indeed, it is doubtful if any significant degree of through working, if any at all, ever took place on a regular basis, and it appears that in its short life, working was more or less confined from and to either end.

The Caledonian WTT of 1886 shows Train No. 30, the 3.25pm West Calder to Edinburgh, being required to work the loop branch, obviously from the Limefield Junction side but with no further detail. This same WTT contains the special instruction, viz:

LOOP LINE FROM LIMEFIELD JUNCTION TO STONEYBURN SIDING – this loop line will be used by Young's Paraffin, Light & Mineral Oil Company's Engine from 6.00am to 9.00am, and no Caledonian Engine must be allowed to enter the Loop Line during these hours.

By 1904, the Edinburgh Dalry Road engine working train No. 2, the 5.53am Gorgie to West Calder and Woodmuir Junction, had a long day, shunting Limefield Junction on the outward journey to put off Hermand traffic and then working Loganlea Colliery from Woodmuir; then, on the return leg, being required to service Hermand Oil Works and brick works before arriving back at Gorgie at about 4.35pm. By this time, however, the loop line was certainly closed to through traffic. The two main oil works standing on the line of route were gone by 1903 and would have offered little, if any, traffic after 1900. The Engineer's report (and drawing) referred to above suggests abandonment by about 1902 and the author thinks that this suggestion is probably accurate.

The Caledonian did, as we have seen, work Loganlea Colliery; earlier, one or more of the Caledonian trip workings also served Foulshiels Colliery at Stoneyburn which was rail connected to the loop, probably up until the Grouping. Even by 1907, the section between Woodmuir Junction, Loganlea Colliery and Foulshiels Colliery was worked on a shared time basis, with the United Colliery pug having access from 06.00 to 08.30, 10.00 to 11.00, 13.00 to 14.30 and 15.30 until 18.00. The Caledonian engine was allocated 08.30 to 10.00, 11.00 to 13.00, and 14.30 until 3.30. This time sharing was to be perpetuated for many years but with the times changing slightly as necessary. However, whilst the through loop was abandoned by 1900 or thereabouts, the Caledonian engines continued to work between Woodmuir Junction and Cuthill Crossing (probably for Foulshiels Colliery traffic) again on a timesharing basis and the footnote in the Caledonian 1913 WTT instructs that:

These times must be strictly observed by all concerned. During the times set apart for the working of the Caledonian company's engine as above, the line between Loganlea Colliery and Cuthill Crossing is to be kept clear of the United Collieries Company's engine.

From the Permanent Way Inspector's written report referred to above, we know that Foulshiels colliery was still being served by the Caledonian in 1917, all traffic being taken out to Woodmuir Junction. In BR days, coal from Loganlea colliery was worked out to the BR sidings at Woodmuir Junction by the NCB pug, and this continued until the colliery closed in 1959.

Limefield Junction was another point where the access was shared on a Time Arrangement basis, with the Caledonian (LM&SR) having sole access between 09.00 until 06.00 the following day. The Young's pug was limited to access between 06.00 and 09.00 daily. As the number of pits and oil works served on the loop line fell into decline, so the working arrangements were altered. By LM&SR days, the working arrangements revealed that the LM&SR engines had access from 09.00 (08.00 Mondays only) until 17.00. The Scottish Oils Ltd pug had access to Limefield Junction (and Hermand mine) between the hours of 17.00 until 09.00 the following day. Each user was responsible for ensuring that no conflicting movement could take place in the event of the West Calder Loop Line not being cleared at the agreed times.

BR locomotives continued to serve both Hermand No. 4 shale mine and Westwood Oil Works, BR actually moving the trains of shale from the mine to the works on a daily basis. This was undertaken by Dalry Road trips 105 and 112 until closure of the oil works in 1962. The author, as a school boy, had several opportunities to ride in the brake van of this Edinburgh trip working between Hermand No. 4 mine and Westwood Oil Works.

Addiewell Oil Works

The Caledonian Railway's access to Addiewell Oil Works, although the main line paralleled the southern boundary of these works, was to be confined to three lines of railway for loaded outward rail tanks and inward empties (see Map A, pp. 20–21). There was a trailing main line crossover immediately on the west side of the works connection. The siding on the works side of this connection split to form the loops which at the western end came together into a single line which butted end-on with the NBR Addiewell Junction to Addiewell Goods line. The access was controlled from Addiewell Oil Works signal box (NS998625), opened in 1869 and situated on the Up side facing south. The signal box was burned down in 1936 and the rail facility was most likely abandoned at that time, since the NBR provided the main rail services to Addiewell and had the lion's share of traffic in any case.

Mid Calder Junction to Oakbank Oil Works

Interestingly, Mid Calder Junction was somewhat unusual in that the nomenclature of the running lines changed there. On the Holytown deviation (the Mid Calder–Cleland 1869 line), trains running towards Edinburgh were Up direction trains and, conversely, those running towards Glasgow were Down trains. At

Mid Calder Junction, entering the West Coast Main Line meant that Up trains from Holytown suddenly were running in the Down direction, and Up trains from Edinburgh for Holytown then ran forward on the Down line from Mid Calder Junction. This came about by the long established tradition whereby lines leading towards the UK capital, that is, London, were always Up lines, whilst those running from London were the Down direction lines. At places like Mid Calder Junction, this was never a point of confusion amongst railwaymen, but amongst the population at large, and indeed many railway enthusiasts, it has been a matter of ongoing confusion over the years.

Initially, the connection to the Oakbank Oil Works was controlled from Oakbank signal box, situated on the south side of the Down (Holytown) line in the vee between the West Coast Main Line and the Holytown line which came together at Mid Calder Junction, immediately to the east (see Map F, p. 166). This Oakbank box was opened in September 1899 and closed in December 1935, following closure of the oil works in 1932. There was also an Up loop between Oakbank Oil Works and Mid Calder Junction signal box and from this Up loop a short branch line, running due west and parallel to the main (Holytown) line, crossed a minor road before swinging north into Oakbank Oil Works (NT077664). Four sidings lay off this loop line and the Caledonian exchanged all traffic, inwards and outwards, with the Oakbank Oil Company's own pugs, of which they had five over the years. All working between the works and these sidings was carried out by the Oakbank pugs and their drivers and firemen (referred to by name) were specially, and individually, authorised to work within the CR sidings at Mid Calder Junction. It is presumed that, like the sidings at Limefield Junction, Caledonian engines and Oakbank engines would have allocated times in which they could work but no such details have been found.

This line between the loop and the oil works was removed sometime after closure of the oil works and Oakbank signal box. It was reinstated in the 1970s by BR from Mid Calder Junction, using the former road bridge which was still extant, and new sidings were laid into Contentibus bing, one of the two bings created at Oakbank over the years. From here, two loaded trains carrying spent blaes were run through to Glasgow (General Terminus) daily and, in the seven years of the contract with the Scottish road haulage company, W.H. Malcolm Ltd, some 2 million tons of shale were transferred to Glasgow. Access was controlled by Mid Calder Junction signal box and the line was worked as a long siding.

This oil works generated a lot of traffic in the years of operation, and over this period was well served by the Caledonian Railway. In 1907, the working of Hamilton engine No. 12 – the 8.45pm weekdays, 10.10pm Sunday, Strathaven Junction to Edinburgh Crewe Junction goods turn – being booked, on Monday mornings only, to uplift all oil tanks at Oakbank for NBR stations, and take them forward to Camps Junction. Train No. 37 – the 4.40pm Edinburgh to Glasgow (direct) Goods – was booked to uplift traffic from Oakbank with, if there were too many wagons, priority being given to shipment or urgent traffic. Oil traffic for Paisley and the south side was then put off at Benhar Junction. By 1913, Train No. 29 – the 3.50pm SX Gorgie to Glasgow Buchanan Street – was booked to clear all traffic at Oakbank with again the proviso that shipment or urgent traffic was given priority. South side traffic was put off at Mossend. On Saturdays, train No. 29 – the 3.50pm SO Gorgie to Glasgow Buchanan Street – took up the working of traffic out of Oakbank.

Overall, whilst the NBR might have had the lion's share of shale oil traffic, nevertheless investigation of old Caledonian Railway working documents reveal just how much traffic was being generated by this industry.

CAMPS JUNCTION TO RAW QUARRY AND CAMPS GOODS

East of East Calder village, a thick seam of Burdiehouse limestone was discovered under the lands belonging to the Earl of Morton. Quality limestone was in great demand for both agricultural purposes and, more especially, in the iron smelting industries of the lower Clyde valley; recognising the potential for some considerable traffic flow, the directors of the Caledonian Railway moved quickly to obtain authorisation to construct a new branch into the limestone workings. Under the authority of the 1865 Cleland and Midcalder (*sic*) Railway and Branches Act, the company moved,

> *to construct a Branch Railway called Railway No. 7 [in the Act] commencing by a junction with the Company's said main line at or near the level crossing of that line, in the Parish of Kirknewton from the road leading from Kirknewton by Hillhouse towards Hatton House and terminating about eighteen chains northwards from the Farm Steading of Burnhouse in the same Parish.*

East of Mid Calder station (now renamed Kirknewton), at Camps Junction (NT108674) a trailing connection, with two loop sidings, ran west to the north of Kirknewton Mains Farm before swinging north and east of Burnside Farm to Camps Goods station immediately west of the Raw Camps quarry complex (NT118674) (see Map F, p. 167). The branch descended on a slight gradient and terminated at Camps Goods Station, crossing the NBR branch from Uphall Junction at a point some 286 yards from its termination. At Camps Goods, the line split into two sidings, one of which served the through goods shed.

On the approach to Camps Goods, a facing connection diverging to the left served the Raw Camps mine (NT095680) which worked the seams of limestone on behalf of the Coltness Iron Company. This limestone, used as a flux in the iron smelting industry, went

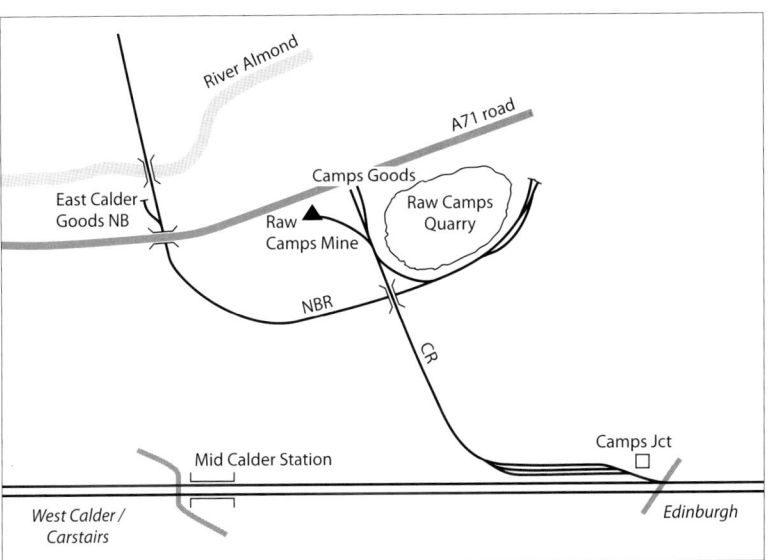

Schematic diagram of Camps Junction and the Camps Branch.

THE SCOTTISH SHALE OIL INDUSTRY & MINERAL RAILWAY LINES

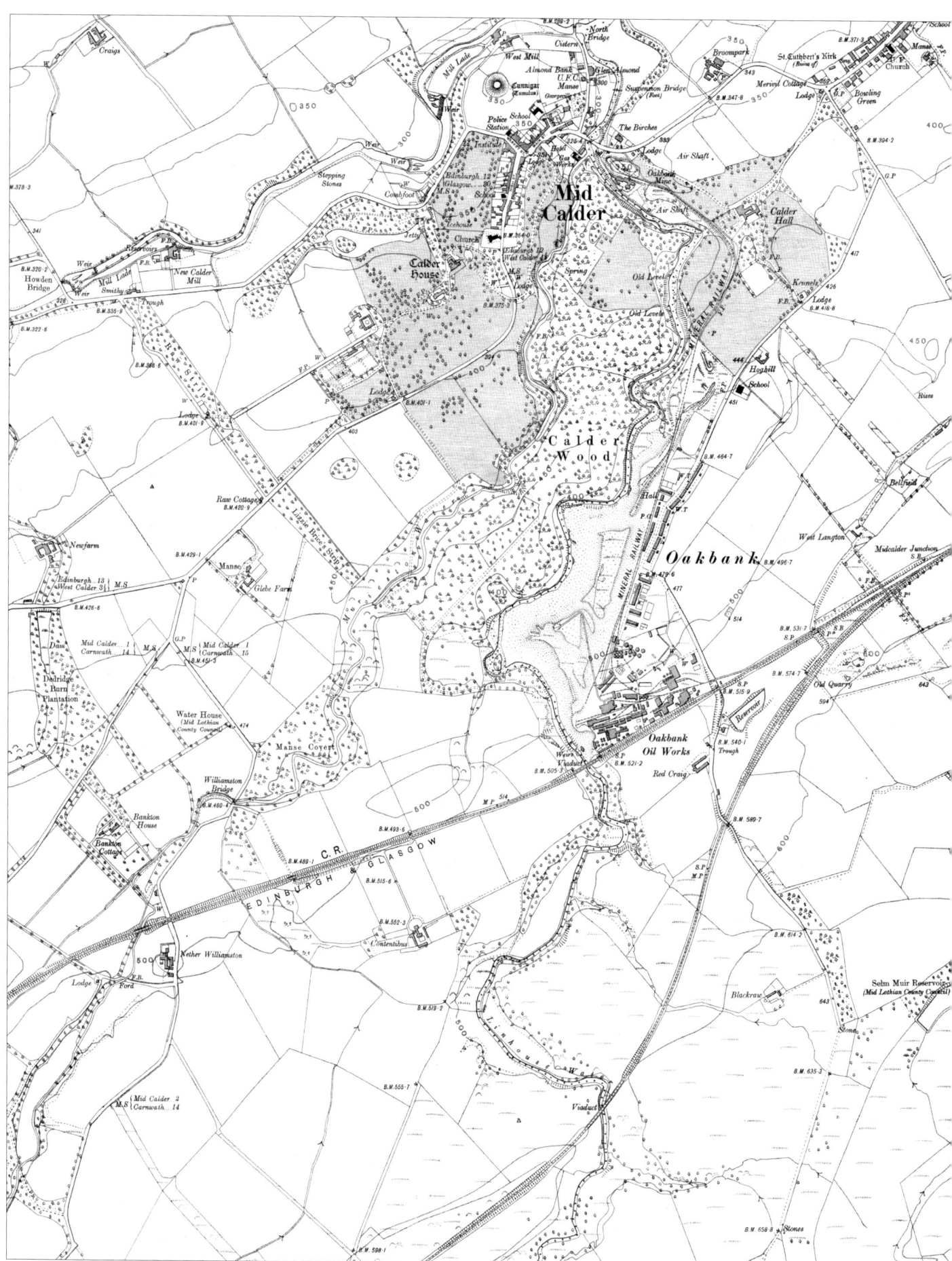

166

Map F: Oakbank Oil Works (1922). From the bottom of the map, the Caledonian Railway Carlisle to Edinburgh main line (1848), now the WCML, bisects the shale reserves in Midlothian as it runs towards Edinburgh. At Mid Calder Junction, the Mid Calder to Cleland chord line (1869) running south-west to north-east converges with the main line of route. To the north of the Cleland line, and rail connected to it, lies Oakbank Oil Works. A mineral line runs due north to access Mid Calder No. 1 mine (also known as Oakbank No. 1). Although not drawn on this map, the approximate route of the Oakbank aerial ropeway can be identified, running west from Oakbank Oil Works across Linhouse Water and Murieston Water, before passing close to Glebe Farm to reach the mine near New Farm.

Further to the east on the main line, the Camps Branch diverges at Camp Junction to run north, terminating at Raw Camps quarry complex and Camps Goods station (CR). (Scale: six inches to one mile, scaled to 70% to fit page.) *Courtesy of the Trustees of the National Library of Scotland*

Oakbank Oil Company 10-ton tank wagon No. 28.
Ed. McKenna collection

Oakbank Oil Company 12-ton tank wagon No. 63.
Ed. McKenna collection

out by rail to Camps Junction, to be worked westwards to the heart of the iron and steel industry in the west of Scotland.

This mine closed in November 1913 and the rails were lifted in 1914. The Caledonian Railway and the Coltness Iron Company both claimed the materials, and since no record of the arrangements under which the siding was constructed could be traced, the materials were shared on an equal basis.

Just after the branch was opened, on the 5th November 1896, one siding was extended northwards to cross the original Mid Calder to Edinburgh road to provide a loading point for limestone at Raw Camps quarry. In the February of 1913, the Caledonian had agreed to provide additional siding accommodation at both Camps Junction and Camps Goods at a cost of £700, but this expenditure proved largely abortive with little or no additional traffic revenue being generated, since the limestone quarrying operation was drastically cut back in that same year. Limestone extraction on a more limited scale continued until 1937, when the Coltness Company purchased the limestone mines at Oxwellmains in East Lothian and abandoned the Raw Camps complex completely.

The abandoned Raw Camps quarry was then used as a refuse disposal point for domestic waste; by 1915 the Caledonian had provided two new sidings, the first for Midlothian County Council (Calders Roads Department) and the other for Edinburgh Corporation for the waste disposal. This siding, known as Camps Rubbish Chute, was worked by internal locomotives owned by Edinburgh Corporation Lighting & Cleansing Department, four different locomotives being employed here over the years. The location was known as Holmes Depot.

At Camps, an exchange facility was formed with the NBR branch from Uphall Junction via East Calder Goods, by means of a 286 yard long chord line. This NBR branch, 1 mile 616 yards in length, opened in November 1866 and was regularly used to transfer crude oil traffic from Tarbrax (owned by the Pumpherston Oil Company) to Pumpherston refinery which was served solely by the NBR (see

EDINBURGH TRAIN ENGINES.

Time of Arrival of Engine at Shed Signal.	Particulars of Work.			
12.35 p.m. 8 8.0 a.m. 8 0	**No. 1—South Leith**—18 in. × 26 in. Tank Engine. Make two trips between Crew Junction and Seafield Yard, do all the necessary shunting at Stations and Sidings on Leith New Lines, and make a third run from Crew Junction to Seafield Sidings with Shipment traffic when required	Morrison Street Crew Junction South Leith Gorgie	... ...	12 45p 1 30 6 0 7 15
	No. 2.			
10.50 a.m.	**No. 3—Camps**—18 in. × 26 in. Goods Engine. Train, Gorgie to Cobbinshaw, calling at Ravelrig Junction, Kaimes Quarry, Camps Station, and Oakbank, Train Oakbank to Westwood Shale Pit, making two runs to Limefield Junction, then work Train to Gorgie, calling at Oakbank, Camps Junction, and Ravelrig Junction	Gorgie Camps Limefield Jct. Gorgie	... ...	11 5a 12 20p 12 50p 2 30 4 0 5 35
	No. 4.			
6.20 a.m. M and T 5.0 a.m. M O T O	**No. 5—Westwood**—18 in. × 26 in. Goods Engine. Train Gorgie to Camps and Oakbank, make up Train at Camps Junction for Breich Sidings and Westwood Pit, and return with Train Oakbank to Gorgie. On Mondays and Tuesdays leaves Gorgie at 5.20 a.m., run to Camps, work Train to Oakbank, lift Train Empty Wagons there for Westwood Shale Mines, work Train of Shale, Westwood to Oakbank, then Empty Wagons, Oakbank to Westwood Shale Mines (Takes forward traffic from Limefield Junction to Westcalder)	Gorgie Midcalder Limefield Junc. Gorgie	... ...	6 40 8 0 8 45 9 40 12 20p 1 0p
10.5 a.m.	**No. 6—Benhar**—18 in. × 26 in. Goods Engine. Leave Gorgie with Train for Kingsknowe, Curriehill, Ravelrig Junction, Newpark, Woodmuir Junction and Benhar; return with Train to Gorgie, calling at Midcalder Junction, Midcalder Station, Camps Junction, Camps Station, and Ravelrig Junction. Delivers traffic from Camps at Curriehill and Kingsknowe. Works Empty Oil Tanks, Midcalder Junction to Camps. Makes second run, Camps to Camps Junction, when necessary	Gorgie Benhar Camps Gorgie	... ...	10 20a 1 0p 7 45p 3 0 3 30 4 55
7.55 a.m.	**No. 7—Loganlea**—18 in. × 26 in. Goods Engine. Leave Edinburgh with Train Empty Wagons for Woodmuir Junction, work traffic to and from Loganlea 2 and 3 Collieries, leave Woodmuir Junction with Train for Crew Junction, calling at Ravelrig Junction, and returning with Train from Granton or Crew Junction to Edinburgh, when required	Edinburgh Woodmuir Crew Junction Edinburgh	... ...	8 10a 9 20a 12 10p 1 15p 2 20 2 35
8.55 a.m.	**No. 8—Fauldhouse**—18 in. × 26 in. Goods Engine. Leave Gorgie with Train for Benhar Junction, calling at Westcalder and Woodmuir Junction. Return Train from Benhar to Edinburgh, calling at Fauldhouse Colliery, Woodmuir Junction, Loganlea No. 2 Colliery, and Oakbank. On Monday, Wednesday, and Friday works Levenseat Quarries	Gorgie (Goods) Benhar Edinburgh	... ...	9 10a 10 10a 11 0 3 30p

CAMPS BRANCH.
WORKED BY TRAIN STAFF ONLY.

	Edinr. MINERAL. No. 2.	O'tairs. MINERAL. No. 5.	Edinr. MINERAL. No. 4.		Edinr. MINERAL. No. 2.	O'tairs. MINERAL. No. 5.	Edinr. MINERAL. No. 4.
	a.m.	a.m.	p.m.		a.m.	p.m.	p.m.
Camps Junction ......dep.	6 50	11 30	3 0	Camps ......dep.	7 30	12 30	4 0
Camps ......arr.	7 0	11 40	3 10	East Quarry {arr. {dep.	... ...	12 35 1 0	... ...
				Camps Junction ......arr.	7 40	1 10	4 10

The top item is an extract from a Caledonian Railway Working Timetable showing the Edinburgh (Dalry Road) engine workings including those working Tarbrax oil traffic. Below it is the entry for the Camps Branch from the same Working Timetable.

p. 186). Indeed, it would appear that there was a regular exchange of traffic at Camps since, in the workings of Carstairs No. 5 engine (in previous years, Carstairs No. 7 engine), it specifically refers to '*delivering all traffic for the NB Line*'.

In 1913 there were three booked turns serving the Camps Branch, two mineral trips (No. 2 and No. 4) from Edinburgh (Dalry Road) shed) and a Carstairs mineral working (No. 5, the Limestone engine); the latter also working the rail tanks of crude oil from Tarbrax which were worked across to Camps exchange sidings for onward transfer to the NBR. The Down siding at Camps Junction had the capacity to hold forty-nine wagons. This single line branch was controlled by Camps Junction signal box and worked under One Engine in Steam with Train Staff arrangements. The Carstairs mineral working (target 107) worked to Camps Junction every Monday, Wednesday and Friday into the early 1950s. Camps Goods (Cal) closed on the 1st June 1960.

Oakbank Oil Company 15-ton tank wagon No. 73.
Ed. McKenna collection

Oakbank Oil Company 14-ton tank wagon No. 35.
Ed. McKenna collection

Oakbank Oil Company 40-ton bogie oil tank wagon No. 66.
Ed. McKenna collection

15

THE NORTH BRITISH RAILWAY

Just as for the Caledonian Railway, the shale oil industry was to generate considerable levels of traffic to rail over the years for the North British Railway (NBR). Post-1860, when Young's patent expired, many new oil companies sprang up and new works were constructed. The larger and more productive saw the benefits of rail connection and thus in 1894 the NBR Goods Manager could list agreements with the following oil companies, covering both the running of private owner wagons or traffic agreements, all these companies by this time being rail connected. The companies were:

Young's Paraffin, Light & Mineral Oil Company	(1888)
Broxburn Oil Company	(1886)
James Ross & Company	(1882)
Linlithgow Oil Company	(1885)
Burntisland Oil Company	(1886)
West Lothian Oil Company	(1884)

The Broxburn Oil Company had 58 wagons registered with the NBR in 1882, by 1899 it had 30 wagons registered for main line running and 156 (mineral) wagons on private lines, in 1917 it had 63 wagons registered.

The Uphall Mineral Oil Company had 51 wagons registered in 1882 but were then taken over by Young's Paraffin, Light & Mineral Oil Company who, in 1899, registered 60 wagons for their Uphall operation.

Young's Paraffin, Light & Mineral Oil Company had 55 wagons on the register in 1882, by 1905 (including Uphall) it had 93 wagons registered, by 1917 this had increased to 181 wagons. This was in addition to a NBR 'pool' of 16-ton coal wagons, actual number not known, which had been allocated to Uphall Junction for shale traffic (see the following section on Uphall Junction).

The Pumpherston Oil Company had registered 64 wagons by 1899 and by 1917 had 128 wagons on the register.

It must be noted in passing that the wagons registered with the NBR in no way reflected the size of the actual wagon fleet in each oil company, since they all had a significant number of privately-owned wagons for internal use which never strayed onto NBR-owned lines.

The NBR had by far and away the biggest share of the traffic associated with the Scottish shale industry. Three main lines traversed the shale fields, the first being the Edinburgh & Glasgow (E&G) main line. This line swung northwards to cross over the Almond Viaduct at Bathgate Junction and passed, in tunnel, under Winchburgh village. It ran north-west across the rich shale fields, running out of them just before Linlithgow station. From the same Bathgate Junction, the Edinburgh to Glasgow via Bathgate main line (Edinburgh & Bathgate Railway) diverged to the left and cut straight across the main shale area running through the heart of same at Uphall. This line ran longitudinally along the shale fields, only leaving just before Bathgate was reached. The third line which crossed the shale fields was the later Edinburgh (Saughton Junction) to the Forth Bridge line (1890) which skirted the north-eastern edge of the shale fields. Two other less important NBR lines served the industry, the first being the Queensferry Junction to South Queensferry via Kirkliston Branch, the other the Bathgate to Morningside Branch.

Rectangular oil tank No. 12, of some antiquity, belonging to the Broxburn Oil Company. This tank carries a registration plate (circular) on the left centre of the solebar indicating that it can run on NBR metals.

Ed. McKenna collection

Map G: Bathgate (1922). Map shows the railways in and around Bathgate. From the upper left, the 1851 Monklands Railway Boghead Branch runs south through Bathgate connecting with the W&C line at Bathgate North Junction. A short distance beyond this junction, the Monklands line again diverges to the right at Polkemmet Junction. (Scale: six inches to one mile.) *Courtesy of the Trustees of the National Library of Scotland*

Bathgate-based ex-NBR Class J36 No. 65282 arrives at East Whitburn from Addiewell with a respectable freight train in tow on the 20th October 1960.
Courtesy W.S. Sellar

The shale industry brought with it increased levels of traffic to rail, with the transportation of raw shale, crude oil and the refined products. Along with coal, ironstone and fireclay mining and foundries, it meant that Bathgate Yard, in the heart of these activities, was to become the largest (at the time) marshalling and sorting yard in Scotland, being over the years, rail connected to at least fifty pits and mines, plus all the various oil works in the county. Bathgate shed provided most of the motive power over this period, although Edinburgh Haymarket shed also provided engines for the Dalmeny/Queensferry area and for Broxburn Branch from Broxburn Junction. As stated, of all the NBR locations in the shale areas, Uphall Junction was to be the main hub for rail access to the major oil works, with Broxburn Junction being only a close second.

Bathgate Upper passenger station with, in right-hand background, the transverse roofs of Bathgate engine sheds.
Bill Lynn collection

Linlithgow to Bridgend Oil Works (Linlithgow Oil Company)

This was a short mineral railway constructed and owned by the Linlithgow Oil Company Ltd. The company was established in 1884 and, at some point between that year and 1889, this branch was provided under an Agreement between the NBR and the oil company, at the company's expense. A connection with the E&G main line was made with the Down main line at a point immediately to the east of Linlithgow station. This railway was constructed primarily to facilitate the transportation of inward rail tanks loaded with crude oil for refining, Bridgend Oil Works carrying out this refining on an agency basis for crude from both Hermand Oil Works and Holmes Oil Works; it must be assumed that the line was also to provide a route whereby these refined products could reach the market place. Clear evidence still exists on the ground as to the route it took.

The railway connected the oil works at Bridgend (also known as Champfleurie Oil Works) (NT039759) to the shale pits at Ochiltree Nos 4 and 6 (NT036755) close by. From Bridgend Works it then ran north-east to cross over the A9 (now B9080) road between Champfleurie House and Bridgend Farm by a bridge (NT042762), the earthworks of which are still clearly visible.

The line swung north, then west to pass immediately behind Champfleurie House and crossed under the minor road leading from the B9080 to Philpstoun (NT028768). The bridge can still be clearly seen although all signs of the actual solum on either side of the bridge arch have been obliterated by more recent agricultural activity.

At some point mid-field, west of this bridge, the line then swung north-west to cross over the Union Canal via a skewed bridge (NT023772), again the abutments of which are still very evident. The line then swung back west to run parallel to, and at a slightly higher level than, the NBR Edinburgh to Glasgow main line. The embankment can still be seen to the north of Wilcoxhome Farm and at a point immediately on the east side of the bridge giving access to West Lothian County Cricket Club grounds in Linlithgow (NT012772). This line joined the main line by a trailing connection.

This line provided evidence of the convoluted routing of rail tanks carrying crude oil, since Hermand Oil Works (on the West Calder Loop) was served by the CR, and Holmes Oil Works by the NBR at Uphall. This line was worked by a privately-owned pug, all traffic, inwards and outwards, being put off at the exchange facility at Linlithgow. The line was lifted soon after closure of the Bridgend Works in 1902. This exchange facility with the NBR was worked by the Linlithgow Pilot, a NBR goods engine which was booked to shunt all sidings including Philpstoun Oil Works between the hours of 08.00 and 18.00 daily. This engine was, in all probability, a Polmont-based engine.

A view approaching Addiewell Junction from Whitburn on the 31st October 1962, taken from the fireman's side of the footplate of Bathgate-based Class J36 No. 65261.
Courtesy W.S. Sellar

Interestingly, this oil company apparently had private sidings for their use at Dundee Tay Bridge, since the Minutes of the NBR Director's meeting mention, in 1894, that the NBR had agreed to extend these sidings at the request of the Linlithgow Oil Company.

Philpstoun Oil Works

In 1885 an agreement was reached between the NBR and James Ross & Company – who had established an oil works at Philpstoun, adjacent to the E&GR main line, on the south side – whereby the NBR laid in a connection and sidings to the then new oil works close by Philpstoun station. The cost of this work was £1,660, of which James Ross & Co. were allowed a deduction of £460 and paid only £1,200. Philpstoun Oil Works was served by a complex of internal mineral lines from the various pits and mines from whence the raw shale was sourced, but was also unique in that from the Philpstoun Nos 1 and 6 mines, shale was delivered into the works by an underground haulageway.

These rail sidings, like all private sidings in the area, were served by the Linlithgow (Polmont) pilot engine on a daily basis.

Bathgate to Addiewell Goods and Addiewell Oil Works

In 1851, the Wilsontown, Morningside & Coltness Railway was extended eastwards from Longridge terminus by the E&GR, which had absorbed the WM&CR in the previous year, finally making an end-on connection with the Edinburgh & Bathgate Railway at Bathgate. On this new line, at what was to become Chemical Works signal box, a connection provided access to Young's new Bathgate (Boghead) Oil Works. The Monklands Railway, not to be outdone, in 1855 constructed a new line running southwards from Blackston Junction on the former Slamannan Railway to Cowdenhead (the Monklands Boghead Branch), cutting through Bathgate and joining the WM&CR line at Bathgate West Junction. Using the latter's metals for a short distance, the Boghead Branch then diverged at what was to become Polkemmet Junction, the branch then continuing westwards, passing to the south of Armadale and terminating at Cowdenhead Colliery (later Woodend Junction). This Boghead Branch itself served many collieries, and a further branch from Armadale Junction was provided to access collieries, foundries, fireclay and brickworks and also the short-lived Armadale Oil Works (Scott's).

At Polkemmet Junction, the WM&CR line ran south-eastwards. The solum has now been cut by the M8 motorway but is still clearly visible in certain stretches between Bathgate and East Whitburn. At the latter location it passed under the A705 road (and site of former East Whitburn passenger station) (NS965652) and continued southwards across farming land to a point on the minor road from the B792 to Longridge. Another branch was later provided immediately south of Whitburn station, running due eastwards to serve Whitrigg colliery (The Lady of the Dales pit) in around 1901 and also served some four other smaller coal pits in the early days.

Addiewell Junction signal box lay on the main line (NS972637) adjacent to, and controlling the level crossing over, the B792 road on the site of former Foulshiels station, and beyond which the main line and the Addiewell Branch diverged. The Morningside line swung westwards, whilst the new Addiewell line, laid around 1863

A rail tour special arriving back at Addiewell Junction from the Addiewell Branch hauled tender first by ex-NBR Class J37 No. 64569 on 19th June 1962. The signalman, Bob Meikle, is out on the token exchange stool ready to receive the token from the fireman. Despite being in rather a rural backwater, a small crowd have gathered at the level crossing gates to witness the event. The operating wheel for the level crossing gates can be seen in the signal box to the left of the box name board.
Courtesy W.S. Sellar

as a separate single line for the new Addiewell Oil Works then under construction, swung away to the south. This latter branch served Foulshiels colliery (United Collieries) (NS977634), then continued southwards to cross the B7015 road on the level, east of Stoneyburn (see Map A, p. 20); it was later crossed by the aforesaid Caledonian West Calder Loop, again on the level, at Cuthill Crossing; it then passed over Breich Water on a viaduct and under the minor road which ran through the Addiewell Rows (Livingstone Street) adjacent to the parish church, lying on the east side of the post-war housing development at Addiebrownhill (NS991626). Here it swung east and entered the complex of exchange sidings on the west side of Addiewell Oil Works and Addiewell Goods.

Electric Token Working was employed between Chemical Works signal box at Bathgate, Whitburn signal box and Addiewell Junction signal box. From Addiewell Junction to Addiewell Goods, the single line was worked under One Engine in Steam with Train Staff, the signalman at Addiewell Junction being custodian of same. Rear-end assistance was also authorised between Bathgate Upper and Addiewell Junction but with the assisting locomotive not coupled to rear of train. This line was worked by trip trains working from Bathgate Upper Yard, with motive power being supplied by Bathgate shed. In the final days of the shale oil industry, Bathgate still had two booked trips, Bathgate No. 2 engine and Bathgate No. 8 engine, serving Addiewell Oil Works on a daily basis.

As an interesting aside, in September 1877 the NBR laid in an additional siding at Addiewell Goods for use by local traders, on ground owned by Young's Paraffin, Light & Mineral Oil Company. The latter granted this ground rent free '*in consideration of the NBR undertaking to convey the Directors of the Oil Company, three or four times a year free of charge, in a saloon, between Bathgate and Addiewell Works*'.

The NBR laid in further sidings at Addiewell Goods in March 1899, again on Young's land, this time at a nominal rent of £1 per annum. A later move to have a passenger train service established between Addiewell and Bathgate came to nothing.

BATHGATE (POLKEMMET JUNCTION) TO THE TORBANE PITS AND ARMADALE

Whilst lying entirely outside the oil shale fields, it is only proper that mention is made of this railway branch since it had strong associations with the shale oil industry. The Boghead Branch of the Monklands Railway, on leaving the WM&CR at Polkemmet Junction, proceeded to run due west. At Boghead Junction (NS958674), connections were laid in on either side of the branch, the line to the south serving, in total, seven Torbane pits. This branch split, the first leg serving (and terminating at) Torbane No. 2 pit (coal and ironstone), whilst the second leg served Torbanehill No. 6, Torbanehill No. 3, Torbanehill No. 2 and Torbanehill No. 5, terminating at Torbanehill No. 8 (NS957662). It was from these pits that the canneloid coal (Torbanite) used in the Boghead Oil Works was produced. Internal tramways connected all

Bathgate Class J36 No. 65261 arrives at Addiewell Goods with a mixed freight train from Bathgate including a large number of empty 16-ton steel mineral wagons. Foulshiels Colliery was closed by this date so empties are for Scottish Oils use only. In the right background (behind locomotive smokebox) can be seen wagons standing on the oil works sidings and, behind that, the gable ends of the Addiewell 'double rows' of oil workers' houses.
Courtesy W.S. Sellar

A view of Livingston station with Livingston signal box under bridge (left). The connection to Deans Oil Works is just around the curve to the left. The view is looking towards Bathgate East.
Bill Lynn collection

the remaining Torbane pits to the railway system and a separate tramway connected Torbanehill No. 2 and No. 3 pits to a private siding on the WM&CR branch line at Inchcross.

This mineral line was recovered after closure of the pits before the end of the nineteenth century but was relaid again at a later date, presumably by the NBR, to serve two pits which had been redeveloped, Torbanehill No. 7 becoming Northrigg No. 1 pit and Torbane No. 2 becoming Northrigg No. 2. The line carried on due south to the site of Torbanehill No. 8 pit where Drum fireclay pits Nos 1 & 2 had been developed.

To the north side of the line at Boghead Junction, the connection to the north side of the line served Bathville Brick & Fireclay Works.

The Boghead Branch itself continued due west. At a point near Trees Farm (NS945674) a further connection was laid in to the north side (Armadale Junction) providing another sub-branch line serving the Armadale area. This line itself fanned out beyond the junction and a line running to the east served Bathville, Etna and Atlas brick works, plus, at a later time, two steel foundries. One leg ran north-eastwards crossing the present B708 road, past the site of the earlier Hardhill No. 1 pit, to serve Hopetoun Nos 8 & 9 pits which lay close by the present Armadale Stadium. The main leg swung round westwards behind Armadale to serve Bathville (Scott's) Oil Works before crossing South Street on the level to serve Armadale No. 15 pit (the Buttress), No. 16 pit and No. 17 pit (all centred around NS930680).

An interesting point about this branch is that, immediately before crossing South Street in Armadale, a loading bank was provided by the NBR. To here, ironstone 'char', left after the burning of the coal content out of the ironstone and a valuable commodity for iron works, was conveyed by road using horse-drawn carts from pits without a rail connection for onward transport by rail to the iron works. The condition of roads in those days was, to say the least, less than satisfactory for heavy horse-drawn carts and thus plateways –

described as '*a track of broad plate rails, each being one yard long and eight inches wide, laid on heavy wooden sleepers*' – were laid along what was to become Armadale Main Street (west) and up the hill in South Street to ease the journey of the cartloads of char to the aforesaid loading bank.

A curving connection from this line ran in a sweeping curve to the north-east, crossing the East Main Street in the town and serving Barbauchlaw Brickworks and the new Barbauchlaw pit (No. 5) (NS943688) which worked until 1952.

FAULDHOUSE CROFTHEAD TO LEVENSEAT AND HANDAXWOOD OIL WORKS

As described above, in 1845 the Wilsontown, Morningside & Coltness Railway Company had laid a new line of railway from Morningside (Lanarks), across the North Lanarkshire moors, to a terminus at Longridge just inside the West Lothian county boundary. This line was built mainly to access the rich limestone, coal and ironstone seams situated around Shotts, Fauldhouse and Levenseat, the latter location having reserves of ironstone and limestone but no coal. Passenger stations were provided at Davie's Dyke, Headlesscross, Fauldhouse, Crofthead and Longridge, and a passenger train service, connecting with a horse-drawn coach service at Longridge, was later operated over this line. Longridge station was actually some distance south of Longridge village, the highest point in West Lothian, and consisted of a single-faced platform only (NS954614). Three through trains between Glasgow and Longridge were in operation daily by 1846.

The coach service was started by the Edinburgh stage coach proprietor, John Croall, on the 16th May 1846. He ran his coach *Luck's All* from Edinburgh via East Calder, Mid Calder and West Calder, to connect with the passenger trains for Glasgow at Longridge station.

The E&GR, which had just absorbed the WM&C, extended this branch line eastwards in 1851 to connect with the original Edinburgh & Bathgate Railway at Bathgate.

There was a short-lived oil works at Handaxwood (Midlothian) near Fauldhouse in the early 1860s (NS941603) which was served by a short branch (or long siding). Accessed from the aforesaid NBR Bathgate to Morningside line at Fauldhouse (Crofthead) by a facing (looking east) connection at Levenseat Branch Junction signal box (NS935604), it served both Levenseat limestone quarry and works. Both the North British and the Caledonian railway companies had their own sidings at this oil works, but Handaxwood Oil Works and the rail connection to it were gone by 1903. The NBR Levenseat Branch crossed over the Caledonian Railway's Edinburgh to Glasgow line (at NS943599) to access the Levenseat sand and limestone workings (see Map E, p. 161). Two chord lines from the Caledonian main line formed a triangular junction with this NBR branch, on both the north and south sides of the Caledonian main line. These Caledonian connections were controlled from Levenseat Junction signal box. The joint access facility created by this arrangement serving the limestone quarry and works was worked under One Engine in Steam conditions, each railway company being allocated an access time – the NBR having access rights between 12.00 and 13.00 daily, and the Caledonian working the complex between 13.00 and 14.00 daily. The Caledonian connection to the north gave that company its own access to the short-lived Handaxwood Oil Works.

After closure of the limestone quarry, Levenseat sand quarry continued to be worked well into the 1950s by a daily trip working from Bathgate over the former NBR branch, taking out wagon-loads of sand which were then worked forward from Bathgate Yard by a Bathgate to Thornton freight working running via Kirkliston, being destined for Pettycur Bottle Works in Fife.

A further NBR branch running north up onto Fauldhouse Moor from Crofthead served numerous small collieries and also the short-lived Benhar (Caledonian) Oil Works, this latter works also being rail connected to the Caledonian Railway's branch from Benhar Junction.

BATHGATE (STARLAW) TO COUSLAND (SEAFIELD) OIL WORKS

In 1877, under an agreement between the NBR and the Seafield Patent Fuel Company dated 1877, a branch line, initially known as the West Calder Branch (strangely, since it ran nowhere near West Calder) struck off southwards from the Edinburgh to Bathgate railway between Livingston Station and Bathgate at a point near Starlaw (NS992679), via a trailing junction to serve the company's Seafield Patent Oil Works (see Map H, pp. 180–81). This junction was later to become Bathgate East Junction. It is thought that the curving layout of track was adopted to avoid taking the railway directly across Easter Inch Moss, a very wet and unstable piece of ground – where peat extraction was carried out on a commercial scale by the Midland Moss Litter Company – and quite unsuitable for the provision of a railway thereon without considerable expense. Instead, the railway skirted the edge of the Moss, but to access Seafield Oil Works a reverse facility was required. Financial problems meant that the branch line was not to be permanently opened at that time, however, and the agreement remained in draft form only. In 1883, the newly-formed Bathgate Oil Company took over the works but this proved equally unsuccessful and it was not until the Pumpherston Oil Company took control that the branch was actually opened for traffic. This branch line diverged to the south and ran to the edge of Easter Inch Moss, as described, towards Blackburn. Access to this branch was controlled by Bathgate East Junction signal box. By means of a reverse junction, the branch then headed east to access Seafield Oil Works (NT006665). An extension of the branch was carried out in 1891 and additional siding accommodation provided in 1909. Today, much of the solum of this line is a public walkway, although it has also been bisected by the M8 motorway; Seafield bing, now landscaped, is known as Seafield Law and is a leisure site.

There was for some period of time, and certainly into the 1950s, an extensive light narrow gauge railway network running across the Moss. Peat extraction (for horticultural purposes) from the Moss was quite a thriving business and peat was conveyed by small wagons from extraction site to the processing area. A small petrol driven tractor provided the motive power.

BATHGATE TO DEANS AND BOGHALL MINES

In 1872, agreement was reached between the Uphall Oil Company and the NBR for a private siding to be provided to the north of the railway between Bathgate and Livingston to serve the Boghall mines and Deans No. 7 pit, to be known as Boghall new siding. The costs of installation were split between the NBR and the Uphall Oil Company, with the NBR paying £343 15/11d and the oil company £441 2/5d with an additional £45 5/5d in labour costs. The oil company also paid 5% per annum on cost of materials and had to maintain the siding to the satisfaction of the NBR's engineer. Access was controlled by a small signal box, Boghall New Siding. Traffic ceased in August 1883 and the signalman was withdrawn. The siding and connection were officially closed in June 1885 and lifted soon after.

BATHGATE TO DEANS OIL WORKS

Under an agreement between the West Lothian Oil Company and the NBR dated 1883, sidings were installed immediately to the west of Livingston Station on the Up Bathgate to Edinburgh line. A trailing connection (NT020681) with an associated siding of some 225 yards in length was laid in, giving rail access to Deans Oil Works, controlled from Livingston signal box. This work, including signalling work, amounted to £573 8s, of which the oil company paid £461 4s, the remainder being written off by the NBR. The oil works closed in 1891 but were taken over and reopened by the Pumpherston Oil Company in 1894.

Over the years, additional sidings and connections were provided. When additional sidings were laid in 1897, Pumpherston Oil Company undertook to pay 4% per annum on the value of the material used to construct a new weighs siding costing £1,625 11/-, whilst the NBR bore the cost of a second siding, estimated at £438 17/6d. In 1888, a further siding was laid in by the NBR at a cost of £200 3/3d, this amount being paid by the oil company in five instalments, spread over five years. Further sidings and connections were provided in 1908 and 1910, again on a percentage of value of material basis.

After the Second World War and the closure of the oil works, the land was sold to the MOD who established an Engineer Park (324 Engineer Park R.E.) close by this connection and this was, until closure, served by rail. Much of the heavy engineering equipment used by the Corps of Royal Engineers was stored here and moved in and out as required, much of it under 'Examine Load' conditions. In 1975, when large quantities of blaes were required for a major engineering project – the western extension of the M8 motorway – a contract was arranged whereby the spent shale in Deans bing would be moved by rail to Shieldhall, in Glasgow. Using a new connection in this same location, three new sidings were driven into the heart of the bing and here W. Griffiths & Son, haulage contractors, loaded the shale to rail. Bathgate depot ran nine loaded trains each day from Deans to Shieldhall and nine empty trains back, carrying, in the eighteen months of the contract, a total of 1.5 million tons of shale and removing Deans bing almost completely. These sidings were worked as a private siding controlled by an electrically-released ground frame (released from Bathgate Central signal box) with the facility to shut trains in.

In the latter days, after closure of the oil works and signal box at Livingston, wrong direction working was authorised over the Up line between Deans siding and Bathgate Yard.

Uphall Junction

Uphall Junction was, amongst all other connections to the NBR rail system, by far the most important and a key location for connecting several major oil works situated in the heart of the shale industry to rail. It was to be the major hub of railway operation for the Scottish shale oil industry and from where several branches diverged. In addition, to augment the oil companies own privately-owned mineral wagons registered to run on NBR metals, the NBR had a pool of 16-ton wooden, end-door mineral wagons specially stencilled 'Shale Traffic Only' and 'Return to Uphall', a clear indication of just how much mined shale for the retorts passed through Uphall. The bulk of this shale traffic emanated from the Ingliston and Newliston pits, all of which were owned by Young's Paraffin, Light & Mineral Oil Company, who had also taken over Uphall Oil Works.

Even in the 1930s, Bathgate shed had a heavy commitment for mineral and general goods working, and had, in addition to the booked train working diagrams, some thirteen 'conditional' turns spread over 24 hours, working to Control Orders and available for any traffic demands required by the shale and coal industries. Two Bathgate Junction (Ratho) pilot engines were also provided on a daily basis by Bathgate shed, Turn 1G (L30) 06.45, and 2G (L30) 24.00, for general shunting and assisting duties.

The booked turns serving Uphall were:
- Up trains Nos 43 and 46 (Bathgate turns) booked to work the Camps Branch daily.
- Up trains Nos 62 and 69 booked to call at Uphall to work all oil traffic forward for East Coast destinations daily.
- Up train No. 63 booked to uplift all oil traffic for destinations north of the Forth Bridge and work forward to Dalmeny (North) Junction on a daily basis.
- Down train No. 61 transferred oil tanks between Broxburn Oil Works and Deans Oil Works daily.

Right into the days of the nationalised British Railways, the remaining oil works – and in particular, Pumpherston Oil Works – generated a considerable amount of both inwards and outwards rail-borne traffic, with Uphall Junction remaining a key location. Indeed, Edinburgh St. Margaret's MPD, not normally involved in train workings in this area, was, post war, allocated one of the afternoon trains, either 62 or 69, where the engine and crew worked a full train of loaded tank wagons from Uphall (Pumpherston

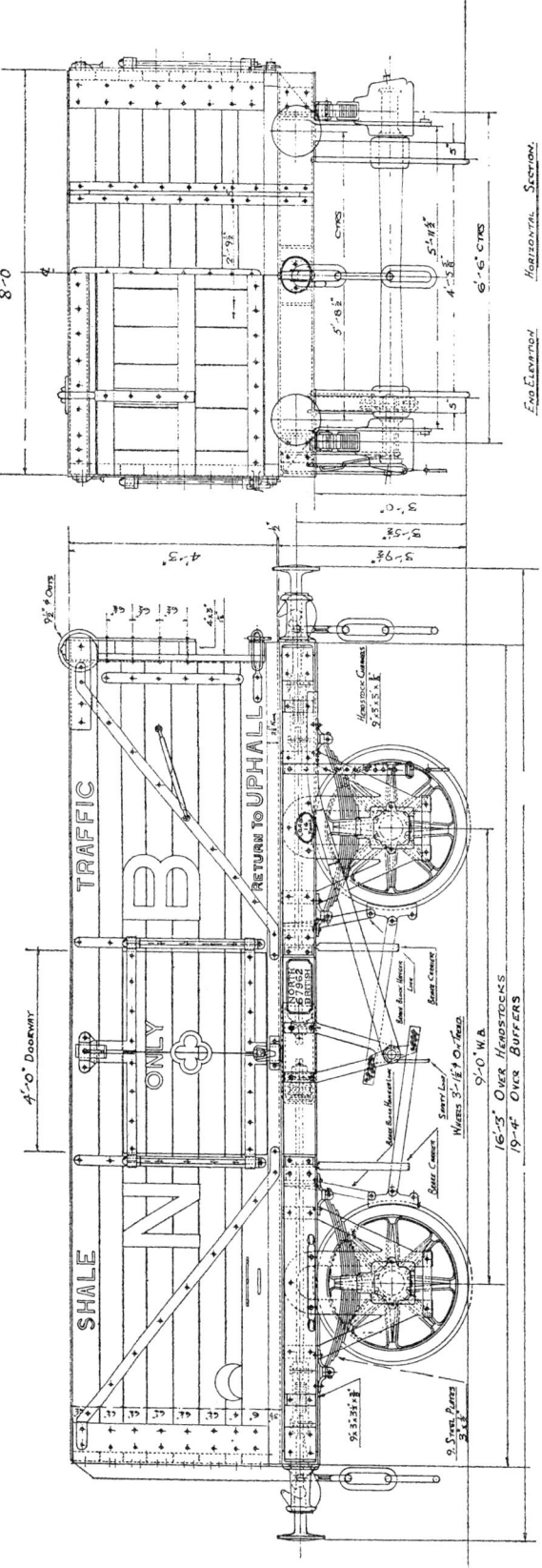

Drawing of a North British Railway 16-ton, end door coal wagon, showing the wagon stencilled 'Shale Traffic Only' and 'Return to Uphall'. Scale 7mm:1ft.

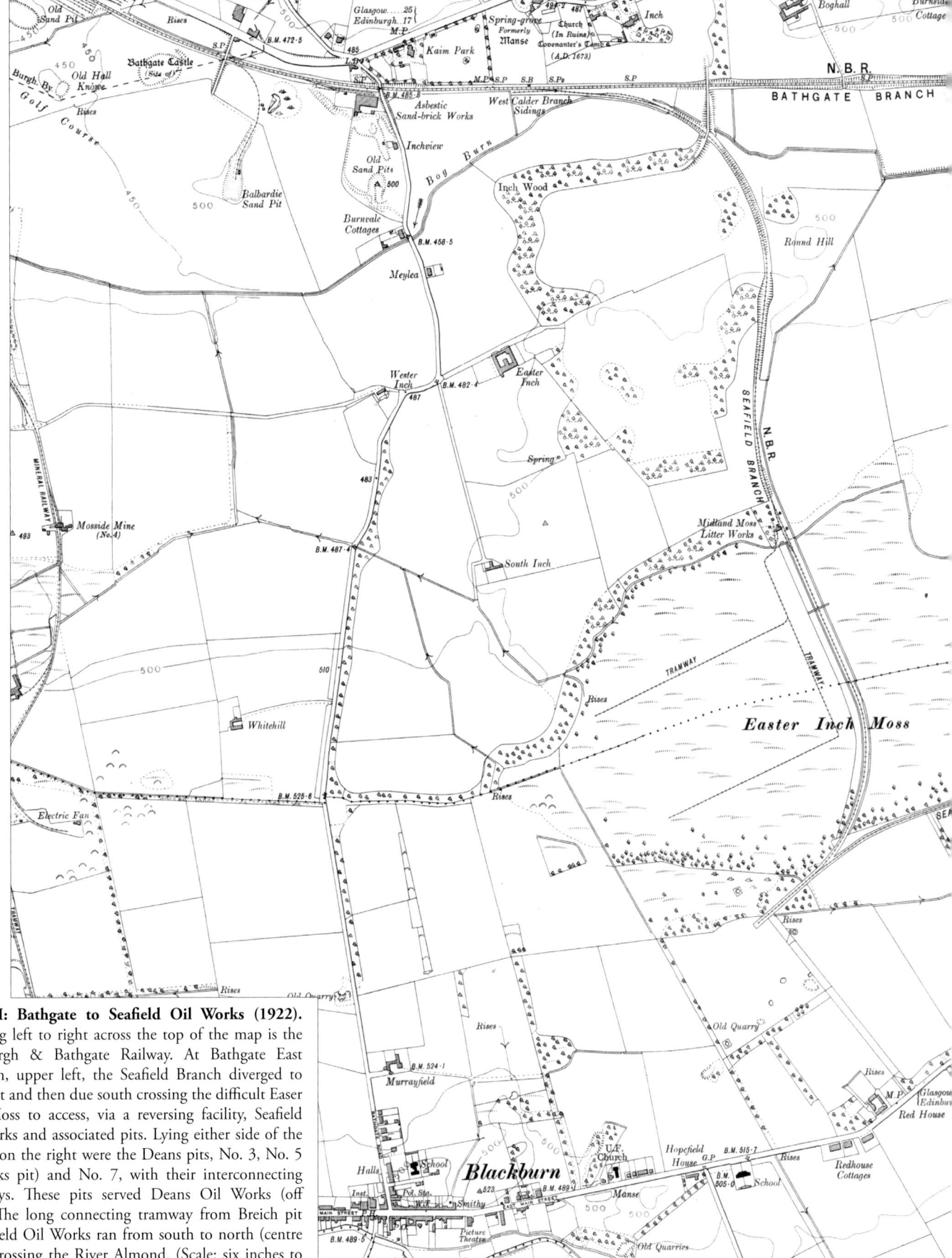

Map H: Bathgate to Seafield Oil Works (1922).
Running left to right across the top of the map is the Edinburgh & Bathgate Railway. At Bathgate East Junction, upper left, the Seafield Branch diverged to the right and then due south crossing the difficult Easer Inch Moss to access, via a reversing facility, Seafield Oil Works and associated pits. Lying either side of the E&BR on the right were the Deans pits, No. 3, No. 5 (Barracks pit) and No. 7, with their interconnecting tramways. These pits served Deans Oil Works (off map). The long connecting tramway from Breich pit to Seafield Oil Works ran from south to north (centre right) crossing the River Almond. (Scale: six inches to one mile, scaled to 85% to fit page.)
Courtesy of the Trustees of the National Library of Scotland

THE NORTH BRITISH RAILWAY

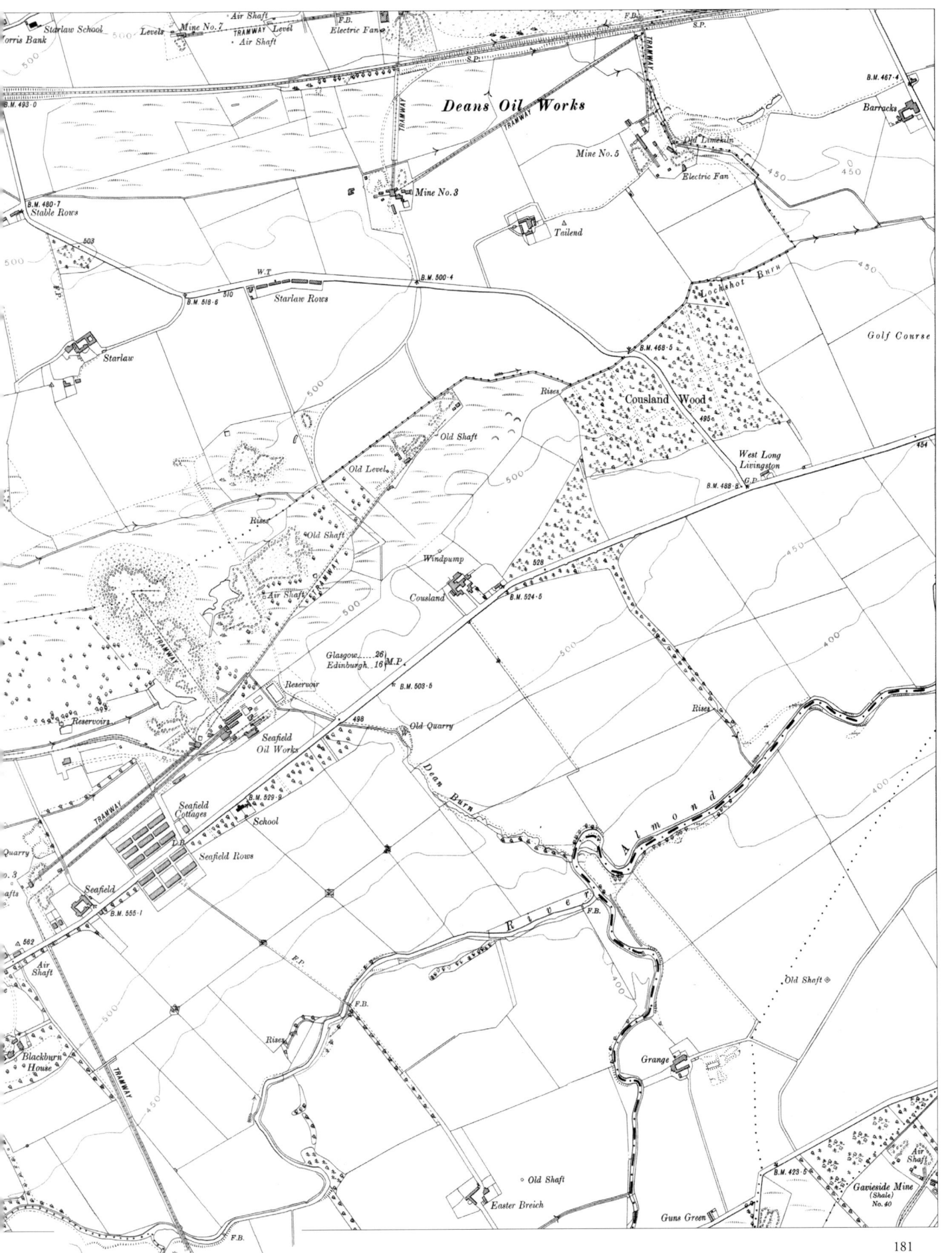

Works) to Portobello Yard for onward working south via the east coast main line, and regularly turned out a Class 'V2', 2-6-2 6MT locomotive for this purpose. In BR days, and pre-Hazchem codes, specially preprinted, double-sided wagon labels were supplied to Pumpherston refinery (and Grangemouth refinery) for outwards loaded oil-related traffic, the reverse of the label giving instructions for the return destination of the discharged tank wagons. Occasionally, things went wrong: in the early 1950s, Pumpherston were supplied with labels from British Railways showing both originating station and return destination as Pumpherston Oil Works siding (Eastern Region BR) Uphall!

UPHALL JUNCTION TO HOPETOUN OIL WORKS

Under an agreement between the Broxburn Oil Company and the NBR dated 1888, under which it was stated that '*the Railway Company shall form and maintain at their own expense, sufficient connections and points to work the traffic*', a branch line was built to give access not only to the Uphall Oil Works immediately adjacent to the main line at Uphall Station but it was also extended to serve the Middleton Hall complex immediately on the south side of Uphall village and, thereafter, it ran on to serve Hopetoun Oil Works at Faucheldean (see diagram, p. 183).

At Uphall Junction, a trailing connection from the Up main line in to a headshunt immediately behind the Up platform of Uphall Station gave access to Uphall Oil Works (NT065708) and a line swung north to run towards Uphall village. At a point some 295 yards south of the village and immediately after crossing over the Edinburgh to Glasgow main road, later the infamous A8, by a low bridge, this line divided, a right hand connection giving access to the Middleton Hall complex sidings, the place which was to become the administration centre and maintenance workshops for Scottish Oils Ltd. Immediately beyond this point, the single line continued, with a short siding giving access to Uphall Goods station (NT058716). The left hand line crossed the main street in Uphall on the level (NT057716) and ran west to service Forkneuk pit (NT054720).

The level crossing was known as Castlehill level crossing and was operated from Castlehill signal box, a small box sitting just on the south side of the main road; the name of the box reflected the site nearby of Strathbrock Castle. The actual site of the level crossing is now a new car dealership. Swinging north by west, the line served a shale pit (closed before 1902) (NT050725) before turning to run north-east past the site of Binny quarry, crossing the minor road from the B8046 to West Binny on the level. Again turning east, it passed under the B8046 road at Ecclesmachan (NT059734) and ran to the south of Kirklands cottages where a siding led off via a trailing connection, before turning south-east in to Hopetoun Oil Works (NT084742). A connection to the right swung south and east to connect into Albyn Oil Works.

The NBR worked only as far as Uphall Goods, the bulk of the working between pits/mines and the oil works being conducted with privately owned internal wagons and engines. The sidings on the Up side at Uphall provided the major traffic exchange facility for both inward and outwards traffic. Bathgate shed supplied the bulk of the engine power although one of the earlier Haymarket Broxburn pilots also provided a service to and from the sidings at Uphall.

UPHALL (DRUMSHORELAND) TO HOLYGATE GOODS, BROXBURN AND ALBYN OIL WORKS

An agreement was reached between the Broxburn Oil Company and the NBR in 1886 to have a branch line built from Drumshoreland to serve both Broxburn (Stewartfield) Oil Works and Albyn Oil Works, north of the village of Broxburn.

The first clause stated:

The Railway Company agreed to construct a Branch Railway from the Edinburgh and Bathgate line at Drumshoreland station to the Broxburn oil works with a depot at Broxburn as shown on the plan.

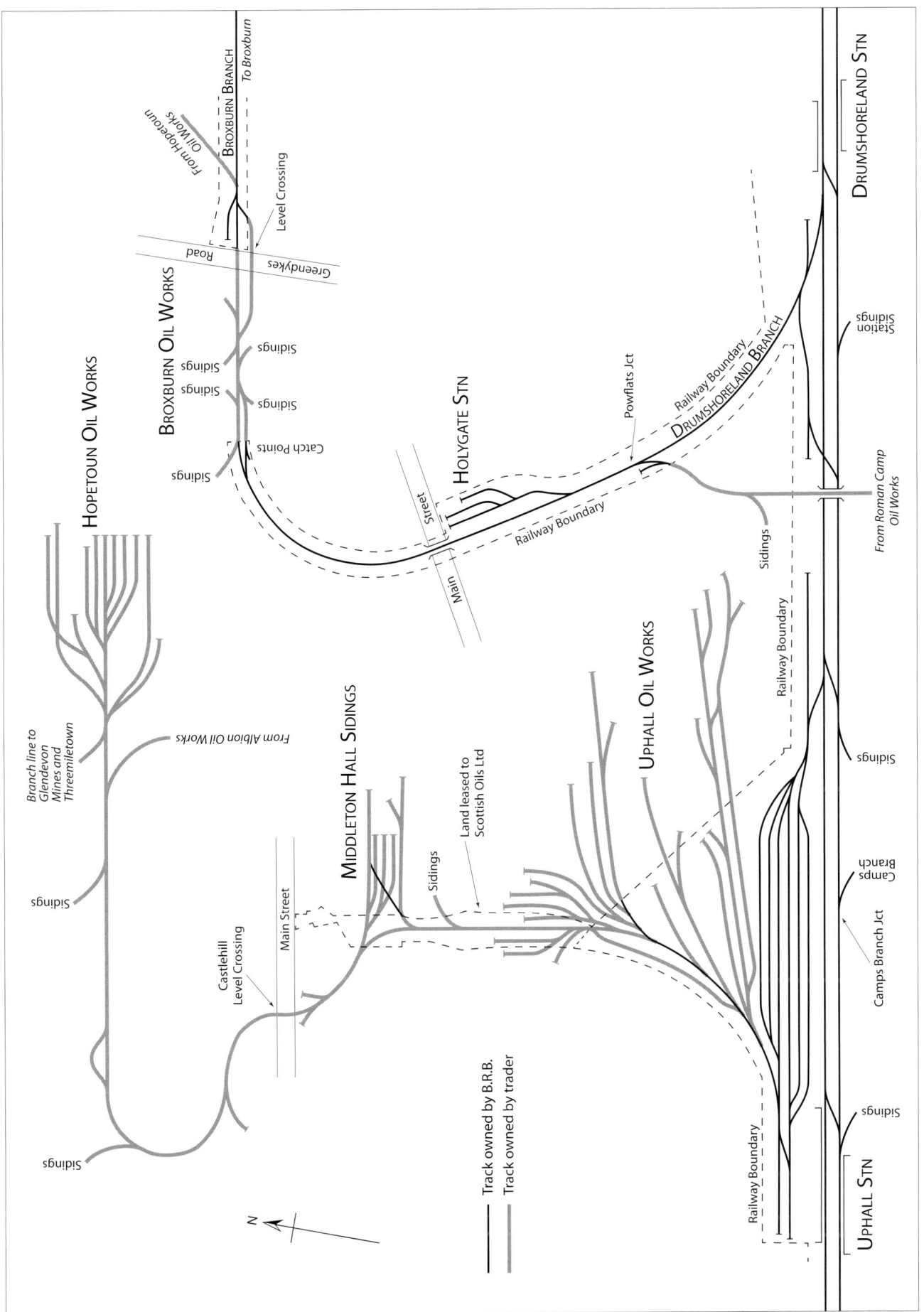

A schematic diagram of the branch lines to the north of the main line between Uphall to Drumshoreland. On the left is Uphall to Uphall Oil Works, Middleton Hall and Hopetoun Oil Works; on the right the Drumshoreland Branch, Holygate Goods and the Broxburn Oil Works Branch. Based on extracts from the NBR/L&NER Engineer's private siding agreement book. The heavy lines were coloured in the original to indicate ownership. (Not to scale.)

A later clause, the fifth, stated:

> The Oil Company shall be entitled subject to the Regulations of the Railway Company to make junctions by means of Branch Railways between the works or pits made or to be made and the said Branch Railway and Railway Company shall form and maintain at their own expense sufficient connections and points to work this traffic.

The branch was duly constructed. At Drumshoreland, the Broxburn Branch was accessed via a trailing connection in the Up direction of travel (NT085076). Additional rail facilities including sidings were provided at Drumshoreland in connection with this work, at a cost of £151 to the Broxburn Oil Company.

This branch ran due north until a point south of the town where a facing connection led into the depot at Broxburn, as mentioned in the agreement, later to be known as Broxburn Holygate Goods station (NT076721). Bypassing Holygate to the left, and crossing over the A899 road by a bridge at Broxburn west-end, the Drumshoreland Branch serviced Broxburn (Greendykes) Oil Works (NT083729) via both training and facing connections. The internal works line then crossed Greendykes Road (B8020), again on the level, and connected with the NBR Broxburn Junction Branch line servicing the Albyn Oil Works complex (NT089731).

The branch from Drumshoreland to Holygate Depot and Broxburn Oil Works was worked under One Engine in Steam Arrangements with Train Staff, and worked by NBR engines from

More recognisable than the Broxburn Oil Company tank wagon shown earlier, this is a 10-ton tank wagon belonging to the same company.

Ed. McKenna collection

A more modern 14-ton capacity Broxburn Oil Company tank wagon, No. 1098, constructed for the conveyance of sulphuric acid only.

Ed. McKenna collection

Bathgate shed (Down train No. 61) to the point where Staff Working finished and Yard Working commenced. NBR locomotives were prohibited from crossing the level crossing at Greendykes Road.

The Haymarket shed Broxburn pilot serviced Albyn Oil Works and also worked the branch up to the limit of NBR working at Greendykes Road level crossing. An internal line worked by company 'pugs' connected Broxburn Oil Works with the South Greendykes and Hayscraigs shale workings.

UPHALL (DRUMSHORELAND) TO ROMAN CAMP OIL WORKS

Under the conditions contained in clause five referred to above, in 1891 the Broxburn Oil Company advised the NBR that they wished to exercise the rights so granted and construct a branch line from the Drumshoreland to Broxburn line via a new trailing connection to be located near Powflats Farm. The branch line was to then run off due south and cross over the Edinburgh & Bathgate Railway at a point just over ½ mile east of Uphall Junction, serving the oil works located at Roman Camp (NT075704), lying on the south side of the E&BR line near Drumshoreland. This junction was to be known as Powflats Junction (NT078712) (see diagrams, pp. 183, 187). The branch was a wholly mineral line. Apart from serving the Roman Camp complex, a single line also ran through the works complex in a southerly direction, then swinging round to the left it connected with the Uphall Junction to Camps Branch (see below). The Broxburn Oil Company paid the NBR £106 10/5d for this connection with the Camps Branch and a further £5 for the gate protecting the internal line. The works were completed in March 1891. It is understood that the NBR served Roman Camp Works via this latter connection from the Camps Branch, and that all traffic passing between Roman Camp and Broxburn Oil Works was worked solely by internal pug engines, via Powflats Junction.

ABOVE: Drumshoreland station looking east towards Edinburgh. *Bill Lynn collection*

BELOW: Broxburn Holygate Goods being shunted by a Bathgate-based EE 350hp diesel shunter on 21st November 1964. *Courtesy W.S. Sellar*

Uphall Junction to East Calder Goods and Camps Raw Quarry

The NBR obtained an Authorisation Act in 1866, under the NB Camps, etc. Branches Act, Railway No. 1, to construct a new branch line diverging from the Edinburgh and Bathgate line at Uphall Junction via a trailing connection from the Down main line. Initially the branch was to run to East Calder Goods station. Completed under an order dated 3rd June 1867, the branch was opened for traffic on the 7th June 1867.

In addition, before the branch reached the West Lothian/Midlothian county march (the Caw Burn), a new connection was laid in to connect with the Roman Camp Oil Works complex (referred to above) and a further connection at some later time to Pumpherston No. 4 shale mine, whilst two new rail connections were laid in within the county of Midlothian, serving, in order, Pumpherston Oil Works and Pumpherston No. 5 mine (all referred to below).

From the trailing connection in the main line, the branch proceeded south-eastwards on a falling gradient, to cross the River Almond on a stone viaduct (NT086684). Immediately north of the B7015 road, at the eastern end of East Calder village, lay East Calder Goods station served by a trailing connection (NT091681). A permanent speed restriction of 5mph was imposed over the Almondell Viaduct.

However, limestone was being worked at Raw Camps quarry by the Coltness Iron Company, and the Caledonian Railway had already gained a foothold in this operation with their new branch from Kirknewton (Camps Junction) to Camps Goods (Cal) which had been opened for goods traffic since the 5th November 1866 (see diagrams, pp. 165, 187). The Coltness Iron Company had approached the NBR to extend the East Calder Branch into the quarry complex. Under an Order dated 1st November 1868, the line was duly extended southwards, crossing under the present B7015 road and swinging to the east, it passed underneath the Caledonian Camps Branch and terminated at the south-east corner of the quarry in two sidings some 266 yards in length. A run-round facility was installed, with a short loop and a loading bank provided at the buffer stops, and a siding serviced the Burnhouse brick works. A trailing connection ran round to the north side of the quarry complex and, in this direction of travel, another trailing connection gave access to the Caledonian branch and goods station and provided an exchange facility. Here, crude oil in rail tanks from Tarbrax, which had been worked in by the Caledonian, was handed over to the NBR for onwards transmission to Pumpherston Oil Refinery.

Several changes took place to the NBR track layout at Camps over the years. In 1893, one siding was extended by 30 yards for the use of Edinburgh Cleansing Department in disposing of street cleaning waste; this was associated with the changes to the Caledonian Railway trackwork referred to at page 166. This was later rendered redundant and was lifted in February 1901. In February 1903, the Coltness Iron Company advised the NBR board that they had no further use for the loops at Camps and these were lifted out in September 1903.

The MOD at some later time purchased land on the north side of the former quarry, immediately adjacent to the B7015 road, and established an army camp there. A long (private) siding was laid in from the NBR line, running parallel to the Caledonian line, before swinging east to serve this camp. The author can remember the rows of Nissen huts and the railway here. The camp buildings at this location were later cleared and a new MOD storage facility built on the site, certainly lasting into the 1960s.

This branch was worked under One Engine in Steam with Train Staff arrangements with the Signalman at Uphall Junction being custodian of the No. 1 Train Staff. Ground frames controlled the connections to East Calder Goods and also Camps sidings, released by Annett key which was attached to the train staff. East Calder Goods closed in January, 1958.

Uphall Junction to Pumpherston No. 4 Mine

In 1901, at the behest of the Pumpherston Oil Company, the NBR laid in yet another a connection diverging to the right just below the connection to Roman Camp, to provide a railway to serve Pumpherston No. 4 mine (NT079691), the cost of £92 being borne by the oil company.

The branch continued straight ahead and, on crossing the Caw Burn (the West Lothian/Midlothian county march), a facing connection gave access to Pumpherston Oil Works sidings and complex. The connection was controlled by a ground frame released by an Annett key.

Uphall Junction to Pumpherston Oil Works

In 1884, the NBR, as a consequence of an agreement with the Pumpherston Oil Company, agreed to a proposal that the haulageway between Pumpherston Nos 1 & 2 mines could cross the Camps Branch on the level, as a temporary expedient. This was a strange concession and details of how it worked and was controlled, in terms of confliction, are not known. Nos 1 & 2 mines did not close until 1907, but it is thought that this situation with the tramway had long been regularised, possibly by an overpass, two of which were provided across the Camps Branch at this location.

In 1890, an agreement was reached between the Pumpherston Oil Company and the NBR to have a series of new sidings laid adjacent to, and on the right-hand side of, the Camps Branch to serve as exchange sidings for, and provide better rail access to and from, Pumpherston Oil Works (NT077694). The main line connection was laid in as a facing connection providing the facilities requested immediately on the south side of the road bridge. This connection was controlled by a ground frame released by an Annett key. Latterly, the connection was to become hand points. The NBR and successors worked inwards and outwards traffic to and from these exchange sidings, the works' internal pugs working therefrom.

Uphall Junction to Pumpherston No. 5 Mine

In 1899, another approach from the Pumpherston Oil Company to the NBR led to a further new connection being laid in, this to provide rail access to Pumpherston No. 5 mine (NT069699), at a cost of £369 to the oil company. The siding connection was worked by a ground frame released by an Annett key.

All traffic from the three locations listed above was collected from the exchange sidings by the NBR engines working the branch, the shale being dropped off at the oil works sidings. The provision of empty wagons was also worked by the NBR.

THE NORTH BRITISH RAILWAY

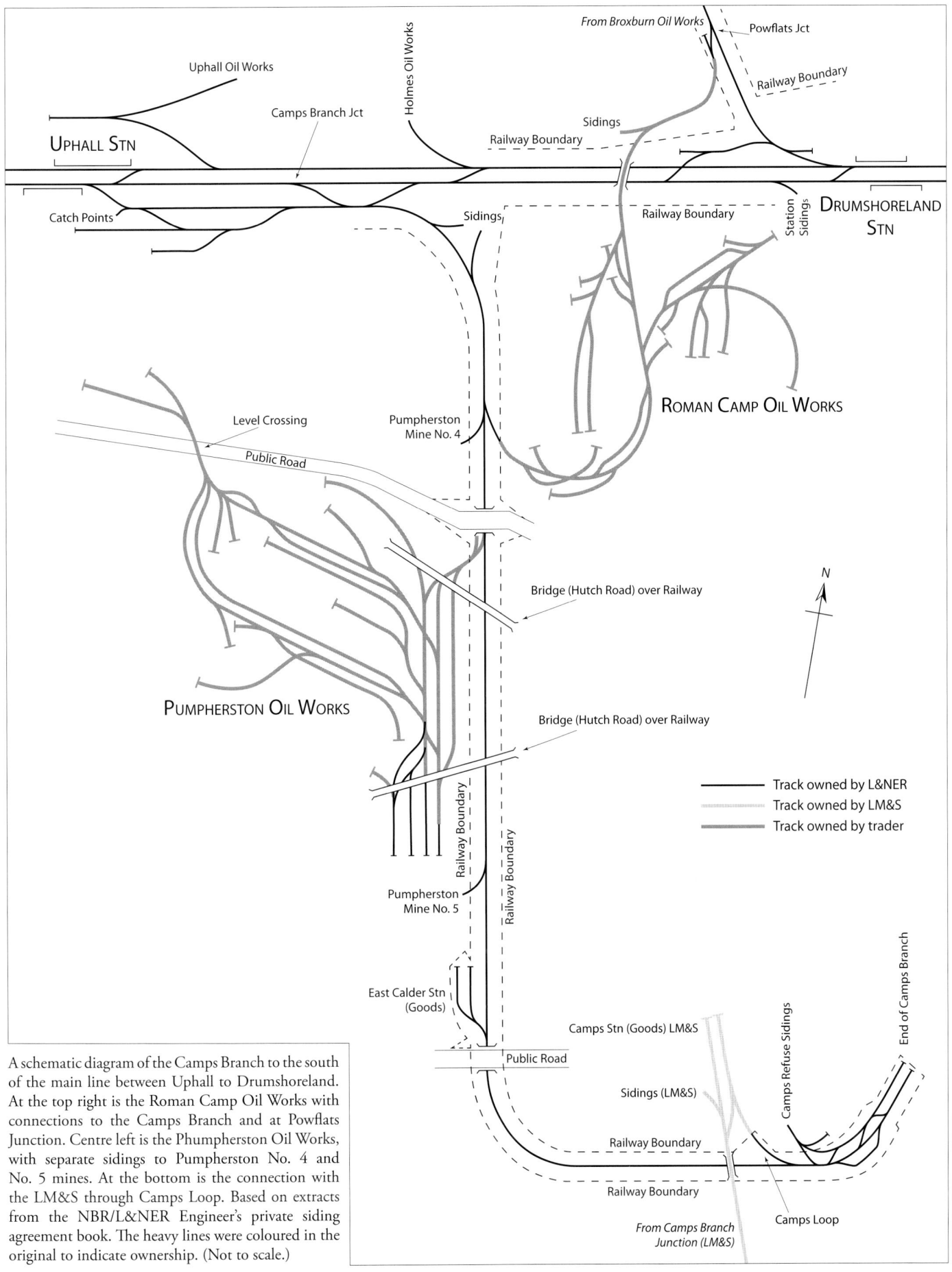

A schematic diagram of the Camps Branch to the south of the main line between Uphall to Drumshoreland. At the top right is the Roman Camp Oil Works with connections to the Camps Branch and at Powflats Junction. Centre left is the Phumpherston Oil Works, with separate sidings to Pumpherston No. 4 and No. 5 mines. At the bottom is the connection with the LM&S through Camps Loop. Based on extracts from the NBR/L&NER Engineer's private siding agreement book. The heavy lines were coloured in the original to indicate ownership. (Not to scale.)

Uphall Junction to Holmes Oil Works

Between Uphall Junction signal box and Drumshoreland, a trailing connection from the Up main line led to a short branch line serving Holmes Oil Works. It is believed that this connection was controlled from a ground frame released by the signalman at Uphall Junction. Holmes Oil Works had a short life and was gone by 1900.

Newliston Junction to Newliston No. 29 Pit

Under an agreement between the NBR and Young's Paraffin, Light & Mineral Oil Company dated 21st November 1882, a branch line was constructed via a facing connection from the Up main Edinburgh to Bathgate line between Clifton siding and Drumshoreland, to serve Newliston No. 29 pit (NT108725), close by the village of Newbridge (see Map C, pp. 28–29).

At the junction (NT011711) there was a loop, with a headshunt on the east side, provided on the Up side of the line. Access to and from the branch was controlled from a new signal box, Newliston Junction, which was located on the Down side of the main line. About mid-way along the loop, a branch line diverged via a facing connection in the Up direction of running, and turned due north for a short distance before swinging north-east to run between Westerton Rows and Stanley House. The branch passed under the Almond Valley viaduct (the 36-arch viaduct) carrying the E&G main line across the River Almond, and serviced Newliston No. 29 shale pit via a headshunt and reversing movement. This branch was worked under One Engine in Steam arrangements with Train Staff (a No. 3 staff) and the custodian was the signalman at Newliston Junction. Since the branch line lay on a heavy rising gradient from the pit up to Newliston Junction, there was also an assisting authority given to permit a banking engine to assist mineral trains (shale) from Newliston Pit sidings to Uphall Junction. Photographic evidence reveals that breakaways with loaded wagons were not unknown, with the resultant chaos!

The line was constructed by Young's with the NBR laying in the main line connection and charging a rent of 6% per annum thereafter. The oil company and L&NER removed the line and all connections in March 1939 upon closure of the pit. The signal box,

Pumpherston Oil Company 8-ton tank wagon. *Ed. McKenna collection*

Pumpherston Oil Company 12-ton tank wagon, No. 120. *Ed. McKenna collection*

Pumpherston Oil Company 10-ton twin tank wagon.
Ed. McKenna collection

Pumpherston Oil Company 20-ton tank wagon with steel frames, No. 109.
Ed. McKenna collection

Pumpherston Oil Company 12-ton tank wagon with steel frames, No. 145.
Ed. McKenna collection

however, remained in use as an intermediate block post between Bathgate Junction and Drumshoreland signal boxes until closure in September 1940, being replaced by a ground frame at that time.

Shale mined at Newliston No. 29 was worked by the NBR to Uphall Junction (for Uphall Oil Works) most likely in some of the NBR pool wagons already described; it is probable that Bathgate engines were employed in this movement, although it is also possible that the Haymarket Broxburn pilot, Ratho pilot or, indeed, the two Bathgate Junction pilot engines that were involved in the transfer of shale traffic from the Ingliston mines, may also have shunted this branch. By 1926, however, it is known that shale from Newliston was worked to Uphall by the return working of the 09.15 Bathgate to Broxburn trip.

There is a first-hand account of a firing trip on a Haymarket freight working in around 1910,* from, as the author described it, a shale oil refinery lying '*mid-way up a mountain*'. The engine was identified as an NBR Drummond 0-6-0 Class 'E', latterly Class J31. On leaving the oil works sidings, the train, loose-coupled as was the norm in these times, and with the locomotive running tender-first, ran away out of control on the steep descending gradient towards a junction with a main line.

---

* *Enginemen Elite* by the late Norman McKillop (Toram Beg), Ian Allan, 1958. McKillop was an engine driver at Edinburgh Haymarket.

Trying to identify this location has exercised many minds over the years. It is likely, and indeed almost certain, that Uphall Junction and the Uphall Oil Works sidings at that same location – the highest point on NBR shale railway lines in West Lothian (120m or 394ft ASL), albeit scarcely 'mid-way up a mountain' – were the central location for this tale. The intermediate signal box, at which a warning note was thrown out, was almost certainly Newliston Junction signal box and the junction ahead was, of course, Bathgate Junction on the E&G main line. This is an interesting, if somewhat embellished, account, which appears to confirm that the Haymarket out-based Ratho engine was at that particular time employed on an 'as required' basis on the Bathgate line when oil traffic was heavy.

Broxburn Junction to Broxburn Albyn Oil Works

Prior to 1848, in the early years of the Edinburgh & Glasgow Railway, a siding was laid in at Broomhouse (so named after a nearby farm) at that company's expense, lying between Ratho and Winchburgh on the north side of the 36-arch viaduct. The siding was 70 yards in length and was accessed via a trailing connection from the then Up main line (E&GR nomenclature and later to be reclassified as the Down line), with a crossover to the other running line. The purpose of this siding is most likely to have been agricultural at that time.

A mishap on the Newliston Branch line. This line was constructed on a heavy gradient leading from Newliston No. 29 pit to the junction with the Edinburgh–Bathgate main line. It would appear that wagons have broken away and run back into quite a dramatic derailment. Here the NBR steam crane from Edinburgh St. Margaret's shed is at work clearing the line. The wagons have been badly damaged and, no doubt, there will be some track damage to repair after rerailing has been completed.

Courtesy of Grangemouth Heritage Trust

ABOVE: Another view of the St. Margaret's steam crane clearing the debris after the derailment on the Newliston Branch. The breakdown train engine is an ex-NBR Class 'C' 0-6-0 freight engine, classed J36 by the L&NER. In the background is the 36-arch viaduct carrying the main Edinburgh to Glasgow railway across the wide valley of the River Almond.
Courtesy of Grangemouth Heritage Trust

BELOW: A further view of the site of the derailment with wagons derailed, badly damaged and much shale spilled around.
Courtesy of Grangemouth Heritage Trust

RIGHT AND BELOW: Young's Paraffin, Light & Mineral Oil Company 10-ton wooden mineral wagon, No. 224, fitted with dumb buffers. Built by R.Y. Pickering of Wishaw.
Ed. McKenna collection

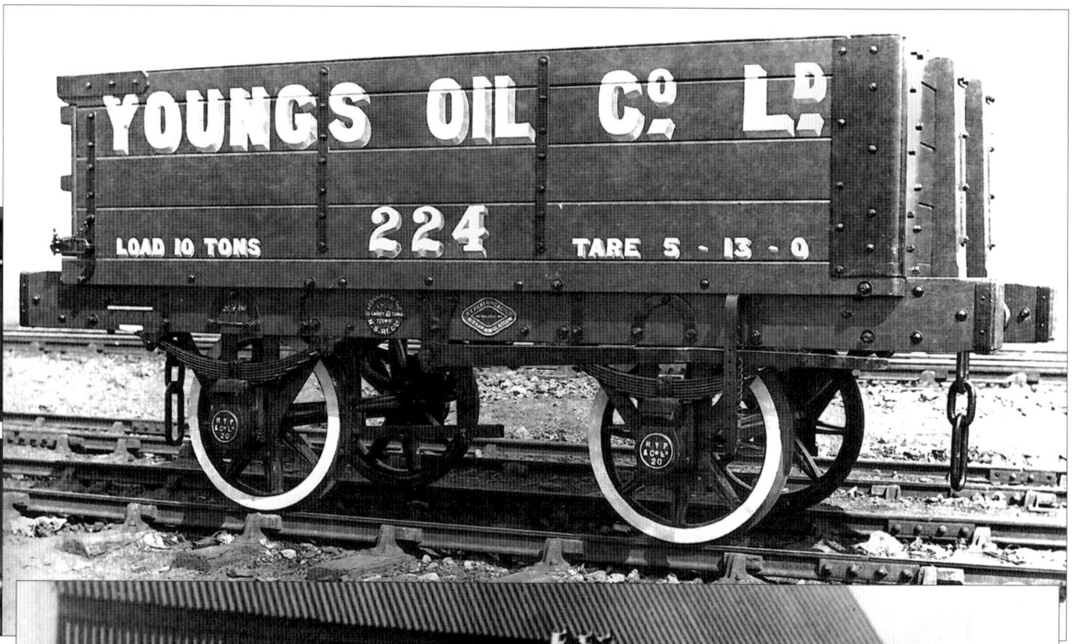

RIGHT: Young's Paraffin, Light & Mineral Oil Company (Uphall) 10-ton tank wagon.
Ed. McKenna collection

By 1866 the Glasgow Oil Company had been formed and established oil works to the north of Broxburn, known as the Albyn Oil Works. This oil works was operational by 1868 and Robert Bell, the Chairman, had a railway constructed from the Albyn Works to the E&G main line (by this time under the ownership of the NBR) at the aforesaid Broomhouse siding. A connection was made into this siding under an Order dated July 1869 and the NBR contributed £150 towards making the connection and providing a further siding. The Albyn Oil Works closed in 1873, but Bell had gone on to set up the Broxburn Oil Company which took over the closed Albyn Works in 1884. With the likelihood of an increase in traffic, this Broxburn Branch was purchased in its entirety by the NBR in May 1884 and in June 1884 a further shunting siding was laid in, the cost of the material being borne by the NBR with the Broxburn Oil Company meeting all the labour costs. A signal box (block post) had been provided at this location (NT098736) and was duly named Broxburn Junction (see Map C, p. 29).

This branch line left the E&G main line at Broxburn Junction and turned south-west to cross over the Union Canal (NT091732) and into Albyn Oil Works. As stated above, it made an end-on connection with the Drumshoreland Branch serving Broxburn (Greendykes) Oil Works at Greendykes Road but, most likely in NBR days and certainly in L&NER days, locomotives of these companies were prohibited from crossing Greendykes Road level crossing (NT083729); thus Albyn was served from the Broxburn Junction end, whilst Holygate and Broxburn Oil Works were served from the Drumshoreland end (see diagram, p. 183). Haymarket shed had traditionally supplied motive power for working the Broxburn Branch and this was perpetuated until final closure.

The line was worked under One Engine in Steam with Train Staff arrangements with the staff being held in Broxburn Junction signal box. However, at the point some 100 yards south of where the Union Canal passed under the line, and right through to Greendykes level crossing, Train Staff working finished and Yard Working (5mph) arrangements applied. A propelling authority was given for trains and lifts of wagons in either direction between Broxburn Junction and Greendykes level crossing, with the authority for trains going onto the branch to propel without a brake van leading. Interestingly, the NBR and L&NER, in Appendices and Working Timetables, always referred to the oil works as the Albion Oil Works.

This branch was, from the outset, to be the domain of the two daily Haymarket engine trip workings, always known thereafter as the Broxburn pilots. It is known that prior to 1910, two Haymarket engines (Wheatley/Drummond Class 'E' 0-6-0s and their crews)

RATHO, SOUTH QUEENSFERRY GOODS, AND DALMENY, via KIRKLISTON.										
DOWN TRAINS.	Distance from Queensferry Junction.	**WEEK-DAYS.**								
		1	2	3	...	4	5	6	6a	
		Goods	Goods	Goods	...	Goods	Goods	Goods	Min.	8.15 a.m. S O (4.3 p.m. S X), South Queensferry Goods on return to Haymarket run to Gorgie with North and Glasgow traffic.
CLASS		D	D	D		D	D	D	B	
Departs from		Haymarket 6 45	Haymarket 7 6a	...	...	South Queensferry 10 30	Haymarket Goods 3 52	Haymarket p.m. 4 51	Bathgate (Up.) 6p40 S O 7 0 S X	Nos. 1, 2, 5, 9 and 12.—‡ Call at Ratho Low Platform for Parcel traffic.
	M. C.	S O a.m.	S X a.m.	S X a.m.		S O a.m.	S X p.m.	S X p.m.	p.m.	No. 6a.—Leaves off Brake Van and lifts Malt Extract Traffic at Kirkliston for Dundee and North thereof. To be assisted from Kirkliston to Dalmeny South by Bathgate Junction No. 1 Pilot.
—Queensferry Junction dep.	... ...	7 9	7 25	...	...	...	4 10	5 10	7 55	
Ratho (Low Platform) „	0 40	‡	‡	......	......	...	‡	......	......	
Kirkliston „	2 10	8 30	8 45	......	......	...	5 27	5 20	8 26	
—Dalmeny Junction „	4 5	9 8	9 20	11 0	......	11 0	6 0	...	8 36	
—Dalmeny (Passenger) arr.	4 68	...	...	11 5	......	11 5	...	...	8 38	
Arrives at		South Queensferry 9 16	South Queensferry 9 28	Haymarket Goods 1 28	...	South Queensferry 1 3	South Queensferry 6 10	Queensferry Jct. 7 0	Dundee 1 30	No. 7.—‡ Calls at Dalmeny South to leave off only, and at Kirkliston to pick up Brake Van.
UP TRAINS.	Distance from Dalmeny (Pass.).	7	8	9	10	12	13	14	15	
		Goods	Goods	Goods	Goods	Goods	Goods	Goods	Goods	No. 13.—Worked by 12.35 S X O.P. Edinburgh to Dalmeny. Traffic on Wednesdays conveyed by 2.17 p.m. W O, O.P. Dalmeny to Edinburgh.
CLASS		B	D	D	D	D	D	D	D	
Departs from		Dundee a.m. 2 30	South Queensferry a m. 10 30	South Queensferry a.m. 10 30	South Queensferry a.m. 10 30	South Queensferry p.m. 1 30	South Queensferry p.m. 2 10	Edinburgh p.m. 3 56	South Queensferry p.m. 7 25	
	M. C.	S X a.m.	S X a.m.	S O a.m.	S O a.m.	W S X p.m.	S X p.m.	S X p.m.		No. 14.—‡ Thereafter to orders of Control office.
—Dalmeny (Passenger) dep.	... ...	5 56	...	11 33	11 35	...	...	...	...	
—Dalmeny Junction „	0 63	6†10	10 38	11 55	11 57	2 0	2 14	......	8 ‡ 5	No. 15.—‡ Runs via Turnhouse.
Kirkliston „	2 58	6‡35	...	12 35	12 35	3 0	...	6 50	...	
—Ratho (Low Platform) „	4 28	......	......	‡	‡	‡	......	......	......	
—Queensferry Junction arr.	4 68	6 45	...	12 49	12 49	3 12	...	7 0	......	
Arrives at		Bathgate (Up.) 8 25	Dalmeny Pass. 11 5	Haymarket Goods 1 3	Haymarket Goods 1 3	Haymarket Goods 3 32	Edinburgh 2 30	‡	Haymarket Goods 8 30	

Extract from the L&NER Working Timetable for 1938 detailing the working of the South Queensferry, Kirkliston and Ratho Branch.

were outstationed at Ratho for this working and also for the South Queensferry Goods – which at the time served the other shale workings on the Newliston and Queensferry branches – since they were, up to that time, twelve-hour shift workings. With the coming of the eight-hour day in 1919, the Broxburn pilot became a double-shifted pilot, as did the South Queensferry pilot, and the locomotives allocated to each were stabled at Haymarket and both workings started and terminated in Haymarket Yard. The two Haymarket engines – originally NBR Class 'E' (J31) 0-6-0s and later NBR Holmes Class 'C' (J36) 0-6-0s – covered both jobs and involved four crews. After 1910, however, it was Bathgate shed which was to provide and crew the two pilot engines at Ratho until closure of the Ingliston and Newliston pits.

Broxburn West Junction to Niddry Castle Oil Works

Immediately beyond Broxburn Junction on the NBR E&G main line, on the west (Down) side, was Broxburn West Junction (NT093744) where a facing connection gave access to Niddry Castle Oil Works (or, as referred to in the NBR/L&NER Appendices, Niddrie Castle). In the 1922 6-inch OS maps, a signal box is shown as being provided here at Broxburn East Junction (see top edge of Map C, p. 29). The connection was, however, latterly to be controlled by a ground frame, electrically released from Broxburn Junction signal box. This actual connecting line was worked as a long private siding into the oil works proper. As in the case of the Broxburn Branch, the Haymarket shed Broxburn pilot worked this long siding. Winchburgh golf club now occupies this site.

Queensferry Junction to South Queensferry and Port Edgar

At Queensferry Junction on the E&G main line, at 7 miles 54 chains from Edinburgh Waverley, the branch line to South Queensferry and Port Edgar diverged to the right via a double junction with the main line, which then formed a branch loop facility which converged to become a single line of railway. This line was authorised in 1863 under the E&GR regime and opened to traffic in March 1868, after the E&GR had amalgamated with the NBR. This line of route was to be used as a freight route for through trains from such places as Bathgate to Thornton Yard in Fife, well into the 1950s, in order to conserve line capacity in the west side of Edinburgh as far as Dalmeny Junction.

Queensferry Junction to Ingliston Pits

The single line was level for the first 200 yards after leaving the main line, but at that point it commenced a descent on a gradient of 1 in 70 for some 530 yards before levelling off at Ratho Lower station. From there the line continued to fall on a gradient of 1 in 70, passing under the main Edinburgh to Newbridge road. At 455 yards from Ratho Lower, a short siding was laid in by the NBR for the use of Hallyards Farm. At the instigation of Young's Paraffin, Light & Mineral Oil Company, at a point just north of Newbridge (NT126733) and some 264 yards beyond the Hallyards siding, a facing connection led off to the right to run north-east, keeping to the south of Hallyards and running parallel to the River Almond, to serve Ingliston Nos 36 & 37 shale pit (NT136735). It was noted in the Minutes of the NBR Director's Meeting in November 1893 that an agreement had been reached with Young's Paraffin, Light & Mineral Oil Company for the construction of this branch line to Ingliston pit, despite the line having actually being opened for traffic in September 1893.

This branch was originally extended for a further 1 mile 66 yards to serve Ingliston No. 33 pit, which lay a mere 660 yards west of what would become the Saughton Junction to Forth Bridge railway line and, later still, the Edinburgh to Aberdeen main line. No. 33 pit was closed in May 1897 and the branch was lifted back to Nos 36 & 37 pit by Young's, the material recovered being used at another, undeclared, location. This connection was controlled by a ground frame released by the Train Staff and was worked under One Engine in Steam arrangements; the line was never signalled.

Initially, this branch line was worked by Young's internal pug engine with the exchange with the NBR occurring at the junction, but in January 1904 the NBR took over all working of the branch, an arrangement that continued until mining of shale ceased in May 1927. In 1926 there was a daily 08.45 Ratho, Ingliston and Uphall shale working and, as the Rosshill mines at Dalmeny were being run down, shale from the Ingliston pits was also worked to Dalmeny Oil Works. Upon closure, the L&NER moved quickly to have the complete branch lifted out and all trackwork had been removed by the end of that year. The shale pits, the site of same, and even the original A9 road, have long disappeared under the 'new' main runway (06/24) and surrounding area at Edinburgh Airport.

Ingliston Pits to Almondhill Paraffin Works, Dalmeny Oil Works, South Queensferry and Port Edgar

At Kirkliston, a siding was extended across the main road to serve a distillery complex; the distillery (Glen Forth Distillery) had been established in 1795 and eventually became part of the DCL Group before closure in 1920. Thereafter a maltings was maintained on the site (Scotmalt Ltd) producing yeast and malt extracts, and both activities provided much inwards and outwards traffic to rail over the years. To the north-east of Kirkliston, there is evidence (on an early OS map) of a private siding having been provided at what was known as the Dundas Oil Company's Almondhill Paraffin Works (NT133750) via a trailing connection on this branch, just to the Dalmeny side of Kirkliston station. This oil works was short-lived, closing in 1870. Here the retorting of shale (in horizontal retorts) won by opencast mining on the Dundas Estates was carried out. Details are hard to come by, but again there is clear evidence of the opencast site on the estate being connected to the Almondhill Works by a narrow gauge tramway. The route of this tramway was largely swept away by the construction of the Winchburgh Junction to Dalmeny Junction NBR chord line upon the opening of the Forth Bridge. The Dundas Oil Company failed

Extract from the L&NER Working Timetable for 1938 detailing the working of the double-shifted Broxburn Pilot. The engine was supplied by Edinburgh Haymarket shed.

in 1870 and was subsequently taken over by the Dalmeny Oil Company in 1871.

Between the site of Almondhill Works and Dalmeny South Junction, a connection was later to be installed to provide access to the Royal Elizabeth Yard, a Royal Navy Victualling Establishment. From this point, the South Queensferry Branch had, prior to the Forth Bridge being built, proceeded to run via Dalmeny before dropping down on a steep gradient to swing westwards into the town proper. Later (in 1878) the branch was to be extended to serve the Royal Navy Establishment at Port Edgar.

With the construction of the Forth Bridge (see Map I, p. 196) and the provision of a new direct main line of railway from Saughton Junction to the bridge proper in 1890, the Kirkliston Branch then joined this new line at Dalmeny South Junction, running over the new metals for a short distance and crossing over the new main line via a series of crossover connections, before it regained its original route down to Port Edgar at Dalmeny North Junction. The original station at Dalmeny (NT143774) was replaced by a new Dalmeny station located on the new main line nearer to the entry to the bridge. A facing connection from the (new) Up main line led to a short series of sidings serving the Dalmeny Oil Works.

On leaving Dalmeny North Junction, the branch continued due north on a steeply falling gradient, down to South Queensferry and Port Edgar, passing underneath the south viaduct of the Forth Bridge. A passenger service operated over the branch until 1930. Passenger stations were provided at Port Edgar, South Queensferry,

BROXBURN RAILWAY.

	Distance from Ratho.	
	M.	C.
DOWN —Ratho	...	...
—Broxburn Junction	2	51
Albion Oil Works	3	31

This Railway is worked by the *Ratho and Broxburn Pilot (Trip and Shunting)*.

CAMPS BRANCH.

		Distance from Uphall Station.	
		M.	C.
DOWN	—Uphall Junction ... ...dep.	0	8
	Roman Camp ... ... ,,	0	50
	Pumpherston { Oil Works Siding ,,	1	14
	{ No. 5 Mine Siding ,,	1	45
	East Calder ... ... ,,	2	48
	Camps Exchange Siding ... ,,	3	37
	Camps Siding ... ...arr.	3	52
	End of Branch ... ... ,,	3	59

Worked by Camps Branch Train—See pages 36 and 38.

HOLYGATE AND BROXBURN OIL WORKS BRANCH.

WEEK-DAYS.	Distance from Drumshoreland.		1 Min. D	2 Min. C	WEEK-DAYS.	3 Min. D	4 Min. C
CLASS					CLASS		
Departs from ... ... {			Bathgate (Up.) a.m. 6 15	Bathgate (Up.) p.m. 2 35	Departs from ... ... {	Bathgate (Up.) a.m. 6 15	Bathgate (Up.) p.m. 2 35
	M.	C.	a.m.	p.m.		a.m.	p.m.
Drumshoreland ... ...dep.	...	...	7 50	3 30	Broxburn Oil Works ... ...dep.	8 50	4 55
Holygate ... ... ,,	1	16	8 25	4 20	Holygate ... ... ,,	......	5 2
Broxburn Oil Works ... ...arr.	1	53	8 35	4 30	Drumshoreland ... ...arr.	9 0	5 20
Arrives at ... ... {			Bathgate Jct. 9 35	Bathgate (Up.) 9 40	Arrives at ... ... {	Bathgate Jct. 9 35	Bathgate (Up.) 9 40

ABOVE: Extract from the L&NER Working Timetable for 1938 detailing the working of the Broxburn Railway, the Camps Branch and the Holygate & Broxburn Oil Works Branch.

BELOW: Photograph showing Ratho station and signal box on the E&G main line. In NBR days this was the stabling point for both the Broxburn Pilot engine (Edinburgh Haymarket engine) and the two Ratho pilots (Bathgate engines). Bill Lynn collection

THE SCOTTISH SHALE OIL INDUSTRY & MINERAL RAILWAY LINES

Map I: Dalmeny (1922). This map shows the new railway layout at Dalmeny after the opening of the Forth Railway Bridge in 1890.

The original Edinburgh & Glasgow Railway Queensferry Junction to Port Edgar railway runs northwards from Kirkliston (bottom left) passing the site of the former Almondhill Paraffin Works (indicated by the spoil heaps marked on the map) and originally ran through old Dalmeny before dropping away leftwards to access South Queensferry and Port Edgar.

For the opening of the Forth Bridge, a new line of railway had been constructed to give a direct connection from Edinburgh to the bridge (lower right). The route of this new line bisected the original railway, requiring a junction at Dalmeny South where the E&GR line joined the new main line. A short distance to the north another new junction, Dalmeny North, was provided for the NBR Winchburgh to Dalmeny chord; it was also at this point that the E&GR branch diverged once more to the right to regain its original route, now passing under the approach viaduct to the Forth Bridge. At Dalmeny North Junction, rail connections gave access to Dalmeny Oil Works. (Scale: six inches to one mile, scaled to 50% to fit page.) *Courtesy of the Trustees of the National Library of Scotland*

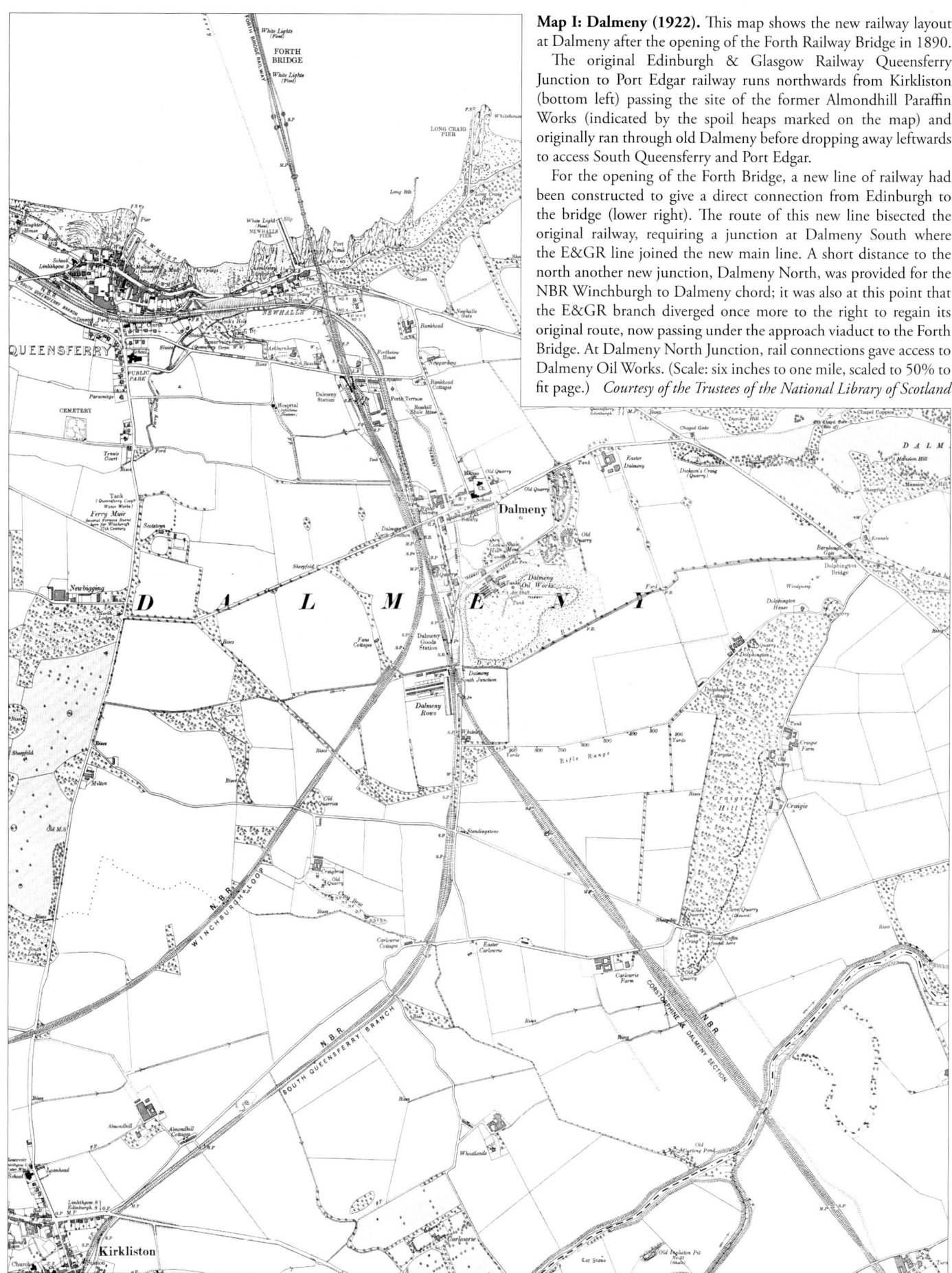

Scottish Oil Agency 14-ton tank wagon, No. 468. *Ed. McKenna collection*

two stations at Newhalls on the east side of the town at different times – Newhalls 1 (1870–78) and Newhalls 2 (1878–1929) – and Dalmeny (old) (closed 1890). South Queensferry was closed to passenger traffic in January 1929, and in September 1930 all passenger services were withdrawn between Ratho Lower and Kirkliston. This line was to carry much, if not all, of the steel required for the construction of the Forth Bridge and was connected to the internal sidings of the bridge workshops located close by, and to the rear of, the Hawes Inn.

This branch line served activities other than the shale oil industry; at South Queensferry the blending plant, and home of VAT 69 Scotch whisky, had been established by W. Sanderson & Co. (later to be part of the DCL Group) which was again a source of considerable rail traffic. The blending plant was to see the closure of the line, latterly maintained for freight only, in 1967, before itself closed in 1984.

EDINBURGH TO LOANHEAD AND STRAITON OIL WORKS (PENTLAND)

On the 20th June 1870, the Edinburgh, Loanhead & Roslin Railway Act 1870 received Royal Assent and the company was empowered to construct:

a line of railway (Railway No. 1), six miles, one furlong and 32 chains in length commencing by a junction with the NBR in the Parish of Newton, and County of   , near the Millerhill station, and terminating in the Parish of Lasswade, and County of Edinburgh, at a point in the field called Longpark, twelve yards or thereby from the northern fence of the road from Roslin to Auchindinny, and sixty yards or thereby from the fence forming the south-western boundary of the grounds of Roslin Free Church manse.

Scottish Oil Agency 14-ton tank wagon, No. 518. *Ed. McKenna collection*

The line was duly constructed and whilst the intention was to continue the line to Penicuik, it was initially to be terminated at Roslin. The NBR had agreed to work the line for a period of thirty years from the date of opening for 45 per cent share of gross revenue, but had also been given the powers to purchase the line within five years of opening. The line was (partly) opened on the 6th November 1873 and opened fully as far as Roslin in August 1874. By 1877 the line had been amalgamated into the NBR and the NBR Sidings Register for that year shows a private siding connected to this branch as the Straiton Oil and Lime Works siding. Shale oil working had pre-dated the opening of the Loanhead Railway, mining having started in 1866 with a small oil works being established at Straiton (NT272660). This operation was financially shaky from the outset and by 1882 the plant, now in poor order, was purchased by the Midlothian Oil Company. The Straiton operation was then bought by the Clippens Oil Company of Paisley in 1884, who at the same time obtained leases at nearby Pentland and set about building completely new oil works to replace the existing plant. The oil produced at Straiton was initially transported by rail to Clippens' original oil refinery at Paisley, but in 1893 the Paisley refinery was closed and all activity was concentrated at the Pentland Oil Works.

Thus the line serving the former Straiton Oil Company's, and later the Clippens Oil Company's, works at Straiton near Loanhead in Midlothian, led from the Loanhead Branch at a point between Gilmerton and Loanhead, roughly at NT284669. The actual location has long been swallowed up by the Edinburgh A720 City By-pass road. This branch ran south for a short distance from the connection on the Loanhead Branch, passing between Loanhead to the north, and Straiton to the south, and was worked by the Clippens Oil Works' pugs as a private mineral line. At the branch connection, new exchange sidings were provided in 1877, controlled by Straiton Sidings signal box, which was not a block post but more of an elevated ground frame. The line served the refinery lying to the east of what is now the A701 road, the actual oil works lying immediately to the west (NT269660)

In the early days of the operation, to transport the crude oil produced at Straiton, empty oil tanks were delivered to the sidings at Straiton by the first goods train of the day on the branch, the 07.40 ex Millerhill; from there they were into the works by Clippens' own pug. Loaded outwards traffic including oil tanks were uplifted by the Down goods at around 13.00 and worked to Millerhill, from whence the loaded tanks were worked forward and eventually handed over to the CR at Coatbridge Whifflet for onward working to the Clippens Oil Works at Linwood, near Paisley.

KINGHORN TO BINNEND OIL WORKS AND MINES (BURNTISLAND OIL COMPANY)

From Kinghorn station yard, this line, running west, passed under Kinghorn High Street by a tunnel constructed on the 'cut and cover' principle, of around 208 yards in length. The railway then ran, still in cutting, into the Links Park where it reversed direction. The reversing point was some 361 yards west of the High Street tunnel mouth. The line then climbed away east to north on a rising gradient of 1 in 40 for 1,076 yards, to service the new candle works which had been established in 1887 on the site of an old flax mill adjacent to Kinghorn loch. From here, the railway swung round and ran westwards for another 2,290 yards to the oil works, on a continuous rising gradient of 1 in 80/1 in 250, before falling down towards the oil works, some 264 yards further on. The overall length of the branch was 2 miles 700 yards and sidings were provided at Kinghorn station yard, Links Park reversing point, the candle works and Binnend Nos 1 and 2 mines. In 1892, a new mine was opened at nearby Grangehill (Binnend Nos 3 & 4) and a short extension to the railway was created to service this mine. The line was never worked by the NBR as such, with the 1891 census recording an engine driver, fireman and platelayer as residents in Binnend village. A Barclay 0-4-0 saddle tank engine, recorded by the Industrial Railways handbook as No. 241, was sold to the company in 1887 and it is likely that this was the works engine for the duration.

Scottish Oil Agency 14-ton tank wagon, No. 544, built by Hurst Nelson Ltd.
Ed. McKenna collection

16

INTERNAL STANDARD GAUGE MINERAL LINES

These wholly private mineral lines were constructed and operated by the various oil companies using their own locomotives and private-owner wagons. Constructed mainly on an 'as and when required' basis, these railway lines crossed public roads at apparent random and, since they were not statutory railways, had no Parliamentary Act under which the level crossings were regulated. Level crossings were thus constructed under local agreements with the various local roads authorities (District Committees and County Councils) and with written assurances being given regarding fencing, gating and means of safe operation. Mostly, the gates, where provided, were operated by the trainmen (shunters or brakesmen) riding with the train. Where no gates were provided, the shunters merely acted as handsignalmen, and it has to be said that such was the limited volume of road traffic then existing, that such level crossings, normally open to road, imposed little or no inconvenience to road users. Even in the last days of operation there was little or no real disruption and, more importantly, no record of any significant accident. Propelling or pulling, these internal trains were worked efficiently and in relative safety.

Addiewell Oil Works to Burngrange, Baads Mine and Fraser Pit

From the oil works complex at Addiewell, one of the largest oil works in the industry, a mineral railway ran through the works complex and swung southwards, being later crossed by the Caledonian Railway's Cleland–Mid Calder line of 1869, and running some 650 yards to the east of Addiewell Station (see Map A, pp. 20–21). There is some evidence to suggest that immediately after passing under the Caledonian main line, in the earliest days a spur swung due west to access earlier Muirhall pits No. 1 and No. 2 lying just to the north of the A71 road. These mines were closed at a date unknown, but had gone before 1907. The main line of route, however, then crossed both the A71 (NT002622) and A704 (NT003621) roads on the level; the level crossings were protected for road/rail movements by hand-operated (trainman worked) level crossing gates. Immediately on the south side of the A704 road, the line split, the right-hand spur connecting the oil works with the newer Muirhall mines. A newer right-hand (west) spur went on to serve Baads No. 22 mine and, at a later date, both the Fraser shale pit (NT004619) and Baads No. 42 colliery (also known as West Mains, NT001610), the coal from which was used in the gas works at Addiewell. The eastern extension of the line ran round behind West Calder cemetery and accessed, in the earlier days, Burngrange No. 39 pit until it closed in 1912 and, later, the newer Burngrange Nos 1 & 2 pit (NT012625), located just to the south side of the former No. 39. The shale (Dunnet) from Burngrange was the main supply for the retorts at Addiewell.

The extension of the line to serve Baads colliery was built and open by 1903. This date can be pinpointed with some accuracy since there is correspondence dated that year between the farmer, one James Rennie of West Mains Farm, and Young's Paraffin, Light & Mineral Oil Company, regarding compensation for ground given up for the railway and fields divided by same. The sum claimed was £27, but, after some protracted correspondence, Rennie finally accepted £17 and an annual supply of as much coal as he could use at a cost of 7/- (35 pence) a ton. The spur to serve Burngrange was re-opened when Burngrange Nos 1 & 2 went into production in the 1930s.

All working over these lines was carried out by Addiewell-based pugs and shale was carried in privately-owned 12-ton wooden wagons.

Addiewell Oil Works to the Gavieside and Polbeth Pits

This route was known to the Addiewell pug drivers as the 'Khyber Pass' (see Map A, pp. 20–21 and Map B, p. 25).

Constructed soon after the opening of the oil works in 1865, this mineral line left the Addiewell complex to the north-east side of the works, immediately crossing the B792 road by a gated level crossing (NT001629) and, on a falling gradient, ran parallel to, and on the south side of, Breich Water. After crossing over an old drove road and right of way by a stone bridge, it passed behind Tennant's March and Mossend Rows. In the early days, a connection swung northwards to cross the Breich Water (NT014636) and round into the Westwood complex. Initially this branch accessed the original Westwood shale pits Nos 12 and 13 via this bridge, but when the pits were closed in 1896 this connection was abandoned. The two Westwood pits were thereafter serviced by a connection which had been laid in by the Caledonian Railway from their West Calder Loop, until closure by 1887. From 1942, access to the new Westwood oil works, refinery and Westwood pit was via the same Caledonian West Calder Loop and the connection to the Westwood complex therefrom.

Meanwhile, the route of this mineral line continued south of Breich Water, to tap the very rich shale workings in and around Polbeth, passing under the secondary road from Mossend to Easter Breich (NT018636). Immediately after this bridge, a chord diverging to the left connected with the Caledonian West Calder Loop, whilst the mineral line continuing ahead crossed under the aforesaid loop (NT021639). The line then ran parallel to a minor road leading to Polbeth, accessing the Gavieside shale mines, Gavieside No. 11 (NT020642), Gavieside No. 27 (NT027646) and Limefield No. 32 pit (NT033649). A facing connection gave access to a spur line which ran north behind the policies of Briestonhill House before turning north-east to serve Gavieside No. 40 pit (NT024653) and the former Gavieside Oil Works until closure. A trailing connection on this line (NT028647) led into another spur line passing the site of an earlier Polbeth No. 26 pit (closed) to Polbeth No. 20 pit (NT024638), Polbeth No. 21 pit (NT025640) and the new Polbeth No. 26 pit, immediately adjacent to the A71 road (NT026639). The winding house of No. 26 pit survived closure in 1947 and was integrated into the new Polbeth community centre built in the 1950s.

Shale from Westwood pit was conveyed to Addiewell for processing until post-1941, when the new Westwood Oil Works were opened; thereafter, the shale was retorted there and, in a reversal of roles, the crude oil being produced at Addiewell was transported to Westwood for refining.

PHILPSTOUN OIL WORKS TO THE PHILPSTOUN AND OCHILTREE MINES

This was another very short railway linking the oil works with the shale working at Philpstoun No. 4 mine (NT045751) and Philpstoun No. 7 mine (NT045755) at the east side of Bridgend Rows. From the pits, it ran east at first, gradually swinging due north to cross the B9080 road at Burnside level crossing (NT056757) between Gateside Farm and Burnside (Threemiletown). It continued northwards to cross over the Union Canal at Fairniehill via bridge No. 37 (NT054768) and into the oil works proper. At some time after the failure of Bridgend Oil Works in 1902, James Ross & Co., the proprietors of Philpstoun Oil Works, took over the mines which had served this former Bridgend Oil Works. They extended this railway to cross the road at the north end of Bridgend Rows (NT043755) near to where the new primary school stands today, to access the other Ochiltree mines, Nos 5 and 6.

HOPETOUN OIL WORKS TO VARIOUS SHALE MINES

From Hopetoun Oil Works, the branch line which had come in from the west from Uphall via Ecclesmachan, had formed a junction with another mineral line (at NT079741) and this latter line then continued northward past Glendevon (Hopetoun) No. 5 mine. The line diverged at a facing junction (NT076746) with the right-hand line running to Glendevon (Hopetoun) No. 6 mine (NT074753), immediately adjacent to the B9080 road (now the site of a filling station and garage). The left-hand spur passed south of Lampindub and Trinlymire farms to serve Hopetoun No. 35 mine, lying immediately east of Redhouse Rows (NT060753). In 1906, a new connection was laid in to serve a new branch line running due north, crossing the B9080 road on the level at, or near, the Threemiletown kennels of the Linlithgowshire & Stirlingshire Hunt (NT061767) and, running adjacent to, and on the right-hand side of, the B8046 road to Philpstoun. The line swung east to service Hopetoun No. 41 (Fawnspark) shale mine just to the south of the Union Canal (NT063765). The original line also ran to the south-west and behind Hopetoun Oil Works, crossing the B8020 road (NT083738) to serve Niddrie shale mine (NT089738).

ALBYN OIL WORKS TO VARIOUS SHALE MINES

Immediately after crossing Greendykes Road, the Uphall to Albyn Oil Works branch diverged (NT084731), with the right-hand route meeting the NBR line from Broxburn Junction (see Map C, pp. 28–29). The left-hand spur accessed the oil works via a trailing connection, but this line, as a purely mineral railway, also continued due north to serve South Greendykes shale mine, before turning due west and crossing the B8020 on the level close by Greendykes cottages (NT080735). The line served Hayscraigs shale mines (NT077737) and continued to meander roughly south-westwards to a point (NT065730) where it divided, the right hand spur continuing to the Carledubs shale mine (NT061729); the left hand spur headed south, crossing the B8046 (NT059724) and serving the shale mine at Crossgreen (NT057723). All train working over this line was carried out by the internal pug engines.

OAKBANK OIL WORKS TO MID CALDER MINES

An internal railway line left Oakbank Oil Works and, paralleling Linhouse Water on the south side of a fairly steep gorge (see Map F, p. 166), ran for about half a mile northwards to Mid Calder village, accessing Oakbank No. 1 mine (NT078674) just south of the road from East Calder. This line was worked by the Oakbank pug engines. The mine worked the Pumpherston shale and closed in 1909.

Oops! One of the Addiewell pugs, possibly No. 9, comes to grief when it is comprehensively derailed within the oil works complex.

Courtesy West Lothian District History Library.

17

OAKBANK OVERHEAD ROPE SYSTEM

It can never be said that the Scottish oil companies were anything other than innovative and unafraid to embrace differing and new technologies. The finest example of this innovation was the Oakbank Overhead Cableway, constructed to convey shale from the New Farm mine near Dedridge to Oakbank Oil Works.

When the principal supplies of shale dried up in 1909 with the closure of Oakbank No. 1 mine (also known as Mid Calder No. 1) at Mid Calder, which had been rail-connected to Oakbank Oil Works, two new mines were opened to mine both the Pumpherston and Dunnet shales. Alderstone No. 3 worked the Pumpherston seam and New Farm Nos 3 & 4 worked the Upper Dunnet shale. The locations of both mines have long disappeared under the spread of Livingston New Town, but the twenty-eight oil company houses (Dedridge Rows) which were constructed by the Oakbank Oil Company are still extant, sitting amongst the modern housing in the area of Livingston which perpetuates the name Dedridge.

Because of the geography of the intervening land between New Farm mine (NT058659 or thereabouts) and Oakbank Oil Works, a mineral railway was quite out of the question (see Map F, p. 166). Thus, in order to transport the shale to the oil works, the Oakbank Oil Company employed German technology, supplied and constructed by J. Pohlig Ltd of Cologne, in the shape of an aerial ropeway from which buckets were suspended. This overhead ropeway ran almost as the crow flies, from New Farm mine at Dedridge to Oakbank Oil Works, passing over what was to become the A71 trunk road from Edinburgh to Ayr, close by the Dedridge Rows, and thereafter crossing over the deep gorges of both Murieston Water and the Linhouse Water *en route*, and conveying the shale in the underslung buckets. During the First World War, all references to the German origin on the ropeway infrastructure were carefully removed, just in case of a protest backlash, but fears were unfounded.

At Oakbank Oil Works the buckets were discharged into a large concrete hopper lying below the operating mechanism and the buckets went round in a loop, being returned empty to the mine. The whole operation proved, on the whole, to be a success, but by 1919 the New Farm mine workings were exhausted and the mines and ropeway closed in the same year.

The aerial ropeway crossing what will later become the A71 Edinburgh to Ayr trunk road, at Dedridge, close by the County Roadman's house. The plaque on the bridge states 'Constructed by J. Pohlig Ltd, Cologne, Germany'.

Two views from the top of the discharging point of the aerial ropeway showing the large shale bunker into which the suspended hoppers of shale were emptied before being sent back empty to the mine. To ease operations, the bunker was built within a bing, with a tramway running horizontally from beneath it to the top of the retorts.

ABOVE: A view of the inside of the shale hopper at Oakbank Oil Works where the overhead buckets from New Farm mine were discharged. The hopper was open-ended to permit the tramway and hutches access and to ease loading of the shale for transfer to the retort benches.

The Oakbank Oil Company's New Farm mine at Dedridge, showing winding house with aerial ropeway. Note the power lines to the winding house on top of the gantry.

Left: Photographed from the Contentibus bing at Oakbank, the Oakbank aerial ropeway can be seen crossing the gorge of the Linhouse Water and running in a straight line towards New Farm mine. The electric power lines are just visible running across the bottom of the gorge to meet the ropeway on the far side, from where they ran on top of the gantries to the winding house.

Right: The aerial ropeway showing bare feeder power wires connected to overhead power lines. The Contentibus bing can be seen in the background.

APPENDIX A

LOCOMOTIVES

The locomotives used by the individual oil companies through to the establishment of the single Scottish Oils Ltd fell into three distinct groups. These were:
- Narrow gauge locomotives used for underground haulage in the shale pits and mines, working on a rail gauge of only 2ft. These were, in the main, 4-wheeled diesel mechanical locomotives, but there were also a number of 4-wheeled battery electric locomotives used underground.
- Narrow gauge locomotives used on surface narrow gauge railways on a 2ft 6ins rail gauge. These were limited to two locations and were, for the time, quite unique in that they were 4-wheeled electric locomotives drawing power from an overhead line system. These locomotives and the system they operated upon have been looked at in some detail at Chapter 13 in this book.
- Standard gauge steam locomotives. By far the most common type used were the 'pugs', the local term used when referring to the 0-4-0 and 0-6-0 outside cylinder steam locomotives which were generally, but not confined to, the saddle tank type. These small tank engines were designed to be self-contained in terms of coal and water supply, but generally ran with a small wooden tender behind to augment the very limited space for coal on the engine. These small saddle tank engines were, in the main, supplied by the noted Scottish industrial locomotive builder, Andrew Barclay & Sons of Kilmarnock in East Ayrshire. The first company, Andrew Barclay & Co., was established in 1859; a sister company, Barclay & Co., was opened in 1871, both companies then working together. The original company failed in 1874, followed by the sister company in 1881 but, in 1884, like the phoenix arising from the ashes, the new company of Andrew Barclay & Sons commenced operation. The Scottish shale oil industry was to prove a good and faithful customer thereafter.

Each company had several pugs working on their premises and over the years there was a considerable exchange of locomotives as oil works closed down and the industry contracted into fewer, but larger and more economical, retorting and refining works.

This list may not be entirely definitive in nature as many records no longer exist and some information is sketchy in the extreme. It is, however, based on best available information.

Broxburn Oil Company's No. 1. Built by Andrew Barclay of Kilmarnock in 1919, works No. 1594. *Hamish Stevenson collection*

THE SCOTTISH SHALE OIL INDUSTRY & MINERAL RAILWAY LINES

ALLOCATION OF STEAM LOCOMOTIVES

BROXBURN OIL COMPANY

Broxburn Works & Middleton Hall

No./Name	Locomotive Type	Built by	Works No.	Date Built	Comments
1	0-4-0 saddle tank with outside cylinders	Andrew Barclay, Kilmarnock	1594	1919	New to Broxburn.
1	0-4-0 saddle tank with outside cylinders	Andrew Barclay, Kilmarnock	Unknown	Unknown	
2	0-4-0 saddle tank with outside cylinders	Andrew Barclay, Kilmarnock	192	1877	
3	0-4-0 saddle tank with outside cylinders	Andrew Barclay, Kilmarnock	205	1879	New to Broxburn and later loaned to BP Grangemouth.
4	0-4-0 saddle tank with outside cylinders	Andrew Barclay, Kilmarnock	251	1882	New to Broxburn.
5	0-4-0 saddle tank with outside cylinders	Andrew Barclay, Kilmarnock	661	1890	New to Broxburn.
6	0-4-0 saddle tank with outside cylinders	Andrew Barclay, Kilmarnock	789	1897	New to Broxburn.
7	0-4-0 saddle tank with outside cylinders	Gibb & Hogg, Airdrie	56	1905	New to Broxburn
8	0-4-0 saddle tank with outside cylinders	Andrew Barclay, Kilmarnock	1152	1908	Transferred from Roman Camp Works, 1955. To P.W. McLellan, Langloan. Re-sold to Dixon's Iron Works 1957.

Broxburn Oil Company's No. 5, built by Andrew Barclay in 1890, works No. 661. Seen here at Uphall in July 1936.   *Hamish Stevenson collection*

Broxburn Oil Company's No. 6 at Uphall in July 1936. Originally built by Andrew Barclay in 1897, works No. 789. New to Broxburn, this engine was then variously transferred to Middleton Hall, Grangemouth and Roman Camp. *Hamish Stevenson collection*

Broxburn Oil Company's No. 6 engine again, seen here at Roman Camp in September 1950, before being withdrawn later that year.
Hamish Stevenson collection

Broxburn Oil Company's No. 1.

Hopetoun

No./Name	Locomotive Type	Built by	Works No.	Date Built	Comments
1	0-4-0 saddle tank with outside cylinders	Andrew Barclay, Kilmarnock	1151	1908	On loan from Young's, Hopetoun Works, 1953. Returned 1956.
3	0-4-0 saddle tank with outside cylinders	Andrew Barclay, Kilmarnock	1960	1929	From Young's, Hopetoun Works. To Motherwell Bridge Co. Uphall, 1964.

Roman Camp Oil Works, Uphall

No./Name	Locomotive Type	Built by	Works No.	Date Built	Comments
6	0-4-0 saddle tank with outside cylinders	Andrew Barclay, Kilmarnock	789	1897	From Broxburn Works, 1924. Returned to Scottish Oils Broxburn Works, 1956.
8	0-4-0 saddle tank with outside cylinders	Andrew Barclay, Kilmarnock	1152	1908	From Young's, Uphall Works, 1936. To Scottish Oils Broxburn Works, 1955.
Hopetoun No. 7	0-4-0 saddle tank with outside cylinders	Grant, Ritchie & Co. Kilmarnock	522	1907	From Young's, Hopetoun Works, 1953. To Scottish Oils Pumpherston Works, 1958.
Hopetoun No. 1	0-4-0 saddle tank with outside cylinders	Andrew Barclay, Kilmarnock	1151	1908	From Broxburn Works, 1956. To Scottish Oils Broxburn Works.

LINLITHGOW OIL COMPANY

Bridgend (Champfleurie) Oil Works

No./Name	Locomotive Type	Built by	Works No.	Date Built	Comments
The Stag	0-4-0 saddle tank with outside cylinders	Barclay's & Co., Kilmarnock	249	1878	New to Bridgend Oil Works. To J. Ross & Co., Philpstoun Oil Works, 1910.
None	0-4-0 saddle tank with outside cylinders	Barclay & Co., Kilmarnock	Unknown	Unknown	
None	0-4-0 saddle tank with outside cylinders	Dick & Stevenson, Airdrie Works	Unknown	Unknown	

LOCOMOTIVES

DALMENY OIL COMPANY

Dalmeny Oil Works

No./Name	Locomotive Type	Built by	Works No.	Date Built	Comments
None	0-4-0 saddle tank with outside cylinders	Dick & Stevenson, Airdrie	Unknown	Unknown	From Young's Hopetoun Works. To Young's, Addiewell Works 1924.
Mabel	0-4-0 saddle tank with outside cylinders	Gibb & Hogg, Victoria Engine Works, Airdrie	Unknown	1902	On loan ex Young's, Uphall Works. Returned to Young's, Uphall Works.

OAKBANK OIL COMPANY

Oakbank Oil Works

No./Name	Locomotive Type	Built by	Works No.	Date Built	Comments
None	0-4-0 saddle tank with outside cylinders	Andrew Barclay, Kilmarnock	110	1871	New to Oakbank.
1	0-4-0 saddle tank with outside cylinders	Barclay's & Co., Kilmarnock	Unknown	1884	Boiler exploded 1902.
1	0-4-0 saddle tank with outside cylinders	Andrew Barclay, Kilmarnock	951	1903	New to Oakbank.
2	0-4-0 saddle tank with outside cylinders	Grant Ritchie & Co., Kilmarnock	698	1920	New to Oakbank. Transferred to Young's, Addiewell, 1931.
3	0-4-0 saddle tank with outside cylinders	North British Loco. Co., Glasgow	5211	1897	New to Oakbank.

J. ROSS & SONS (PHILPSTOUN OIL COMPANY)

Philpstoun Oil Works

No./Name	Locomotive Type	Built by	Works No.	Date Built	Comments
1	0-4-0 saddle tank with outside cylinders	Andrew Barclay, Kilmarnock	278	1885	New to Philpstoun.
2	0-4-0 saddle tank with outside cylinders	Andrew Barclay, Kilmarnock	1028	1905	New to Philpstoun.
2	0-4-0 saddle tank with outside cylinders	Andrew Barclay, Kilkmarnock	1120	1907	New to Philpstoun Oil Works.
3	0-4-0 saddle tank with outside cylinders	Andrew Barclay, Kilmarnock	1960	1929	New to Philpstoun Oil Works. Transferred to Young's Hopetoun Works, 1934.
The Stag	0-4-0 saddle tank with outside cylinders	Barclay's & Co., Kilmarnock	249	1878	Transferred from Champfleurie Oil Works, 1910 and scrapped 1924.
None	0-4-0 saddle tank with outside cylinders	Andrew Barclay, Kilmarnock	61	1867	Transferred from Young's Addiewell Works, 1924 and sold to Shanks & McEwan.

Above: J. Ross & Son's Philpstoun Oil Works No. 2 in very clean condition. This was an Andrew Barclay product, works No. 1120 of 1907.
Hamish Stevenson collection

Hopetoun No. 1 Pug.

Hamish Stevenson Collection

LOCOMOTIVES

Young's Paraffin, Light and Mineral Oil Company

Hopetoun Oil Works & Shale Mine, Winchburgh

No./Name	Locomotive Type	Built by	Works No.	Date Built	Comments
(Hopetoun) No. 1	0-4-0 with outside cylinders	Andrew Barclay, Kilmarnock	1151	1908	New to Hopetoun Oil Works. Transferred to Broxburn Oil Works, 1953.
3	0-4-0 with outside cylinders	Andrew Barclay, Kilmarnock	1960	1929	Transferred from Philpstoun Oil Works, 1934. Transferred to Broxburn Oil Works, 1953.
4	0-4-0 with outside cylinders	Dick & Stevenson, Airdrie	Unknown	Unknown	New to Hopetoun Works. Transferred to Addiewell Oil Works then to Uphall Oil Works. Retired from Uphall and transferred to Dalmeny Oil Company.
6	0-4-0 tank with outside cylinders	Dübs & Co., Glasgow	1437	1881	Transferred from Addiewell Oil Works, 1900. Transferred to Uphall Oil Works.
(Hopetoun) No. 7	0-4-0 saddle tank with outside cylinders	Grant, Ritchie, Kilmarnock	522	1907	Transferred from Uphall Oil Works. Transferred to Roman Camp Oil Works, 1953.

An unidentified Young's Paraffin, Light & Mineral Oil Company locomotive seen here in August 1950. *Hamish Stevenson collection*

Uphall Oil Works & Shale Mines

No./Name	Locomotive Type	Built by	Works No.	Date Built	Comments
1	0-4-0 saddle tank with outside cylinders	Neilson & Co., Glasgow	Unknown	c.1880	Rebuilt 1920.
2	0-4-0 saddle tank with outside cylinders	Dick & Stevenson, Airdrie	Unknown	Unknown	
3	0-4-0 saddle tank with outside cylinders	Dick & Stevenson, Airdrie	Unknown	Unknown	
4	0-4-0 saddle tank with outside cylinders	Dick & Stevenson, Airdrie	Unknown	Unknown	Transferred from Addiewell Oil Works. Transferred to Hopetoun Oil Works.
4	0-4-0 saddle tank with outside cylinders	Andrew Barclay, Kilmarnock	Unknown	Unknown	Transferred from Addiewell Oil Works.
5	0-4-0 saddle tank with outside cylinders	Neilsons, Glasgow	1236	1866	Transferred from Addiewell Oil Works, 1905.
5	0-4-0 saddle tank with outside cylinders	Andrew Barclay, Kilmarnock	84	1869	Transferred from Addiewell Oil Works.
6	0-4-0 tank with outside cylinders	Dübs & Co., Glasgow	1437	1881	Transferred from Hopetoun Works.
7	saddle tank with outside cylinders	Grant, Ritchie, Kilmarnock	522	1907	New to Uphall. Transferred to Hopetoun Oil Works.
(Uphall) No. 8	0-4-0 saddle tank with outside cylinders	Andrew Barclay, Kilmarnock	1152	1908	New to Uphall. Transferred to Roman Camp Works, 1936.
Mabel	0-4-0 saddle tank with outside cylinders	Gibb & Hogg, Airdrie	Unknown	1902	On loan from Gibb & Hogg. Loaned to Dalmeny Oil Company. Returned to Grant & Hogg.

Uphall Oil Works (Young's) No. 8. Built by Andrew Barclay in 1908, works No. 1152. *Hamish Stevenson collection*

Addiewell Oil Works

No./Name	Locomotive Type	Built by	Works No.	Date Built	Comments
1	0-4-0 well tank with outside cylinders	Hawthorn, Leith	Unknown.	1865	New to Addiewell.
1	0-4-0 saddle tank with outside cylinders	Andrew Barclay, Kilmarnock	1153	1908	New to Addiewell. To United Fireclay Co., Armadale, 1957.
2	0-4-0 well tank with outside cylinders	Hawthorn, Leith	Unknown.	1865	New to Addiewell.
2	0-4-0 saddle tank with outside cylinders	Andrew Barclay, Kilmarnock	66	1867	New to Addiewell. To Shanks & McEwan, Glasgow, 1920.
2	0-4-0 saddle tank with outside cylinders	Andrew Barclay, Kilmarnock	1970	1929	New to Addiewell. Scrapped 1960.
3	0-6-0 saddle tank with outside cylinders	North British Locomotive Co.	1236	1866	New to Addiewell. To Uphall Oil Works, 1905.
3	0-4-0 saddle tank with outside cylinders	Andrew Barclay, Kilmarnock	1254	1911	New to Addiewell. To United Fireclay Products Ltd, Armadale, 1957.
4	0-4-0 saddle tank with outside cylinders	Andrew Barclay, Kilmarnock	61	1867	New to Addiewell. To James Ross & Co., Philpstoun Oil Works, 1924.
4	0-4-0 saddle tank with outside cylinders	Dick & Stevenson, Airdrie	Unknown	Unknown	Transferred from Hopetoun Oil Works, date unknown. Transferred to Uphall Oil Works, date unknown. Transferred to Dalmeny Oil Works, date unknown. Retired at Dalmeny, 1924. Disposal unknown.
4	0-4-0 saddle tank with outside cylinders	Grant Ritchie, Kilmarnock	698	1920	Transferred from Oakbank Oil Works, 1931. Transferred to Westwood Oil Works, 1959.
5	0-4-0 saddle tank with outside cylinders	Andrew Barclay, Kilmarnock	84	1869	New to Addiewell. Transferred to Uphall Oil Works.
6	0-4-0 saddle tank with outside cylinders	North British Locomotive Works, Glasgow	1574	1870	
7	0-4-0 saddle tank with outside cylinders	Andrew Barclay, Kilmarnock	137	1873	New to Addiewell.
8	0-4-0 side tank with outside cylinders	Dübs & Co., Glasgow	1098	1878	Originally built for India. New to Addiewell.
9	0-4-0 side tank with outside cylinders	Dübs & Co., Glasgow	1437	1881	Originally built for India. New to Addiewell. To Hopetoun Oil Works, 1900.
9	0-4-0 saddle tank with outside cylinders	Andrew Barclay, Kilmarnock	1373	1914	New to Addiewell. Scrapped 1958.
None	0-6-0 saddle tank with outside cylinders	Peckett & Sons, St. George, Bristol	Unknown	Unknown	On loan ex A. Baird, Hamilton. Returned to A. Baird, Hamilton.
None	0-4-0 saddle tank with outside cylinders	Grant Ritchie, Kilmarnock	Unknown	Unknown	On loan from Gibb & Hogg, Aidrie. Returned to Gibb & Hogg, Airdrie.

Two photographs of Addiewell (Young's) No. 4 at Addiewell. This engine was a Grant Ritchie product of 1920, works No. 698. The top picture is dated May 1956. The small wooden tender is also identified as No. 4 to match the locomotive. *Hamish Stevenson collection*

LOCOMOTIVES

Addiewell Oil Works (Young's) No. 9 seen here on 8th July 1936 at Addiewell. This locomotive was built by Andrew Barclay in 1914, works No. 1373. The engine was scrapped in 1958.

Hamish Stevenson collection

PUMPHERSTON OIL COMPANY

Pumpherston Oil Works

No./Name	Locomotive Type	Built by	Works No.	Date Built	Comments
1	0-4-0 saddle tank with outside cylinders	Andrew Barclay, Kilmarnock	901	1900	New to Pumpherston.
2	0-4-0 saddle tank with outside cylinders	Andrew Barclay, Kilmarnock	1129	1907	New to Pumpherston.
3	0-4-0 saddle tank with outside cylinders	Andrew Barclay, Kilmarnock	1416	1915	New to Pumpherston. To United Fireclay Ltd, Armadale, 1956.
1	0-4-0 saddle tank with outside cylinders	Andrew Barclay, Kilmarnock	2065	1939	New to Pumpherston. Scrapped 1964.
(Hopetoun) No. 7	0-4-0 with outside cylinders	Grant Ritchie & Co. Kilmarnock	522	1907	From Broxburn Oil Co., Roman Camp Oil Works, 1958.

NB. There was one other 0-4-0 steam saddle tank engine of 1880s vintage working at Pumpherston but no details can now be found.

CLIPPENS OIL COMPANY

Straiton Oil Works

No./Name	Locomotive Type	Built by	Works No.	Date Built	Comments
1	0-4-0 saddle tank with outside cylinders	Andrew Barclay, Kilmarnock	778	1897	New to Straiton. To Glasgow Iron & Steel Co.
7	0-4-0 saddle tank with outside cylinders	Andrew Barclay, Kilmarnock	193	1878	New to Straiton. Rebuilt by Shotts Iron Co., 1907. To Shotts Iron Works, 1920.
6	0-4-0 saddle tank with outside cylinders	Barclay's & Co., Kilmarnock	282	1881	From Coats Works, 1920. To Connell, Coatbridge, 1956. Scrapped.

ABOVE: Pumpherston Oil Company No. 1, built by Andrew Barclay in 1900, works No. 901. Note the steel-bodied tender complete with handrail.
Hamish Stevenson collection

BELOW: Pumpherston Oil Company's No. 2 engine. This was another Andrew Barclay product, built in 1907, works No. 1129.
Hamish Stevenson collection

Scottish Oils Ltd

Westwood Oil Works

No./Name	Locomotive Type	Built by	Works No.	Date Built	Comments
5	0-4-0 saddle tank with outside cylinders	Andrew Barclay, Kilmarnock	661	1890	Transferred from Broxburn Oil Co., 1940. Scrapped Connell, Coatbridge, 1957.
4	0-4-0 saddle tank with outside cylinders	Grant Ritchie & Co., Kilmarnock	698	1920	On loan from Young's Addiewell. Returned 1959. Re-loaned 1959 until 1960 and returned to Addiewell. Scrapped on site by Connell, 1963.

Clippens Oil Company No. 1 at Straiton, complete with spark arrestor and in sparkling condition. The engine was built by Andrew Barclay in 1897, works No. 778.
Courtesy Jeff Hurst

Allocation of Diesel Mechanical Locomotives

Roman Camp Shale Mines
(Underground haulageway: 2 feet gauge)

No./Name	Locomotive Type	Built by	Works No.	Date Built	Comments
None	4 wheel diesel mechanical	Ruston & Hornsby, Lincoln	198277	1939	New to Roman Camp. Scrapped.
None	4 wheel diesel mechanical	Ruston & Hornsby, Lincoln	213856	1942	New to Roman Camp. Scrapped.
None	4 wheel diesel mechanical	Ruston & Hornsby, Lincoln	235683	1945	New to Roman Camp. Scrapped.

Philpstoun Shale Mines
(Underground haulageway: 2 feet gauge.)

No./Name	Locomotive Type	Built by	Works No.	Date Built	Comments
None	4 wheel diesel mechanical	Ruston & Hornsby, Lincoln	201975	1940	New to Philpstoun Nos 1 and 6 mines. Scrapped.
None	4 wheel diesel mechanical	Ruston & Hornsby, Lincoln	201983	1940	New to Philpstoun Nos 1 and 6 mines. Scrapped.
None	4 wheel diesel mechanical	Ruston & Hornsby, Lincoln	202004	1940	New to Philpstoun Nos 1 and 6 mines. Scrapped.
None	4 wheel diesel mechanical	Ruston & Hornsby, Lincoln	222073	1943	Transferred to Philpstoun Nos 1 and 6 mines from MOD. Scrapped.

Uphall Mine
(Underground haulageway: 2 feet gauge)

No./Name	Locomotive Type	Built by	Works No.	Date Built	Comments
None	4 wheel diesel mechanical	Ruston & Hornsby, Lincoln	229612	1944	New to Uphall. Disposal unknown.
None	4 wheel diesel mechanical	Ruston & Hornsby, Lincoln	221617	1943	From Admiralty RNAD, Broughton Moor, Cumberland, 1946. To Anglo-Iranian Oil Co. Ltd, 1948.

Burngrange Nos 1 & 2 Pit
(Underground haulageway: 2 feet gauge)

No./Name	Locomotive Type	Built by	Works No.	Date Built	Comments
None	4 wheel diesel mechanical	Ruston & Hornsby, Lincoln	191678	1939	New to Burngrange. Disposal unknown.
None	4 wheel diesel mechanical	Ruston & Hornsby, Lincoln	213858	1942	New to Burngrange. To South Crofty & Eastpool Mines, Cornwall, 1962.
None	4 wheel diesel mechanical	Ruston & Hornsby, Lincoln	252865	1947	New to Burngrange. Disposal unknown.

Fraser Shale Pit
(Underground haulageway: 2 feet gauge)

No./Name	Locomotive Type	Built by	Works No.	Date Built	Comments
None	4 wheel diesel mechanical	Ruston & Hornsby, Lincoln	213857	1942	New to Fraser pit. Disposal unknown.

Westwood Pit
(Underground haulageway: 2 feet gauge)

No./Name	Locomotive Type	Built by	Works No.	Date Built	Comments
None	4 wheel diesel mechanical	Ruston & Hornsby, Lincoln	182143	1936	New to Westwood. Disposal unknown.
None	4 wheel diesel mechanical	Ruston & Hornsby, Lincoln	229613	1944	New to Westwood. Disposal unknown.
None	4 wheel diesel mechanical	Ruston & Hornsby, Lincoln	242888	1946	New to Westwood. Disposal unknown.
None	4 wheel diesel mechanical	Ruston & Hornsby, Lincoln	280867	1949	New to Westwood. Disposal unknown.

LOCOMOTIVES

Westwood Oil Works showing Oakbank Oil Company diesel locomotive No. 2 at work. This 0-4-0 diesel mechanical locomotive was built by Andrew Barclay of Kilmarnock, works No. 412, to the order of Scottish Oils Ltd, and was supplied new to Westwood Oil Works in 1957.

Pumpherston Oil Works
(Standard Gauge)

No./Name	Locomotive Type	Built by	Works No.	Date Built	Comments
1	0-4-0 diesel mechanical	Andrew Barclay, Kilmarnock	341	1940	Transferred from Westwood Oil Works 1963. Sold to Miles Druce Ltd, Leeds.
2	0-4-0 diesel mechanical	Andrew Barclay, Kilmarnock	399	1956	New to Pumpherston. Transferred to BP Refinery, Grangemouth, 1969.
4	0-4-0 diesel mechanical	Hudswell Clarke, Leeds	D697	1950	On loan from BP Refinery, Grangemouth, 1970.
5	0-4-0 diesel mechanical	Hudswell Clarke, Leeds	D699	1951	On loan from BP Refinery, Grangemouth, 1973.

Westwood Oil Works
(Standard gauge)

No./Name	Locomotive Type	Built by	Works No.	Date Built	Comments
None	0-4-0 diesel mechanical	Andrew Barclay, Kilmarnock	341	1940	New to Westwood Works. Transferred to Pumpherston Works, 1963
2	0-4-0 diesel mechanical	Andrew Barclay, Kilmarnock	412	1957	New to Westwood Works. Transferred to BP Grain, Kent, 1963.

Scottish Oils Ltd diesel locomotive No. 4 was delivered new to Pumpherston Oil Works from Hudswell Clarke in 1970, works No. D697. Seen here on loan to Grangemouth Refinery crossing Grangeburn Road with tanks for Grangemouth Docks.
Jim Summers/Hamish Stevenson collection

BELOW: Winchburgh tramway locomotive No. 2 on maintenance duty.

LOCOMOTIVES

The Winchburgh tramway. Here, former Oakbank Oil Company electric locomotive No. 1 stands outside the loco shed at Niddry Castle Oil Works.
R.C. Nelson/Hamish Stevenson collection

ALLOCATION OF ELECTRIC LOCOMOTIVES

WINCHBURGH TRAMWAY
(Surface haulageway: 2 feet 6 inches gauge)

No./Name	Locomotive Type	Built by	Works No.	Date Built	Comments
1	Overhead electric 4 wheeled locomotive	Baldwin Locomotive Works, Philadelphia, USA	20586	1902	New to Winchburgh. Rebuilt in Middleton Works, 1945. To Connell's, Coatbridge. Scrapped.
2	Overhead electric 4 wheeled locomotive	Baldwin Locomotive Works, Philadelphia, USA	20587	1902	New to Winchburgh. To Royal Scottish Museum, Edinburgh, 1961. Transferred to Almond Valley Heritage Museum, Livingston.
3	Overhead electric 4 wheeled locomotive	Westinghouse	None	1907	New to Winchburgh. Scrapped, 1961.
4	Overhead electric 4 wheeled locomotive	English Electric Company	722	1929	New to Winchburgh. Scrapped, 1961.
5	Overhead electric 4 wheeled locomotive	Andrew Barclay, Kilmarnock/ Metropolitan Vickers, Manchester	None	1942	New to Winchburgh. To Connell's, Coatbridge. Scrapped, 1961.
6	Overhead electric 4 wheeled locomotive	Andrew Barclay, Kilmarnock/ Metropolitan Vickers, Manchester	None	1942	New to Winchburgh. To Connell's, Coatbridge. Scrapped, 1961.

THE SCOTTISH SHALE OIL INDUSTRY & MINERAL RAILWAY LINES

HOPETOUN SHALE MINE
(Gauge 2 feet 6 inches)

No./Name	Locomotive Type	Built by	Works No.	Date Built	Comments
None	0-4-0 battery electric	Metro Vickers/ Andrew Barclay, Kilmarnock	Unknown	Unknown	Disposal unknown.
None	0-4-0 battery electric	Unknown	Unknown	Unknown	Disposal unknown.
None	0-4-0 battery electric	Unknown	Unknown	Unknown	Disposal unknown.

RIGHT: Former Oakbank Oil Company electric locomotive No. 3 stands in Niddry Castle Oil Works sidings. *Hamish Stevenson collection*

BELOW: Former Oakbank Oil Company electric locomotive No. 4 stands outside the shed at Niddry Castle Oil Works on the 31st August 1959. *Hamish Stevenson collection*

APPENDIX B

KNOWN OIL COMPANIES AND WORKS POST-1850

It has been asserted that, post 1864, when Young's patent expired, there were soon to spring up over 120 oil companies of varying sizes and activities in their operation. Around half of these were involved in retorting cannel coal or other bituminous coals and thus were not properly part of the shale oil industry. Many of the others were involved in refining and producing specific oil related products such as paraffin, and they either quickly went to the wall or were absorbed or purchased outright by larger concerns. As the pressures on the shale oil industry, caused by imported American oil products, began to bite, further closures and more mergers occurred. The undernoted list contains details of oil works either directly retorting/refining oil from shale or refining crude oil purchased from such works within this period of time.

List (by County) of Known Works Producing Shale Oil or Refining Shale Oil Products During the Life of the Industry

West Lothian (Linlithgowshire)

Albyn Oil Works, Broxburn (Fernie's *later* Glasgow Oil Company)
Almondhill Paraffin Works
Armadale Paraffin Works (Scott's)
Bathgate (Boghead) Chemical Works
Bathville Paraffin Works
Benhar Oil Works (Simpson)
Benhar Paraffin Oil Works
Blackburn Oil Works
Boghall (Starlaw) Oil Works (Meldrum & Co., Uphall Oil Company)
Breich Oil Works (*also* New Hermand *or* Mid Breich) (New Hermand Oil Company)
Broxburh Oil Works (Broxburn Oil Company)
Broxburn Paraffin Works (Steel, *later* Glasgow Oil Company)
Broxburn (Buchan) Paraffin Oil Works (Poynter's: Calder Oil Company)
Broxburn Paraffin Refinery (Greendykes) (Bell)
Broxburn Greendykes Paraffin Oil Works (Bell, *later* Broxburn Oil Company)
Broxburn Stewartfield Shale Oil Works (Steel)
Broxburn Stewartfield Oil Works (Bell)
Chamfleurie Oil Works (Bridgend) (Linlithgow Oil Company)
Dalmeny Oil Works (Pattison, Dalmeny Oil Co.)
Deans Crude Oil Works (West Lothian Crude Oil Co., Pumpherston Oil Company)
Drumcross Paraffin Oil Works (Lester & Wylie)
Drumcross Paraffin Oil Works (Palmer & Paget)
Drumshoreland Paraffin Works (Almondfield) (McLintock, Fraser)
Grangepans (Thirlestane) Paraffin Oil Works (John Nimmo & Son)
Hallfarm Oil Works (Broxburn) (Liddle)
Holmes Oil Works (Holmes Oil Co.)
Hopetoun Oil Works (Fauchledean) (Uphall Mineral Oil Co. Young's)
Niddrie Castle Oil Works (Winchburgh) (Oakbank Oil Company)
Philpstoun Oil Works (J. Ross & Sons)
Roman Camp Oil Works
Roman Camp Paraffin Oil Works, *also known as* Almondfield (Fraser & Fraser)
Roman Camp Oil Works (Drumshoreland Muir)
Seafield Crude Oil Works (Seafield Patent Oil Company, Bathgate Oil Company, Pumpherston Oil Company)
Stankards Paraffin Refinery
Stankards Shale Oil Company
Uphall Oil Works (Peter McLagan, Uphall Oil Co., Young's)
Westwood Oil Works (Scottish Oils)
Westwood Paraffin Oil Works (Stewart's)

Mid Lothian (Edinburgh-shire)

Addiewell Oil Works (Young's Paraffin, Light & Mineral Oil Company)
Bellsquarry Oil Works
Burngrange Shale Oil Works
Cobbinshaw Bog Oil Works (Ferris & Fernie)
Cobbinshaw North Paraffin Oil Works (Mungle, Scottish Mineral Oil & Coal Co.)
Cobbinshaw South Oil Works (South Cobbinshaw Oil Co., South Cobbinshaw Fireclay & Brick Co.)
Dean Oil Works (Charles Handyside), Newtongrange
East Hermand Crude Oil Works (Dunnet's Works)
Edgefield Candle Works & Edin Oil Works (Dickson Candle Co.)
Gavieside Paraffin Works (Fell/West Calder Oil Co.)
Grange Shale Oil Works (Charlesfield) (Raeburn)
Harwood (Hartwood) Paraffin Oil Works (Andrew Walker)
Hermand Shale Oil Works
Levenseat Oil Works (Handaxwood) (Hamilton & Ross)
Magdelene Chemical Works
Oakbank Paraffin Oil Works
Pentland Paraffin Oil Works (Clippens)
Pumpherston Oil Works (Pumpherston Oil Co.)
Pumpherston Shale Oil Works (Buckside)
Straiton Paraffin Oil Works (Taylor's. Straiton Oil Co., Midlothian Oil Co., Clippens)
Whitebog (Eldin) Oil Works (Hawthornden)
Whitehill Shale Oil Works

Lanarkshire

There were known to be some forty-seven oil companies operating in Lanarkshire but for the purposes of this book only two, Tarbrax Oil Works and Lanark (Whitelees) Oil Refinery, have been discussed at length, since both were an integral part of the Lothians shale oil fields, with Tarbrax lying just over the County March with Midlothian, geographically separated by a mere few hundred yards. Many of the other works, especially in the Monklands Parish, were concerned with oil production from canneloid coal. Most of the works listed here were gone prior to 1900.

Arden Oil Works
Auchenheath Oil Works
Avonhead Oil Works
Birkenshaw Oil Works (Shawburn)
Bishop Street Oil Manufactory
Blochairn Chemical Works
Bredishome Oil Works
Calderbank Works
Canalbank Paraffin Oil Works (Robertson)
Crown Point Oil Works (Palmer, Clyde Oil Works)
Drumbrow Oil Works
Drumgray Oil Works
Dryflat Oil Works
Forth & Clyde Paraffin Oil Works
Glenmains Oil Works (Bellsdyke)
Greengairs Oil Works
Hareshaw Oil Works
Hillhouseridge Oil Works (Grey, Benhar Oil Co., Shotts Oil Works)
Kirkmiurhill Oil Works
Kirkwood Oil Works
Lanarkmoor (Whitelees) Oil Manufactory
Loanhead Oil Works (Buchanan)
Lochburn Road Paraffin Refinery
Longrigg Oil Works (Nimmo)
Millburn Oil Works
Nettlehole Oil Works (J. Smith)
Palacecraig Oil Works (Calder Oil Co.)
Port Dundas Paraffin Oil Works (Robertson, Donald & Co.)
Possil Oil Works
Rochsoles Oil Works (A. Law & Co.)
Rochsolloch Oil Works (Airdrie Mineral Oil Company)
Rosebank Oil Works (Clydesdale Chemical Co.)
Roughcraig Oil Works (Clarkston, Spence & Forsyth)
Sheepford Locks Oil Works (Adam & Hamilton)
Shettleston Oil & Chemical Works
Stand Oil Works (J. Struthers & Co.)
Stanrigg Oil Works
Stonehouse Oil Works
Swinehill Oil Works
Tarbrax Oil Works
Vulcan Chemical Works (Townsend)
Wattson Oil Works (J. Ross & Son)
Whitehill Oil Works (Clippens Oil Co.)
Whiterigg Chemical Works
Woodhall Oil Works (Stanrigg Oil Co.)

Fife

The Lothian shale fields extended under the River Forth and into Fife for a short distance.

Binnend Oil & Mineral Works
Inverkeithing Oil Works
Kilrenny Oil Works (Rowatt & Yuill)
Lochgelly Paraffin Oil Works
Methil Paraffin Oil Works
Westfield & Capledrae Oil Works

Aberdeen

Aberdeen Oil Works

Ayrshire

Annick Lodge Oil Works
Craigie Oil Works
Fergushill Oil Works
Hurlford Oil Works
Kilwinning Oil Works Lanemark Oil Works
Pathhead Oil Works

Renfrewshire

Clippens Oil Works
East Fulton Shale Oil Works
Inkermann Shale Oil Works
Linwood Shale Oil Works
Linwood by Blackston Shale Oil Works
Nitshill Paraffin Oil Works (Lesmahgow Coal Company)

Stirlingshire

Ballat Oil Works
Coneypark Oil Works
Falkirk Oil Works
Forthbank Oil Works
Limerigg Oil Works

Roxburgh

Hawick Oil Refinery

APPENDIX C

GLOSSARY OF MINING TERMS

Anticline	An upward fold or arch of rock strata.
Bed	A stratum of shale or other sedimentary rock.
Bing	A tip or heap of shale waste or retorted (burned) shale blaes.
Brush (to)	To remove excess material in order to increase the dimensions of a roadway.
Brushing	Digging out the pavement or taking down the roof to create more headroom in a roadway.
Cage	In a mine or pit shaft, the platform or assembly used to hoist personnel and materials to and from the underground workings.
Coal dust	Fine particles of coal capable of passing through a No. 20 sieve.
Craw Picker	In shale mining, the Inspector who checked the quality of shale at the pit-head.
Creep	The forcing of pillars into the soft bottom (pavement) by the weight of a strong roof.
Cuddie	A weight mounted on wheels, or a loaded hutch used to counter-balance the hutch on a cuddie brae (from 'cuddie' the Scots term for a horse or pony).
Cuddie-Brae	An inclined roadway worked as a self-acting incline, the cuddie acting as a drag on a fully loaded hutch running down.
Cundie	Unfilled space between pack walls; a small roadway or air course.
Cut Chain	A chain used on inclines which may be cut at different places to suit various levels.
Cut Chain Brae	An incline on which cut chains operation is employed.
Depth	Denotes either the vertical depth of a shaft in a pit or the inclined depth of a mine. Generally, when referring to pits, the depth is expressed in fathoms (6 imperial feet).
Dip	Declivity or declination of the strata.
Dip and Rise	The slope or inclination of the strata.
Dip Level	The lowest drift or roadway following the strike of the strata.
Dook	A roadway driven to the dip, visually the main road going to a dip.
Downcast	The shaft down which air currents enter the underground workings.
Drawer	The person who works with the miner at the face to load the shale dislodged by explosives into hutches and takes the loaded hutch(es) from the face to the shaft or terminus of a haulage road.
Endless Rope	A system of haulage where hutches are hauled by being attached to a wire rope always moving in one direction.
Explosive	Black powder (gun powder) packed into paper tubes and ignited in shot holes in the face to dislodge shale.
Face	Exposed area of the shale seam from which shale is being extracted.
Faceman	*See* Miner.
Fireman	A specially-trained, competent miner working under the oversman with responsibility for supervising the ventilation in a working and testing for gas. Originally, in early underground coal workings, a person delegated to enter a gas-filled workplace wearing a protective covering of wet sacking or other heavy cloth, and carrying a lit candle on a long pole with which to ignite (fire) the gas – a dangerous process which saw many fatalities and men badly burned. Also, because of the sacking covering, these firemen were often referred to as 'penitents'.
Fall (Roof Fall)	A mass of roof rock or shale which has fallen in any part of the mine.
Fan	A rotary mechanical or electrical device used to supplement the natural ventilation of the workings by forcing the extraction of stale air.
Fathom	A vertical linear measurement consisting of 6 imperial feet used to describe depth.
Fault	A geological slip between two portions of the strata which have moved relative to each other.
Fault Line (Zone)	A major fault as opposed to a single fracture which could be hundreds of yards in length.
Fill	Any material which is put back in place to provide support to the roof in a working.
Friable	Easy to break or crumble naturally.
Fuse	A chord-like substance with a core of black powder used to ignite the explosive.
Graith	The miners' tools: picks, drill bits, shovels, etc.
Haulage	The transfer of shale from the face by pony, inclined endless-rope or locomotives.
Haulageway	The narrow gauge trackwork on which the loaded and empty hutches run.
Hoist	Drum on which the hoisting rope was wound in engine house as the cage was raised to the surface.
Hoisting	Vertical transportation of shale, materials and men within a shaft.
Hutch	A small four-wheeled wagon or tub for conveying shale, carrying around one ton and running on rails.
Hydro-carbon	A family of chemical compounds containing carbon and hydrogen atoms in various combinations and found especially in fossil strata including shale.
Inbye	In the direction of the working face.
Incline	A roadway towards the rise of the strata along which the shale is transported by mechanical means. The length of the main working incline roads was measured in feet and not fathoms.

In Gaun e'e	An early term for a small drift or mine, starting from the surface of the ground into the seam being worked.
In Situ	In the natural or original position.
Lamp	A generic term covering several different types of illumination used underground. Early lamps consisted of a reservoir holding oil or tallow, a 'stroup' or spout containing the wick and a 'hanger' or hook used to fasten the lamp to the miner's cap. Later, lamps employing calcium carbide and a water supply were used, the action of the water on the calcium carbide producing acetylene gas (C_2H_2) which was burned as a naked flame. The naked flame lamps were superseded by battery electric headlamps. The term also addressed the flame safety lamps such as the Davy lamp.
Lampman	The person employed in the lamp house at each mine/pit to control issue and return of all lamps (other than personal tallow and acetylene lamps), who filled and trimmed safety lamps and ensured electric safety lamp batteries were fully charged whilst not in use.
Level	A nearly horizontal passage or tunnel in a mine which often acted as a drainage adit.
Main Entry	The point at which the mine shaft left the surface of the ground and went underground.
Miner	The person responsible for working the face and extracting the shale by drilling bore holes, placing and firing shots and the general safety fo the working area.
On-Cost	Charges for labour and maintenance additional to the payment for shale production.
Opencast	Working of outcropping shale seams on or near the surface, akin to quarrying. The overburden was removed and, after shale extraction, then replaced.
Outbye	Nearer to the shaft and therefore farthest from the working face.
Outcrop	Place where the seam or strata comes close to the surface.
Overburden	Layers of soil and rock covering the seams of shale.
Oversman	A deputy or under-manager in charge of underground operations.
Pack (or Pack Wall)	A wall or wooden structure erected and filled with waste (fill) to help support the roof where shale is being extracted by the Longwall method.
Pavement	The floor of a seam.
Pillar	*See* Stoop.
Pit Ponies	Small horses employed underground to haul trains of loaded hutches from workplaces along the haulageways to the pit bottom. Replaced in the main by mechanical methods.
Pit Props	*See* Roof Supports.
Placeman	*See* Miner.
Pyrites	Hard, heavy, shiny yellow mineral (FeS_2) found mainly in coal and used in the production of sulphuric acid.
Rib	A thin stratum of rock in the shale seam.
Rise	The upward direction of the strata.
Roof	Stratum above the seam or working.
Roof Fall	*See* Fall.
Roof Support	Posts, jacks, arched girders or wooden props used to support the rock overlying the shale seams.
Room	A working space in stoop and room working.
Seam	The stratum or bed of shale.
Shaft	A vertical hole or pit sunk through the strata for the purpose of finding and mining shale Seams and through which miners, material and mined shale was raised or lowered. Shafts also provided ventilation to underground workings. Generally lined with brick. Depth measured in fathoms.
Shift	The pit worker's day, or the generic term for pit mining teams all underground at the same time. The working day consisted of three eight-hour shifts: • the day shift started early in the morning and finished around midday • the back shift was the afternoon shift finishing near to midnight • the night shift covered the intervening period between the two former shifts.
Sink (to)	Construct a vertical shaft by digging out and shoring up.
Slip	The movement of a geological fault.
Split	A room or end driven through a stoop.
Stemming	Packing the explosives into the bore hole.
Stoop	A broad pillar of shale left unworked and *in situ* to help support the roof. Generally stoops were recovered later during retreat mining.
Straw (or Strae)	A make-do fuse comprising straw filled with black powder.
Subsidence	Gradual sinking or ground movement, but sometimes abrupt collapse of the overburden above worked-out sections of the pit and often a result of retreat mining.
Surface Mine	*See* In Gaun e'e.
Syncline	A downward fold in the strata when it dips downwards from both sides towards the axis. Opposite of Anticline.
Timber	*See* Roof Support.
Trees	The name given by shale miners to the wooden pit props used to support the roof at the working face. Timber props were preferred as they gave an audible warning of roof weighting or other strata movement. *See also* Roof Support.
Tram	A self propelled vehicle capable of hauling a train load of loaded hutches both underground and on the surface.
Tramway	The plateway or railway upon which the trams (trainload of hutches) were moved.
Undercut	To cut out the underside of the seam.
Upcast	The shaft by which the ventilating current of air returned to the atmosphere, the flow often assisted by extraction fans.
Weighting	Fracturing and lowering of the roof as a result of shale extraction.
Wire Rope	A steel rope used in the winding shafts and as continuous haulage system both underground and on surface.
Working Face	Place where the shale was being extracted.

BIBLIOGRAPHY

The Oil Shales of the Lothians, 3rd Edition, HMSO, 1927
Land of the Mountain and Flood: the Geology of Scotland, Scottish Natural Heritage/Birlinn, 2009
Industrial Locomotives of Scotland, Industrial Railway Society, 1976
Brief Description of the Operations of the Scottish Shale Oil Industry Scottish Oils Ltd, Scottish Oils Ltd, 1951
James Barrowman, *A Glossary of Scotch Mining Terms*, 1886
John Butt, *James 'Paraffin' Young*, 1983
John L. Carvel, *The Coltness Iron Company*, 1948
Robert Duncan, *Sons of Vulcan*, Birlinn, 2009
Alastair Findlay, *Shale Voices*, Luath Press, 2002
C. Hamilton Ellis, *The North British Railway*, Ian Allan, 1955
Dr Barbra Harvie, *Oil Shale Bings*, West Lothian Council, 2005
William F. Hendrie, *Discovering West Lothian*, John Donald Publishers, 1986 and 1995
R. Hynd-Brown, *Armadale Past & Present*, 1906
David Kerr B.Sc., *Shale Oil: Scotland*, Privately published by the Author 1994
John H. McKay, 'The Social History of the Scottish Shale Oil Industry', unpublished PhD thesis, Open University, 1984
Miles Oglethorpe, *Scottish Collieries*, RCAHMS, 2006
C. Randall, *IOM Historical Research Report No. TM/90/02*, IOM, 1990
Iltyd I. Redwood, *A Practical Treatise on Mineral Oils and Their By-Products*, 1897
A. Seaton, *IOM Historical Research Report No. TM/85/02 1985*, IOM, 1985
John Thomas, *A Regional History of the Railways of Great Britain (Volume 6)*, David & Charles, 1984
F. Worsdell, *The Tenement – A Way of Life*, W&R Chambers, 1979

West Lothian Local Bio-diversity Plan, West Lothian Council
John Airey's Railway Map of Scotland 1875, National Library of Scotland Map Room
6-inch Ordnance Survey Maps 1910/1920, National Library of Scotland Map Room
UK Oil Shales Past and Possible Future Exploitation, HMSO 1975
Minute Books, North British Railway: 1865–1922, National Archives
Minute Books, Caledonian Railway: 1860–1923, National Archives
Railway & Canal Commission Plans & Maps Produced During Proceedings in Application by Anglo American Oil Co. Ltd against the Caledonian Railway Co. and the North British Railway Co. 1915. Caledonian Railway Company copy
The Scotsman, 1900–1963
The West Lothian Courier, 1880–1980
The Midlothian Advertiser, 1890–1965

Journal of the North British Railway Study Group (various)
The True Line, Journal of the Caledonian Railway Association, 1997–2000

A railway wagon containing shale is being emptied into the crusher hopper by means of a crane. The location is unknown, but this method of discharging wagons caused frequent damage.
Courtesy of Grangemouth Heritage Trust

INDEX

accidents, 63–68, 119
Addiewell
  Goods, 164, 175–76
  Junction, 175–76
  Oil Works, 19, 23, 35, 36, 40, 77, 95, 97, 102, 110, 142, 155, 164, 176, 199, 200
  pits, 22
  village, 118, 123
afterdamp, *see* gases
Albyn Oil Works, Broxburn, 95, 96, 97, 182, 184–85, 192, 200
Alfreton, Derbyshire, 33
Almond, River, 16, 23, 26, 102, 188, 194
Almond Valley, 13, 188
Almond Valley Heritage Centre, 9
Almondell Oil Works, 96
Almondhill Paraffin Works, 194
ammonium sulphate, *see* sulphate of ammonia
animal fats, 33
Armadale, 13, 127
  Junction, 175, 177
  pits, 177
Arthur's Seat, 15
ash of oil shales, 18

Baads Mines, 22, 110, 199
Bass Rock, 15
Bathgate, 13, 118
  East Junction, 178
  locomotives, 173, 176, 178, 179, 190
  Oil and Chemical Works, 34, 95, 97, 142, 175, 176
  Oil Company, 103, 105, 178
  West Junction, 175
  Yard, 163, 173, 176
Bathville (Scott's) Oil Works, 177
Beilby, (Sir) G.T., 77, 103
Bell, R., 35, 95, 96–97, 109, 192
Bellsquarry, 13
Benhar
  Caledonian Oil Works, 159, 178
  Junction, 159
  Paraffin Works, 159

Binnend
  mines, 109, 198
  Rows, 109
Binney, E.W., 4, 95
Binny Craig, 15, 27
  anticline, 27
Blackbrae Oil Works, 22, 110
Blackness shale field, 30
blue oil, 92
Boghead cannell coal, 34
Boghead Junction, 71, 176, 177
Breich,
  pits, 26, 31, 163
  shale field, 23, 31
Breich Water, 13, 110, 162, 176, 199
brick-making (SOL) Pumpherston, 42
Bridgend, 19, 98, 174
  Oil Works, 77, 98, 198, 200
  Rows, 98, 200
British Aluminium Company, 109
British Oil & Candle Company, 103
Broxburn, 13, 118, 125, 182
  growth of population, 118
  Junction, 192, 200
  Oil Company, 95–96, 120, 129, 171, 184–85, 192
  pits, 200
  shale field, 27
  shales, 18, 26, 27, 30, 31, 40, 45, 47, 96, 98, 101, 107
  Works, 179, 184, 192
Burngrange
  accident, 7, 64, 65, 66, 67
  Nos 1 & 2 pit, 40, 52, 57, 199
  No. 39 pit, 23, 199
Burntisland
  Oil Company, 30, 109, 171, 198
  Oil Works, 30, 109
  shale field, 13, 16, 18, 19, 30

Cadell H.M., 15, 16
Calder Fault, *see* faults.
Caledonian Mineral Oil Company, 103, 110
Camilty Gunpowder Mill, 58
Camps
  Branch, 169, 179, 185–86
  Goods station CR, 165, 168, 169, 186
  Junction, 165, 168, 169
  Raw Camps limestone quarry, 165, 168
  Rubbish Chute, 168
  shale, 18, 19, 27, 30
candles, 35, 97, 98
Carstairs locomotives, 158, 159, 169
Champfleurie shale, 18, 30, 99
Champfleurie & Philpstoun shale field, 30
Charlesfield Estate, 110

chemical composition
  oil shale, 18
  shale oil, 18
chokedamp, *see* gases
Chrichton, A.H., 79, 99
Christison, (Sir) R., 33
Clapperton Trough, 26
Clippens Oil Company, 30, 31, 105, 107, 198
coal measures, 15, 16
Cobbinshaw, 16, 17
  limestone and coal, 158
  Loch (reservoir), 19, 158
  North Mining (Bairds), 158
  North Oil Works (Mungles), 110, 157
  shale field, 19
  South Oil Works, 157
Cockburnspath, 13
Cockleroy, 15
Coltness Iron Company, 165, 168, 186
Conacher, H.R.J., 81
cracking process and 'cat' crackers, 88
Crofthead, 160, 178
Cowdenhead, 175
crude oil production
  shale crushing, 75
  shale retorting, 75–86
  see also refining
crude solid paraffin, 92, 94
Crum Brown, A., 18
Curly shale, *see* Pumpherston shale
Cuthill rail crossing, 162, 163, 164, 176

Dalmahoy shale, 18, 30
Dalmeny, 13, 16, 179
  Estate, 104
  Oil Company, 104, 129, 194
  Oil Works, 37, 104, 194
Dalmeny/South Queensferry shale field, 30
Deans
  Oil Works, 40, 103, 105, 179
  pits, 26
  shale field, *see* Livingston
Dechmont Arch, 26

229

Dedridge Rows, 26, 103, 201
detergent, alkyl sulphate, 42
dolerite floats, 27
Drumcross Oil Works, 102, 103
Drumshoreland, 182, 184, 185, 188, 190, 192
Duddingston
   pits, 30, 60, 103, 149, 152
   shale field, 30
Dundas Estate, 104, 194
Dundas Shale Oil Company, 104, 194
Dunnet Shale Groups, 18, 19, 22, 23, 24, 26, 27, 30, 31, 40, 45, 47, 99, 104, 107, 109, 110, 201
Duntarvie Castle, 27
dust from oil shale, 61, 62, 129

East Calder, 177
   Goods station, 168, 186
Easter Inch Moss, 178
Ecclesmachan, 182, 200
   shale field, 27
Edinburgh Dalry Road locomotives, 164, 169
Edinburgh Haymarket locomotives, *see* Haymarket (Edinburgh)
Edinburgh St. Margaret's locomotive shed, *see* St. Margaret's (Edinburgh)
Ennis, L., 131
Excise Duty exemption, 40, 41, 113
explosives, 58
Explosives Act (1853), 58, 59

faults, geological
   Calder, 18, 26
   Ecclesmachan, 27
   Langside Blackburn, 23, 26
   Middleton Hall, 18, 19, 26, 27
   Midhope, 30
   Murieston, 18, 19, 26
   Niddry Castle, 27
   Ochiltree, 18, 27, 30, 149
   Wilsontown, 19
   Winchburgh, 18, 27
Fell, Alexander, 102
Fells shale, 18, 19, 22, 23, 24, 26, 27, 31, 101, 102, 105, 107
firedamp, *see* gases
fish oils, 33
Fivestanks shale, 27
folding of strata in Scottish shale, 15, 17
Forth & Clyde Paraffin Oil Works, 102
fossils, 16
Foulshiels Colliery, 163, 164, 176
fracking, 31, 44

Fraser shale, 18, 30
Fraser pit, 47

gases, presence in mines, 61
gas, natural exploration, *see* fracking
Gateside anticline, 30
Gavieside
   Paraffin Oil Works, 102, 163, 199
   pits, 22, 23, 64, 102, 199
Geikie, (Sir) A., 16
geography of the Scottish shale fields, 13
geological positions, 16
geological time periods, 15, 31
geology of the Scottish shale fields, 15
Glasgow Oil Company, 95, 96–97, 192
Glasgow Scottish Oil Company, 110
Grange Farm Oil Works, 110, 163
Grangemouth BP, 42, 182
green oil, 87, 88
Greendykes Road level crossing, Broxburn, 184, 185, 192

Hamilton locomotives, 165
Handaxwood Oil Works, 31, 110, 160, 178
Harburnhead, 19
Hartwood
   Church, 22
   Estate, 22
   Farm, 22
   House, 22
   mine, 110
   Oil Works, 110
Harwood Water, 22, 163
Haymarket (Edinburgh) locomotives, 173, 185, 190, 193
Helensburgh, 13
Henderson., N.M., 77, 79, 96, 102
Hermand, 23, 99
   mines, 7, 23, 40, 110, 164
   Oil Company, 22, 99–101, 157
   Oil Works (old), 22, 99, 163, 174
   Oil Works (new), 100, 163
   Old (Shuttlehall) 23, 99
   siding, 163
Holmes
   Oil Company, 27, 109
   Oil Works, 98, 109, 174, 188
Holygate Goods Depot, 184
Hopetoun
   anticline, 30
   Oil Works, 40, 98, 182, 200
   mines, 200
Houston Coals, 27
Houston Wood, 26
Howden House, 113
Hurlet Coal, 23

Ingliston/Newliston shale field, 30
Ingliston pits, 30, 98, 179, 194
igneous intrusions and rocks *see* volcanic activity
Iinclined shaft winding, 51

kerogen of oil shale, 15, 18
Kimmerage (Dorset) shales, 9
Kinghorn, 199

tunnel, 198
station, 198
Kingscavil Rows, 98
Kirk, A.C., 77
Kirkliston
   Arch, 30
   station, 194, 197

Lanark Oil Refinery, 103
leisure activities, 131–33, 135–36
Levenseat
   Branch Junction (NBR), 160, 178
   Junction (CR), 160, 178
   limestone, 110, 160, 178
   Oil Company, 110
   sand quarry, 178
   shale, 31, 110
Lillie's coal shale, 31, 107
Limefield
   House, 36
   Junction, 163, 164, 165
limestone, 31, 177
Linhouse Water, 26
Linlithgow, 13, 174
   Oil Company, 30, 98, 100, 157, 171, 174
Linwood Oil Works, 31
Little Harwood *see* Hartwood
Livingston Almond Valley Heritage Centre, 9
Livingston/Deans shale field, 26
Livingston New Town, 13, 136
Livingstone, David, 36
Loanhead, 13, 30, 198
Loganlea Colliery, 163, 164
Longridge, 177

McKay, Dr J., 131
McLagan (MP), Peter, 37, 96, 105
Meldrum, E., 34, 95, 97
methane gas, *see* natural gas
   in shale mines, *see* mining, health & safety
Midlothian Oil Company, 30
Mid Calder, 13, 177
   Junction, 165
   No. 1 mine, 200, 201
   Oil Company, 102–103
   shale field, 26
Middleton Hall, 110, 182
Midlothian Oil Company, 107, 198
mining, health & safety
   1855 Mines Act, 61
   1887 Coal Mining Regulation Act, 61, 69
   1911 Coal Mines Act, 61
   1920 Mining Industry Act, 61
   1945 Mines & Quarries Act, 61

INDEX

Mining Methods of Oil Shale
  longwall, 47
  longwall in steps, 47
  opencast, 45, 49
  retreat or broken working, 49
  stoop & room, 48–49
  stoop & room (modified), 49
MOD 324 Engineers Park (R.E.), 178
Montrose, 13
Moore (JP) Mrs S. 119
Motherwell locomotives, 159, 160
motor spirit
  diesel oil, 40, 94
  petrol, 99
Muirhall pits, 22, 199
Muldron Junction, 160
Mungle shale, 18, 23, 27, 110
Murieston Fault, *see* faults

natural gas occurring in shale, 31
natural oil
  Bilsthorpe, 113
  Dukes Wood, 113
  Eakring, 113
  occurring in shale
    Breich pit, 31
    Westwood pit, 31
New Deeps colliery, 33
New Farm mine, 26, 103, 201
Newliston Junction, 188, 190
Newliston No. 29 pit, 30, 98, 179, 188, 190
Niddrie Castle Oil Works, 30, 40, 41, 98, 99, 103, 149, 193
North British Oil & Candle Company, 103
North Cobbinshaw Oil Works, 110, 157

Oakbank
  Oil Company, 26, 30, 102, 103, 104, 109, 123, 129, 149, 157, 201
  Oil Works, 40, 102, 104, 165, 200, 201
  overhead cableway, 201
  Rows, 129
Ochiltree
  Fault, *see* faults
  pits, 30, 98, 200

Paraffin
  origins, 33
  see also refined products
Pennsylvania USA, discovery of liquid petroleum, 37
Pentland
  Hills, 15
  Oil Works, 31, 198

Philpstoun
  House, 30
  mines, 30, 99, 103, 200
  Oil Works, 37, 40, 49, 99, 175, 200
  Oil Company, *see* J. Ross & Co.
Philpstoun & Champfleurie shale field, 30
pits and mines, equipment
  drilling, 57
  haulage, 52
  lighting, 53, 54
  power supplies, 55, 56
  water control, 59, 60
  winding systems, 51
Playfair, (Prof.) Lyon, 33
Polbeth, 22, 163
  pits, 23, 52, 199
  shale field, 23, 31
Polkemmet
  colliery, 159
  Junction, 175
pollution, 129–30
Polmont locomotives, 174, 175
Port Edgar, 194
Pumpherston, 13
  anticline, 26
  Arch, 26, 27, 30
  group of shales, 18, 19, 30, 45, 105, 109
    Jubilee, 26, 105
    Maybrick, 26
    Curly, 26, 27
    Plain, 26
    Wee, 26
  Oil Works/Refinery, 37, 97, 105, 112, 113, 168, 179, 182, 186
  Oil Company, 19, 26, 101, 103, 105, 123, 129, 168, 178, 179, 186
  pits and mines, 186
  shale field, 26
pyrobitumin, *see* kerogen

Raeburn, John, 97, 110
Raeburn shale, 18, 22, 26, 27, 30, 105
railways
  Caledonian Railway branches
    Addiewell Oil Works Junction, 156, 164
    Benhar Junction to Polkemmet Moor, 159–60
    Camps Junction to Raw Quarry and Camps Goods, 165, 168–69
    Levenseat Junction to Levenseat and Handaxwood Oil Works, 160
    Mid Calder Junction to Oakbank Oil Works, 164
    Muldron Junction to Muldron pit, 160
    Tarbrax Junction to Cobbinshaw mines and Tarbrax Oil Works, 157
    West Calder Loop, 160, 162–64
  classification of, 141
  non standard (Narrow) gauge railways, 143–47
    surface haulageways, 144–45, 147
    tipping of shale waste, 144–45
    see also Winchburgh, tramway
  North British Railway branches
    Bathgate to Addiewell Goods and Addiewell Oil Works, 175–76
    Bathgate to Deans/Boghall mines, 178
    Bathgate to Deans Oil Works, 178–79
    Bathgate to Seafield Oil Works, 178
    Bathgate to Torbane pits and Armadale, 176–77
    Broxburn Junction to Albyn Oil Works, 192–93
    Broxburn West Junction to Niddry Castle Oil Works, 193
    Crofthead to Levenseat quarry and Handaxwood Oil Works, 177–78
    Edinburgh Loanhead and Straiton Oil Works, 197–98
    Hallyards to Kirkliston, Almondhill Paraffin Works, Dameny Oil Works and Port Edgar, 194, 197
    Kinghorn to Binnend Oil Works, 198
    Linlithgow to Bridgend Oil Works, 174–75
    Newliston Junction to Newliston No. 29 pit, 188, 190
    Philpstoun Oil Works, 175
    Queensferry Junction to South Queensferry and Port Edgar, 193
    Queensferry Junction to Ingliston pits, 193–94
    Uphall Junction, 179
    Uphall to East Goods Raw quarry and Camps, 186
    Uphall to Holmes Oil Works, 188
    Uphall to Hopetoun Oil Works, 182
    Uphall to Pumpherston No. 4 mine, 186
    Uphall to Pumpherston No. 5 mine, 186
    Uphall to Pumpherston Oil Works, 186
    Uphall to Roman Camp, 185
    Uphall (Drumshoreland) to Holygate Goods, Broxburn Oil Works and Albyn Oil Works, 182, 184–85
  internal standard gauge mineral railways, 199
    Addiewell Oil Works to Burngrange and Baads mine, 199
    Addiewell Oil Works to Gavieside and Polbeth pits, 199, 200
    Albyn Oil Works to various pits, 200
    Hopetoun Oil Works to various pits, 200
    Oakbank oil Works to Mid Calder No. 1 mine, 200
    Philpstoun Oil Works to Philpstoun and Ochiltree pits, 200
Ratho, 179, 190, 190, 193, 194, 197
Raw Camps Quarry, *see* Camps
Raw limestone pit, *see* Camps

refining of crude oil, 87, 88
  first distillation, 87, 88
  second distillation, 88, 92
Refined Products
  batching oil, 94
  burning oil, crude, 88
  burning oil, heavy, 92
  cleaning oil, 92, 94
  diesel oil, 40, 94
  heavy gas oil, 88, 92
  lamp/power oil, 92, 94
  light gas oil, 2, 94
  lighthouse oil, 92
  lubricating oil, 92, 94, 120
  motor spirit, petrol, 94
  naphtha, 87, 94, 97, 120
  paraffin, 35, 88, 97, 120
  paraffin wax, 35, 92, 93, 97, 120
  residuum, 92, 94
Reichenbach, Carl Ludwig von, 33
retorts
  Beilby retort, 77
  Beilby & Young Pentland retort, 77, 98, 99, 100, 103, 104, 105, 107, 109
  Bryson Pumpherston retort, 79
  description of modern developments, 75
  Henderson retort, 77, 96
  horizontal retorts, 75
  Phipstoun retort, 79, 99
  Westwood retort, 79
Roman Camp
  Oil Works, 40, 96, 185, 186
  pits, 26, 37
  Rows, 129
Rosebery, Lord, 47, 104
Ross, Messrs J. & Co., 27, 30, 37, 99, 129, 171, 175, 200
Rosshill mines, 30, 104

St. Margaret's (Edinburgh) locomotive shed, 179
salt water accompanying natural oil, 31
safety lamps, 61
Scottish Oils Ltd, 40, 41, 99, 110, 112, 113, 118, 121, 129, 182, 201
Scottish Shale Museum, 9

Seafield and Cousland
  shale field, 26
  pits, 26
Seafield Patent Fuel Company, 178
Seafield Patent Oil Works, 103, 178
Seafield Oil Works, 40, 101, 105, 178
Selligue, Alexandre Francois, 33
Simpson, A, 103
Simpson, G, 103
Simpson, Sir James Young, 102
shale bings, list of, 113
shale communities
  education, 130–31
  health, 62
  pay and conditions, 120–21, 123
  see also leisure activities
Shale Miners & Oil Workers Trade Union, 121
shale oil company houses (rows), list of, 122
shale oil industry
  early years, 33, 34, 35
  the growth years, 35, 36, 37
  decline, 37, 40, 41, 42
  renaissance, 42, 43
  locomotives
    battery, 222
    diesel mechanical, 221
    diesel mechanical (underground), 219, 220
    electric, 222
    steam, 205–219
sills and floats, 27
Slamannan Railway Company, 175
South Cobbinshaw Oil Works, 98, 102, 103, 157
South Queensferry, 194, 197
Stankyards, 97, 110
Starlaw Oil Works, 97
Stewartfield Original Paraffin Works, 35, 182
stinkdamp, see gases
Straiton
  Oil Company, 107
  Oil Works, 30, 107, 198
  shale field, 30
Stranraer, 13
sulphate of ammonia, 37, 87, 104, 120
sulphuric acid, 94, 96, 97
surface haulage, 144–45, 147
Sutherland, D., 103
sweating and sweat houses, 92, 94

Tarbrax, 16
  Oil Company, 103, 105
  Oil Works, 37, 103, 168, 186
  pits, 19
Teschenite sills, 31
Thornton, James, 99, 103
Thornton, Thomas, 99
Torbane/Torbanehill pits, 34, 95, 176–77
Torbanehill mineral, see Boghead cannel
Totleywells mine, 30, 103, 149, 152
Under Dunnet shale, see Dunnet

underground haulage, 52, 143–44
Union Canal, 27, 99, 174, 192, 200
Uphall, 13
  Castlehill level crossing, 182
  Goods, 77
  Junction, 168, 173, 179, 182, 186, 188, 190
  Mineral Oil Company, 97, 98, 178
  Oil Works, 40, 95, 97, 105, 179, 182, 190
  village, 118, 182
Uphall Station, 97, 182

vegetable oils, 33
ventilation of mines and pits, 53
vertical pit haulage, 45, 51
volcanic activity, 18, 27, 31, 33, 49
volcanic rock (Dolerite), 27

Walkinshaw Oil Company, 31, 100
wax, see paraffin
West Calder, 7, 22, 36, 95, 97, 99, 118, 125, 163, 177
  oil boring at, 31
  Oil Company, 98, 102, 103
  Oil Works, see Gavieside
  population changes, 118
  shale field, 19, 27
  water, 22, 102, 129, 163
West Lothian Bio-diversity Action Plan – Oil Shale Bings, 113–15
West Lothian Oil Company, 103, 105, 171, 178
West Mains mine see Baads mine No. 42
Westwood
  New Pit, 31, 57, 60, 199
  Nos 12 & 13 pit, 23, 26, 163, 199
  Oil Works, 40, 79, 110, 113, 163, 164, 199, 200
  (Stewart's) Paraffin Works, 110, 163
whitedamp, see gases
Whitequarries, 30, 49, 99, 103, 149, 152
Wilson, J.S.G., 16
Wilsontown Fault, see faults
Winchburgh, 13, 97, 125, 154
  tramway, 149–54
    electric locomotives, 149–50
    passenger coaches, 152
Woodmuir Junction, 160, 164
Woodmuir Colliery, 160
Woolfords, 158

yields
  crude oil, 22, 26, 30, 34, 27, 77
  sulphate of ammonia, 22, 26, 37, 77
Young, Dr J., 119, 129
Young, Sir J., 33–35, 95
Young, James Junior, 36
Young, W., 77, 105
Young's Paraffin, Light & Mineral Oil Company, 36, 95, 97–98, 110, 129, 157, 171, 176, 179, 188, 194, 199

PICTURES IN INDEX: P. 229L, an early view of Greendykes Road in Broxburn showing oil company rows and the chimneys of Broxburn Oil Works to rear; P. 229R, Broxburn Main Street looking east around the mid-1850s; P. 230L, Union Street corner with the Main Street in West Calder at the turn of the twentieth century; P. 230R, the strange and topsy-turvey conglomeration of buildings at the east end of West Calder Main Street known as the Peoples Palace; P. 231L, the original West Calder high School in Hartwood Road, West Calder; P. 231R, West Calder Main Street looking east around the turn of the twentieth century; P. 232, a view of the Oakbank Oil Company's Hopetoun Rows at Winchburgh, built for workers at the Niddrie Castle Oil Works, and still extant and occupied today.